Green Polymeric Nanocomposites

Green Polymeric Nanocomposites

Edited by
Satya Eswari Jujjavarapu and
Krishna Mohan Poluri

CRC Press is an imprint of the
Taylor & Francis Group, an **informa** business

CRC Press
Taylor & Francis Group
6000 Broken Sound Parkway NW, Suite 300
Boca Raton, FL 33487-2742

First issued in paperback 2021

ISBN-13: 978-1-138-48657-7 (hbk)
ISBN-13: 978-1-03-217484-6 (pbk)
DOI: 10.1201/9781351045155

Visit the Taylor & Francis Web site at
http://www.taylorandfrancis.com

and the CRC Press Web site at
http://www.crcpress.com

Dedicated to all our inspirational mentors
and beloved family members

Contents

Preface

"The environment and the economy are really both two sides of the same coin. If we cannot sustain the environment, we cannot sustain ourselves." These words of Dr. Wangari Maathai, 2004 Nobel Peace Prize winner, have indeed become more pertinent in the present day. From 2010 onwards, there has been an expanding inclination toward green materials. Green polymers have been suggested as potential alternatives to petroleum-derived synthetic polymers. Green polymers are either derived from natural sources or are synthesized using green chemistry methods. Either way, they are ecofriendly as they lead to a decrease in waste generation and energy consumption, avoid the use of toxic chemicals, preserve non-renewable resources like petroleum, and do not accumulate in the environment as traditional plastics do. Many consumers have now become aware of these factors, thus paving the way for the overwhelming popularity of these green polymers. For instance, the global green polymer market is expected to reach $51.2 billion by 2023, increasing from $36.0 billion in 2018 at a compound annual growth rate (CAGR) of 7.3%.

The integration of nanotechnology with these green polymers has led to the fabrication of green polymeric nanocomposites with enhanced thermal, mechanical, physicochemical, biodegradable, and biocompatible properties. The nanofillers employed in these composites possess several advantages owing to their large surface-area-to-volume ratio. Therefore, these composites have the amalgamated benefits of both the green polymers and the reinforced nanomaterials. All these factors explain the steeply rising demands of the market for these composites and their incorporation into various sectors. As of today, green polymeric nanocomposites are already being actively utilized in healthcare, the automobile industry, agriculture, paint and ceramics, construction, electronics, textile, and packaging industries. In the medical sector, these composites are currently being employed for the purposes of regenerative and reconstructive medicine. Research is also focused on the fabrication of implants, pills, capsules, skin grafts, drug carriers, etc. from such composites. As far as day-to-day commodities are concerned, these polymer nanocomposites have found a place in several products. Recently, a starch-based polymer nanocomposite has already gained popularity in the manufacture of shopping bags and meat liners.

This book aims to capture the recent developments with regard to such green polymeric nanocomposites. The book has been broadly divided into ten chapters. Chapters 1 to 5, edited by Dr. Satya Eswari Jujjavarapu, delineate the fundamental concepts of these composites. Terms such as green molecules, natural polymers, nanocomposites, etc. are defined and their significance is discussed. Comparative analyses are drawn among various synthesis as well as characterization (quality assessment) techniques for these composites. Electrospinning, being the most common fabrication method owing to several inherent advantages, is discussed at length. Composites comprising either of polymers from animal sources, plant sources, or microbial sources are illustrated thoroughly. Characteristics, applications, and optimization parameters of these composites are reviewed with appropriate examples for each of these different kinds of polymers.

Chapters 6 to 10, edited by Dr. Krishna Mohan Poluri, provide insights into the advanced technical aspects of research and the wide range of applications of these composites. The green polymers often present intrinsic disadvantages in certain cases, such as biomedical applications. Therefore, the various techniques that are employed to improve the polymer matrices of these composites are elucidated with special emphasis on an emerging technique known as non-thermal plasma surface modification. Then, the applications of these composites in diverse areas, namely, tissue engineering, drug delivery, food packaging, catalysis, environmental uses like adsorption of heavy metal ions and air filtration, etc. are portrayed. The polymeric nanocomposites are termed "facsimiled bones" owing to how closely they mimic the bone microstructure. Their present and future implementations in orthopedic uses are presented in detail. Finally, a record of the life-cycle assessment and the future perspective of these composites is depicted in order to outline their impact on human health and the environment. The book concludes by discussing the present and future markets for these composites by underlining their associated limitations and thus providing recommendations and avenues for rectification of these limitations.

Dr. Krishna Mohan Poluri
Dr. Satya Eswari Jujjavarapu

Acknowledgments

We are grateful to all of our authors for their precious contributions towards this book, in order to make it a comprehensive resource covering basic to advanced aspects of "green polymeric nanocomposites". We would also like extend our earnest gratitude to the editorial staff of CRC Press, Mrs. Allison Shaktin and Ms. Camilla Michael for their constant support and encouragement. The technical support provided by Ms. Sharanya Sarkar, Mr. Mukesh Kumar Meher, and Dr. Khushboo Gulati of IIT-Roorkee is greatly appreciated. KMP acknowledges the support of research grants from Department of Biotechnology (DBT), Science and Engineering Board of Department of Science and Technology (SERB-DST), Ministry of Water Resources, Ministry of Human Resource Development – Indian Institute of Technology Roorkee (MHRD-IITR) from Government of India, and BEST Pvt. Ltd. KMP also acknowledges the facilities provided by Mahatma Gandhi Central Library, Department of Biotechnology, Centre for Nanotechnology and Institute Instrumentation Centre at IIT-Roorkee. SEJ acknowledges Mr. Sekharbabu Jujjavarapu and Mr. Chaitanya Kumar Yadagini for constant motivation and support.

Dr. Krishna Mohan Poluri
Dr. Satya Eswari Jujjavarapu

Acknowledgements

We are grateful to all our contributors for their precious contributions towards this book, in order to make it a comprehensive resource [illegible] of [illegible] polymer [illegible] composites. We would also like to [illegible] to the [illegible] of CRC Press, Ms. Aliso[n] Shin and Ms. [illegible] for their constant support and encouragement. The [illegible] support provided by Ms. [illegible] and Dr. Khushboo [illegible] [illegible] acknowledge [illegible] [illegible] Department of [illegible] [illegible] [illegible]

[illegible]
[illegible]

Editor Biographies

Dr. Satya Eswari Jujjavarapu is currently working as an Assistant Professor in the Department of Biotechnology at National Institute of Technology (NIT), Raipur, India. Her fields of specialization include bioinformatics, biotechnology, and process modeling, evolutionary optimization, and artificial intelligence. She has more than 35 publications in SCI/Scopus-indexed journals and 35 proceedings in international and national conferences. Her research contributions have received wide global citations. She has also published six book chapters and four books (currently in press) with international publishers. She is an active member of various organizations and has received various awards.

Dr. Krishna Mohan Poluri is currently working as an Associate Professor in the Biotechnology Department, IIT-Roorkee. He earned his PhD from Tata Institute of Fundamental Research TIFR–Mumbai, and Post-doc from Rutgers University and University of Texas Medical Branch (UTMB–Texas). His areas of expertise are structural biology, protein engineering, biomolecular interactions, glycoimmunology and structure-based design of therapeutics and scaffolds, bionanotechnology and algal biotechnology, etc. He has published 85 publications, including research articles, editorials, books, and book chapters in various reputed international journals. He has authored the book *Protein Engineering Techniques*. Dr. Poluri is also a guest editor/editorial board member and ad hoc reviewer for several international research journals. He has won several awards and fellowships for his research work, most prominently the Young Scientist Award from the Indian Science Congress Association (ISCA), 2009, National Academy of Sciences India (NASI), 2014, and Innovative Young Biotechnologist Award (IYBA), 2013, by DBT.

Contributors

Potla Durthi Chandrasai
Department of Biotechnology
National Institute of Technology
Warangal, India

K. Chandrasekhar
Department of Biotechnology
National Institute of Technology
Warangal, India

Murthy Chavali
Shree Velagapudi Rama Krishna Memorial College (Autonomous)
Nagaram, India
and
PG Department of Chemistry
Dharma Appa Rao College
Nuzvid, India
and
MCETRC
Tenali, India

R.R. Deshmukh
Department of Physics
Institute of Chemical Technology
Mumbai, India

P. Gopinath
Nanobiotechnology Laboratory
Centre for Nanotechnology
Indian Insitute of Technology
Roorkee, India

Khushboo Gulati
Department of Biotechnology
Indian Institute of Technology
Roorkee, India

Archita Gupta
Department of Bioengineering
Birla Institute of Technology
Mesra, India

Enamala Manoj Kumar
Bioserve Biotechnologies (India)
Hyderabad, India

Gopalakrishnan Kumar
Ton Duc Thang University
Ho-Chi-Minh City, Vietnam

Prasun Kumar
Chungbuk National University
Department of Chemical Engineering
Cheongju, South Korea

Mukesh Kumar Meher
Department of Biotechnology
Indian Institute of Technology
Roorkee, India

Sweta Naik
Department of Biotechnology
National Institute of Technology
Raipur, India

Neelamshobha Nirala
Department of Biomedical Engineering
National Institute of Technology
Raipur, India

Padmini Padmanabhan
Department of Bioengineering
Birla Institute of Technology
Mesra, India

P.V.A. Padmanabhan
Research Division of Plasma Processing
Department of Physics
Sri Shakthi Institute of Engineering and Technology
Chinniyampalayam, India

K. Navaneetha Pandiyaraj
Research Division of Plasma Processing
Department of Physics
Sri Shakthi Institute of Engineering and Technology
Chinniyampalayam, India

M.C. Ramkumar
Department of Physics
School of Basic Science
Vels Institute of Science, Technology and Advanced Studies
Chennai, India

Sharanya Sarkar
Department of Biotechnology
Indian Institute of Technology
Roorkee, India

Sneha Singh
Department of Bioengineering
Birla Institute of Technology
Mesra, India

Veena Thakur
Pt. Ravishankar Shukla University
Raipur, India

Anita Tirkey
Department of Biotechnology
National Institute of Technology
Raipur, India

Aditya L Toppo
Department of Biotechnology
National Institute of Technology
Raipur, India

Deepak Kumar Tripathi
Department of Biotechnology
Indian Institute of Technology
Roorkee, India

1 Introduction to Green Molecules, to the Present Situation and to Previous Research on Green Polymer Nanocomposites

Neelamshobha Nirala

CONTENTS

1.1 THE IMPORTANCE OF GREEN MOLECULES

Synthesis of nanomaterials with the desired quality and properties is one of the primary concerns in current nanotechnology. But these synthesized nanomaterials should be environmentally friendly and maintain sustainability. To fulfil this criteria, concepts of green chemistry are incorporated in the synthesis of nanomaterials. Green chemistry involves the utilization of renewable sources, the minimization of waste production and the saving of energy (Jawed & Khan, 2018), thus helping to save the environment from excessive waste created by the accumulation of plastics, which is extremely harmful to living organisms, reduce the financial burden and save non-renewable sources for the future. In nanocomposites, the term "green" defines the materials that are biodegradable and renewable in nature. As a result, synthetic polymers are now being replaced by various degradable polymers (both from natural and synthetic sources) and green synthesis of nanoparticles takes place. The green synthesis of nanoparticles includes the selection of green solvents, nontoxic material as a stabilizer and an eco-friendly benign reducing agent.

1.2 THE IMPORTANCE OF GREEN NANOPARTICLES

In the past few decades, nanoparticles have gained huge interest due to their vast applications in the field, ranging from food packaging to paint and ceramics; from wastewater treatments to bioanalysis; in the agricultural industry; from diagnostic and therapeutic applications to the design of biosensors; in optical imaging, targeted drug delivery systems, tumor detectors and radiotherapy dose enhancers. Day by day its applications are amplified in various fields. Nanoparticles (NanoPs) are available in a range of sizes from 1 to 100 nms. Morphological diversity of NanoPs can be seen in the form of triangles, hexagons, pentagons, cubes, spheres, ellipsoids, nanowires and nanorods (Nadaroglu, Alayli, & Ince, 2017). Nanoparticles can be divided into three types based on their geometry as nanoparticles, nanoplatelets and nanofibers, nanotubes, nanorods or whiskers. Nanoparticles are particles where all three dimensions are in the order of nanometres. Colloidal silica, noble nanometals and metal oxides are examples. Nanoplatelets are where only the thickness is in the nanoscale range. Layered clay minerals, silicic acid and zirconium phosphates are just a few examples. When two dimensions of nanoparticulates are in the nanoscale range they are called nanofibers or tubes, of which carbon nanotubes are an example (Shchipunov, 2012).

NanoPs show improved properties (physical, chemical, optical, electrical and catalytic) and functions compared to their bulk materials because of their nanoscopic size and large surface area to volume ratio. Further, due to its wide applications, NanoP synthesis has been improved by the use of many physical and chemical methods such as plasma arcing, sputtering, laser ablation, spinning, sol–gel processes and so on. Apart from synthesizing NanoPs of various shapes, sizes and characteristics, these physiochemical processes also suffer from many limitations. These methods are highly expensive, labor intensive and require high temperatures, pressure and radiations, highly toxic reductants and stabilizing agents, which are harmful to the environment and living organisms. To overcome these limitations and to produce NanoPs that utilize inexpensive, environmentally benign reducing agents and non-toxic stabilizing agents that are biocompatible, green synthesis of NanoPs has emerged (Green biosynthesis of nanoparticles mechanism and applications, 2013).

Green synthesis of metal nanoparticles means the reduction of metal ions into NanoPs using biological systems. These biological systems include micro-organisms such as bacteria, fungi, yeast, diatoms and plant tissues (Singh et al., 2018). The metabolites present in the biological systems act both as reducing and stabilizing agents (which is required to avoid the agglomeration of NanoPs) for NanoP synthesis. This not only reduces the cost associated with the chemical process but also saves the environment from the toxic effect of chemicals. Biosynthesis of NanoPs is a one-step bio-reduction method which is quite simple, fast, economical and efficient, and can be easily scaled for industrial applications. Green synthesis of NanoPs showed enhanced antimicrobial, antifungal, antioxidant, anticancer and catalytic activity compared to chemical methods (Alananbeh, Refaee, & Qodah, 2017; Gold, Slay, Knackstedt, & Gaharwar, 2018; Vishwanatha et al., 2018). Among the various available methods of biosynthesis of NanoPs, utilization of plant extracts are comparatively more favorable. Cost of maintenance and cultivation is much less in plants, as waste disposal requires less effort and there is no need to take care of numerous parameters, as in the culture of bacteria. Also, the therapeutic effects of plant extracts are transferred into nanoparticles. NanoP synthesis can use the whole plant but it also has certain limitations: the shape and size of NanoPs may change depending upon their localization in plants, and the proficient extraction, isolation and purification of NanoPs from the whole plant is difficult, whereas phytochemicals from plant extract can reduce metal ions much quicker than bacteria and fungi, which need a longer incubation time. NanoPs synthesized by plant extracts contain a functionalized surface that can attach with organic ligands (Makarov et al., 2014; Green biosynthesis of nanoparticles mechanism and applications, 2013).

1.3 THE IMPORTANCE OF NATURAL POLYMERS

Natural polymers are polymers obtained from natural resources. Human beings have learned to use these from the very beginning of evolution for different purposes such as shelter, food and clothing. During the Second World War, due to the scarcity of natural polymers and their inherent limitations, synthetic polymers gained major interest. Synthetic polymers are produced by human beings using petroleum products. The use of various chemical methods helps in the economical production and tailoring of synthetic polymers with required properties. The resistance of synthetic polymers to various modes of degradation and vast petroleum reserves further increased its applications. However, the same advantages of synthetic polymers with overuse created many threats to the environment and living organisms. Further, the increasing costs of petroleum and its limited resources, energy demands, chemical toxicity, ingestion by marine life and aggregation of plastic wastes means that the recyclability of synthetic plastics is still in the developmental stage and their non-degradability demands the replacement of synthetic polymers by natural polymers as much as possible.

Natural polymers are degradable by various enzymes and hence it does not produce any adverse effects on the environment. Both the raw materials needed for the production of natural polymers as well as its production costs are economical. They are biocompatible, non-toxic and maintain CO_2 neutrality. Due to the natural origin of natural polymers, they are safe and free from any major side effects. Ease of availability, low organic salt content and no regional limitations on industrial production are other advantages of natural polymers. Polymers are called green when they exhibit properties such as eco-friendly, biodegradable and renewable (Jacob John & Thomas, 2012; Olatunji, 2016; Pillai, 2010).

1.4 THE IMPORTANCE OF NANOCOMPOSITES

Composites are the combination of two or more different materials into a single substance with a distinct interface between them. These physically and chemically different materials form two different phases, called the matrix phase and the reinforcement phase. The purpose of reinforcement material is to bear the load whereas the matrix phase is responsible for binding together and distributing the load among the reinforcement material. Composites usually show superior properties to their composing materials. On the basis of material used as matrix, composites are broadly classified as metal matrix composites, ceramics matrix composites and polymer matrix composites. So, in polymer matrix composites, a polymer is used as a matrix (Shekar & Ramachandra, 2018). In this chapter we will focus our study on polymer matrix composites. Composites are called nanocomposites (NCs) when at least one dimension of the reinforcement material is in the range of 1 to 100 nm. So in simple words, a combination of reinforcement elements as nanofillers (e.g. nanoparticles) with matrix is called a nanocomposite. These nanofillers can be of three types based on their morphology, as shown in Figure 1.1 (Fu, Sun, Huang, Li, & Hu, 2019). Nanocomposites offer many advantages, such as industrial significance in designing different structures and materials with better properties and flexibilities, less interfacial defects due to the size of nanoparticles and large volume to surface area. But the direct ingestion of nanoparticles by higher organisms and their entry into

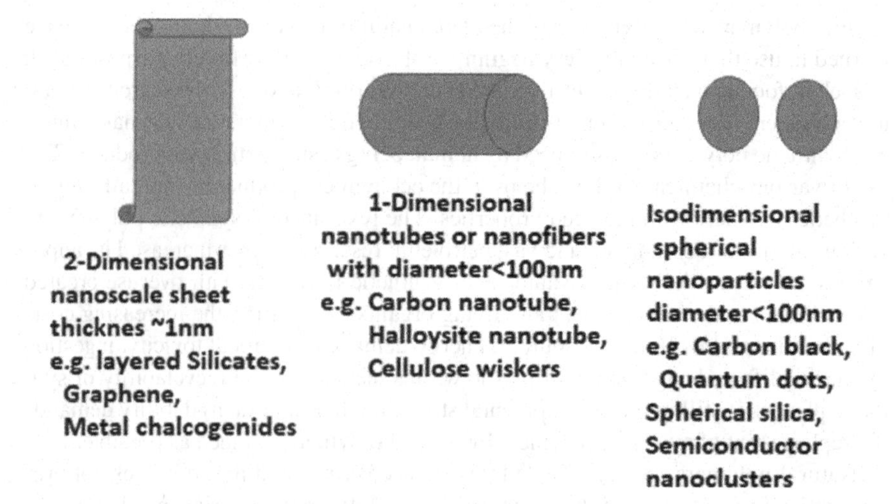

FIGURE 1.1 Schematic diagram of various types of nanofillers used as reinforcement phase.

the aquatic ecosystem creates a need for other composites, called bionanocomposites. Bionanocomposites are made up of biopolymers and its reinforcement materials include inorganic solids like metal nanoparticles, hydroxyapatite and clay (Arora, Bhatia, & Attri, 2018). Biopolymers have huge availability, they are biodegradable and renewable in nature and show solubility in polar solvents.

Nanocomposites are called green nanocomposites if they tend to be renewable and biodegradable. Most of the natural polymers (like cellulose, starch, lignin, poly-lactic acid and so on), as well as a few synthetic polymers, are degradable in nature but are not renewable in nature. As a result, many natural and degradable polymers have been studied for green nanocomposites. There is a very slight difference between the bionanocomposites and the green nanocomposites, as the former include any natural or biopolymers, and the latter are fixed for degradable natural polymers. They are often used as synonyms for each other. The properties of bionanocomposites depend upon their processes of fabrication, the properties of individual comprising components, the degree of mixing of two phases, types of filler materials and their orientations, size, shape, volume fraction and characteristics of nanoparticles, type of adhesion and the nature of the interphase developed at the matrix interface (Lone & Akter, 2018).

Overall, green polymer nanocomposites (GPNCs) are eco-friendly, cost-effective, offer excellent mechanical properties, innate nontoxic properties, antimicrobial properties, biocompatibility and biodegradability. Due to these properties, the applications of GPNCs have expanded to building blocks, development of electronic devices, self-cleaning applications, food packaging and biomedical applications, which further includes wound dressing, vaccinations, drug delivery systems, tissue engineering and many more (Bhawani, Bhat, Ahmad, & Ibrahim, 2018).

1.5 TYPES OF NATURAL POLYMERS AND THEIR PROPERTIES

Natural polymers (NPs) are a class of polymers obtained from living organisms such as plants, animals or microbial sources. They are also referred to as biopolymers as nature provides them during the growth cycles of living organisms (Ghanbarzadeh & Almasi, 2018). Due to their bio-origin they offer many favorable properties such as being biodegradable, renewable, biocompatible and non-toxic, safe and devoid of side effects. Moreover, their ease of availability and cost-effective production makes them economical. They offer wide applications in biomedical engineering, the pharmaceutical industry (as drug formulation excipients), the food industry, as biomaterial, hydrogel wound dressing, in the textile industries, as boiler fuel for the paper and pulp industry, as a binder, a dispersant, sequestering agents, food additives, drug delivery systems, bone cement and scaffolding for bone-tissue engineering and many others. There are various ways to categorize natural polymers such as on the basis of their sources of origin: they are broadly divided into plant-based NPs (examples are cellulose, starch, pectin etc.); animal-based NP (chitosan, collagen, fibrin, silk etc.); and NPs of microbial origin (xanthan, dextran or alginate) (Ghanbarzadeh & Almasi, 2018; Doppalapudi et al., 2015). NPs can also be segregated into six classes on the basis of their chemical composition, namely: polysaccharides, polypeptides, polynucleotides, polyesters, polyisoprenes and lignin (Olatunji, 2016). We have listed some of the commonly found NPs with their properties in Table 1.1.

TABLE 1.1
Types of Natural Polymers, Their Origin and Properties

Name of Polymer	Origin	Properties
Cellulose (Costa, Almeida, Vinhas, & Sarubbo, 2017; Kaushik, Sharma, & Ramakrishna, 2016; Menon, Selvakumar, Kumar, & Ramakrishna, 2017; Ruocco, Costantini, Guariniello, & Costantini, 2016; Vishakha, Kishor, & Sudha, 2012)	It is the most abundantly found polysaccharide obtained from the cell walls of plants, whereas its other sources include algae (Cladophorah and Microdyction), bacteria (genus gluconacetobacter) and marine animals (Ascite family commonly called tunicates).	Insoluble in water, crystalline in structure, high molecular weight, poor thermal stability, high tensile strength, infusible and insolubility in most solvents, non-compatibility with hydrophobic polymers, hydrophilic in nature, and broad chemical modifying capacity.
Hemicellulose (Ghanbarzadeh & Almasi, 2018; Hu, Du, & Zhang, 2017; Muchlisyam & Harahap, 2015; Olatunji, 2016)	It is also a polysaccharide obtained from plants.	Amorphous branched structure, less mechanical strength than cellulose, hydrophilic, soluble in alkali medium and easily hydrolysed in acid.
Lignin (Agrawal, Kaushik, & Biswas, 2014; Ralph, 1998)	It is the most abundant aromatic polymer obtained from plants.	Complex structure, hydrophobic, insoluble in most solvents, resistant to microbial degradation, optically inactive, it provides biomechanical strength, it has a combustion heat of 26.6KJ/g.
Starch (Ghanbarzadeh & Almasi, 2018; Olatunji, 2016; Doppalapudi et al., 2015; Vishakha et al., 2012)	It is a polysaccharide obtained from plants.	It is an odourless and white tasteless powder in its pure form, has rapid enzymatic degradation, and strong hydrophilic and low mechanical strength.
Pectin (Olatunji, 2016; Vishakha et al., 2012)	It is a complex anionic heteropolysaccharide obtained from plants.	Resistant to enzymatic attack of upper GI tract.
Inulin (Mensink, Frijlink, Van Der Voort Maarschalk, & Hinrichs, 2015)	It is a fructan-type polysaccharide obtained from plants like chicory, Jerusalem Artichoke, garlic, onion etc.	Similar to pectin it is also resistant to upper gastrointestinal tract enzymatic attack.
Glucomannan (Behera & Ray, 2016)	It is a polysaccharide obtained from the tuber of Amorphophallus konjac K. Koch plant.	High viscosity, soluble in water and has high water-holding capacity.

(*Continued*)

TABLE 1.1 (CONTINUED)
Types of Natural Polymers, Their Origin and Properties

Name of Polymer	Origin	Properties
Chitin (Elieh-Ali-Komi & Hamblin, 2016; Islam, Bhuiyan, & Islam, 2017; Rinaudo, 2006)	It is the second most abundant mucopolysaccharide available in nature. It is mainly found in the exoskeleton of marine crustaceans (crab and shrimp), mollusks, and insects along cell walls of fungi.	It is crystalline in structure, insoluble in most of the solvents, has gel-forming properties, is inelastic, has affinity for proteins and antibacterial properties, is biocompatible, nontoxic, inert in the gastrointestinal tract of mammals, and has an ability to absorb heavy metal ions.
Agar (Vishakha et al., 2012; Rhein-Knudsen, Ale, & Meyer, 2015)	Hydrocolloid polysaccharide extracted from agarophyte members of the Rhodophyta.	It has an excellent, unique and reversible gelling property, where it can be converted into gel form by heating up to 32 to 43°C and does not melt below 85°C.
Guar gum (Doppalapudi et al., 2015)	Plant-based polysaccharide.	High molecular weight and viscosity, gelling property, insoluble in organic solvents.
Alovera gel (Doppalapudi et al., 2015)	Plant-based polysaccharide.	Promotes wound healing, offers antifungal activity, anticancer, anti-inflammatory, hypoglycaemic effects, immunomodulatory and gastro-protective properties.
Silk fibroin (Doppalapudi et al., 2015)	Fibrous protein obtained from the larvae of insects.	Exhibits slow biodegradability, chemical and elasticity properties biocompatibility and good mechanical strength.
Alginates (microbial) (Doppalapudi et al., 2015; Olatunji, 2016)	Anionic polysaccharide found in brown seaweed and marine algae.	High molecular weight, non-toxic polymers. They form water-soluble salts with low molecular weight amines, monovalent cations and quaternary ammonium compounds, Used as a stabilizing agent, voscosifier, drug carrier and as binding agent.
Carageenane	Polysaccharide extracted from edible red seaweeds of Rhodophycea class	They form water soluble salts and highly viscous solutions.

(Continued)

TABLE 1.1 (CONTINUED)
Types of Natural Polymers, Their Origin and Properties

Name of Polymer	Origin	Properties
Albumin (Doppalapudi et al., 2015)	Family of globular proteins.	Non-toxic, non-immunogenic, experiences heat denaturation, not glycosylated, soluble in water and moderately soluble in concentrated salt solutions.
Collagen (Doppalapudi et al., 2015)	It is one of the most abundant mammalian proteins.	Weak antigenicity, biocompatibility, formation of fibres through self-aggregation and cross-linking, poor mechanical strength.
Gelatin (Doppalapudi et al., 2015)	A protein acquired by hydrolysis of collagen, derived from the skin, white connective tissue, and bones of animals.	Water-soluble, viscous, wide isoelectric range, physicochemical properties are dependent on the method of extraction.
Polyhydroxyalkanoates (Doppalapudi et al., 2015)	Polyesters originated from bacteria.	Piezoelectric properties, biocompatibility and biodegradability by simple hydrolysis of ester bonds in aerobic conditions.
Fibrin (Doppalapudi et al., 2015)	Non-globular plasma protein obtained from fibrinogen.	High adhesive and tensile strengths, biocompatibility and resorption, ability to improve cellular interactions.
Poly-γ-Glutamate (Doppalapudi et al., 2015)	Anionic, biodegradable homo-polyamide produced by microbial fermentation.	Water-soluble.
Polyphosphate (Doppalapudi et al., 2015)	Polyanhydrides extracted from many species of bacteria, yeast, and plant and animal cells.	Acts as a substitute for adenosine triphosphate (ATP) in various kinase reactions, plays a role in cellular regulation, and also participates in heavy metal detoxification based on metal chelating property.
Poly-ε-caprolactones (PCL) (Ghanbarzadeh & Almasi, 2018)	Aliphatic polyester.	Soluble in a wide range of organic solvents, semi-crystalline, low glass transition temperature of -62°C, low melting temperature of 57°C and high thermal stability.

Natural polysaccharides have shown significant applications in drug delivery and pharmaceutical industries because of their chemical structure. Sometimes they show water solubility by forming multiple hydrogen bonds, whereas they become water-insoluble by forming intermolecular hydrogen bonds with each other to produce a crystal structure. Plant polysaccharides have gained new interest due to their biodegradable and biocompatible natures. The algae-based polysaccharides, due to their extraordinary gelling and thickening properties, are the main hydrocolloids used as texturing agents for food and non-food applications. Mammalian polysaccharides have excellent mucoadhesive capacity, which may help in the enhancement of the absorption of drugs and proteins via mucosal tissues. Even the microbial polysaccharides provide a more economic alternative source to the current polymer-exploiting industries (Bhatia, 2016).

1.6 WHY WE COMBINE NATURAL POLYMERS WITH GREEN NANOPARTICLES

In the past few decades, polymer nanocomposites have been able to attract lots of research attention due to their light weight, great strength, stiffness and ease of fabrication. Natural polymers are rich renewable resources, biocompatible and also offer many environmental benefits such as being biodegradable and helping to reduce greenhouse gas emissions, dependency on fossil fuels, the deposition of harmful solid wastes and the discharge of waste associated with polymer production and the surge in employment in the agriculture sector (Shekar & Ramachandra, 2018). Nevertheless, although natural polymers have a range of physical properties suitable for wide applications, they have limited potential to replace synthetic polymers. They suffer from weak mechanical properties and microbial contamination, lack properties required for material fabrication and have batch to batch variation, high gas and water permeability and a low heat-degradation temperature. Natural polymers may contain functional groups that usually do not produce toxic effects but may generate undesirable immunological effects. There comes the need of nanoparticles whose addition into the polymer matrix not only improves its strength but also affects many physical properties. Nanocomposites can accrue the benefits of nanoparticles only when they are efficiently dispersed in the polymer matrix. Nanoparticles possess a large surface area that provides sufficient contact for effective interactions between the two phases (matrix and reinforcement), which further increases with the reduction in dimensions. But achievement of homogenous distribution is not so easy, as nanoparticles have a strong tendency of agglomeration due to their large surface and high surface energy (Shchipunov, 2012). Overall, nanoparticles allow the tuning of the structure, function and properties of nanocomposites. Even a single nanoparticle can be added to create a specific functionality lacking in natural polymers. Different combinations of natural polymers and nanosized particles are experimented with to achieve the desired properties for wider applications.

In the next chapters we will talk about various green sources for the production of green polymeric nanocomposites. Various green routes are sought to produce nanocomposites, and their varied applications and limitations need to be overcome for industrial production. Different methods are used to improve their surface properties by using green technology and biodegradation of nanocomposites and their future aspects.

1.7 THE PRESENT SCENARIO AND PREVIOUS RESEARCH ON GREEN POLYMER NANOCOMPOSITES

Green synthesis of polymer nanocomposites not only helps in improving their properties but also provides a path which is environmentally friendly, cost-effective, biodegradable and presents wide application in biomedical engineering. We have included a few examples where green synthesis was used to improve the qualities of either nanoparticles or polymer nanocompositions (PNCs). The detected presence of heavy metal ions such as lead, mercury and cadmium in the water, even in trace amounts, is considered to be environmental pollution. To assess the quality of real water samples Muthivhi el al. prepared a colorimetric sensor using a simple and cost-effective green synthesis of silver nanoparticles, which eliminates the requirement of time-consuming procedures and skilled persons in performing analytical techniques. They utilized gelatin as a non-toxic capping agent to avoid agglomeration and maltose as a reducing agent. They reported the minimum concentration detection of mercury ion as 10^{-12} M by silver nanoparticles (Muthivhi, Parani, May, & Oluwafemi, 2018). Another example is the sensing of aqueous ammonia concentration in liquid phase at room temperature, which is also toxic to living organisms. They prepared an optical sensor from the green synthesis of gold nanoparticles by using guar gum as a reducing and capping agent. They reported that their prepared colloidal solution is highly stable and able to detect ammonium in ppb (parts per billion) levels (Pandey, Goswami, & Nanda, 2013). They eliminated the need for high temperatures and sophisticated instruments for the detection of ammonia. Controlled synthesis of magnetic nanoparticles has also gained interest due to its wide applications in magnetic storage devices, magnetic recorders, MRI contrast enhancement agents, sensors and others. A one-pot green synthesis of high-purity magnetite nanocrystal from sodium alginate and $FeCl_3$ in alkaline solution was reported (Gao et al., 2008). A nontoxic nanocomposite which is water-soluble was prepared from silver nanoparticles in a polymer matrix, which showed excellent antimicrobial activity against several strains of gram positive and gram negative bacteria as well as fungi (even though the antimicrobial property of NCs toward bacteria is two to three times stronger than fungi). They included poly(1-vinyl-1,2,4-triazole), a nontoxic water-soluble polymer, as the stabilizing matrix, and glucose and dimethyl sulfoxide (DMSO) as the reducing agents. Here DMSO also works as a good solvent for PVT and facilitates the reaction by providing homogenous distribution of AgNPs into polymer matrix (Prozorova et al., 2014).

Targeted drug delivery systems is a growing area of research, especially in cases of cancer, to overcome the limitations associated with conventional drug release, such as uncontrolled release of a drug, poor specificity, nonspecific distribution, low bioavailability and rapid clearance, which ultimately retard the therapeutic efficiency of drugs. Yiu and his colleagues introduced a protocol to assemble a complex nanocomposite for a highly specific drug delivery system with combined magnetic and antibody targeting properties. To achieve these dual-targeted mechanisms they formed a multifunctional porous material using mesoporous molecular sieve SBA-16 as temples, which was impregnated by magnetic iron oxide, functionalized by amine groups and finally tagged by rabbit immunoglobin G in order to establish a generic binding mechanism for antibodies into the nanocomposite (Yiu, Niu, Biermans, Van Tendeloo, & Rosseinsky, 2010).

Some polymers, such as poly(N-isopropylacrylamide) and PEO-PPO-PEO Triblock Copolymers (Pluronic) are temperature-sensitive in nature. That is, they show a sharp change in their aqueous solution properties with respect to temperature and thus allow controlled, triggered release of drugs (He & Alexandridis, 2017). Recently, Lombardo et al. have reviewed the recent advances of smart nanocarriers obtained from organic and inorganic nanostructured materials for efficient transport and controlled release of drug molecules. They explained the major factors that strongly affect the design of nanocarriers, their emerging approaches, limitations and challenges as an efficient drug delivery system (Lombardo, Kiselev, & Caccamo, 2019).

Many polymers, such as polypyrrole, polyacetylene, polyaniline and polythiophene, exhibit both optical and electrical properties comparable to metals and semiconductors but still offer ease of fabrication and good processability compared to metals. They are called conducting polymers. The suitability of well-known conducting polymers for biomedical applications was reviewed by Kaur, Adhikari, Cass, Bown, & Gunatillake (2015). Conducting polymers offer many applications in biomedical engineering, such as neural simulation and sensing, medical implants, bioimaging, biosensing and tissue engineering but still suffer from biocompatibility and poor solubility in common organic solvents. To overcome this major limitation, two approaches have come forward. First the blending of conducting polymer with biodegradable polymer (Chen, Yu, Guo, Ma, & Yin, 2018) and second, the green synthesis of conducting polymers. Good reviews about conducting polymers, their synthesis, properties and biomedical application can be found in Balint, Cassidy, & Cartmell (2014) and Kenry and Liu (2018).

Another area of exploration of nanocomposites which are commonly used in drug delivery systems and tissue engineering is that of nanocomposite hydrogels. Hydrogels are porous network structures with swelling/dwelling behavior, cell adhesion and excellent biocompatibility but lacking in mechanical properties and multifunctionality. Biosynthesis of hydrogels from naturally abundant polymer (cellulose and chitin), their properties and applications were reviewed in Shen, Shamshina, Berton, Gurau, & Rogers (2015). By combining hydrogels with various organic and inorganic nanofillers, there is improvement in their properties, such as mechanical strength, tunable cell adhesion, biodegradability and self-healing (Song, Li, Wang, Liao, & Zhang, 2015). Recently, hydrogels were prepared using natural polymers (guar gum) to increase their water retention properties for agricultural application (Abdel-Raouf, El-Saeed, Zaki, & Al-Sabagh, 2018). A green synthesis of hydrogel was achieved even with bacterial cellulose derived from acetic acid bacteria, where the water retention of hydrogel was shown to be dependent on the total solid content of bacterial cellulose (Chaiyasat et al., 2018).

1.8 COMPARATIVE STUDY OF THE SYNTHESIS PROCESS OF PNCs

Polymer nanocomposites offer many advantages, such as being light weight and fire retardant, having improved thermal, mechanical and barrier properties, and exhibiting excellent flammability properties (Dantas de Oliveira & Augusto Gonçalves Beatrice, 2019). Improvement in the properties of PNCs depends upon many parameters, which include size, shape, type of filler materials and their orientations, the degree of mixing of filler incorporated, their concentration, shape and size, the nature of biopolymers, the interaction of fillers with polymer matrix and also on the synthesis process. Based on the degree of dispersion of nano-sized layer fillers on polymer matrix NCs they can be divided into three classes: microcomposite or phase-separated (polymers are not able to penetrate inside the silicate layer), intercalated (when several polymer chains seep between interlayers) and exfoliated (polymers are extensively penetrated such that delamination of the layer structure occurs) (Liu et al., 2006). In general, the synthesis process can be divided into three approaches. They are in-situ polymerization, melt intercalation and exfoliation adsorption.

1.8.1 In-Situ Polymerization

In this method, first the nanofillers or layered silicate are soaked in the liquid monomer solution, so that low molecular weight monomer leaks in between the layers of silicate. As a homogenous mixture is formed, an initiator is added and it is exposed to suitable conditions such as radiation or high temperature so that polymerization can take place. To increase the crosslinking of the polymer chain, curing of the polymer needs a suitable catalyst, but that may modify the physical properties as well as affect the functionality of produced NCs. This method is the most viable and promising approach, as it provides efficient dispersion of fillers as well as enhances their interfacial adhesion with polymer matrix. The many advantages of this method include straightforward and carry-out processing and thermodynamic compatibility at the matrix reinforcement interface, an important method for the synthesis of insoluble and thermally unstable polymer nanocomposites which is not suitable by solution or melt processing. This method also provides fine tuning of the synthesized polymers through the proper selection of input materials and their synthesis condition. However, it is difficult to control intragallery polymerization (Camargo, Satyanarayana, & Wypych, 2009). Polymerization results in intercalated and exfoliated NCs. This method usually involves the synthesis of composites based on thermoplastic and thermoset.

1.8.2 Melt Intercalation

This method involves annealing the polymer matrix above the softening temperature of polymer, then mechanical mixing of polymer with an appropriate modified nanofiller and finally kneading the composites to achieve uniform distribution. This method is applied only for thermoplastics such as wheat gluten, whey, gelatin, soy and PHA. In the case of starch, which is not a true thermoplastic in its native state, plasticizers are added at temperatures of 90 to 180°C under shear stress. This is an environmentally friendly method, as it does not require any toxic solvent. Moreover, this method is compatible with industrial processes such as polymer extrusion, blending processes and injection molding, which makes it economical. One major limitation of this method is the use of high temperatures and pressure, which can damage the surface modification of fillers, and pressure which may degrade the polymer during extrusion. In the case of clay as filler, fine tuning of layered silicate surface chemistry is important for compatibility of fillers with polymer. Since the proper dispersion of filler particles depends upon polar interaction of polymers and filler, polar compatibilizer can be added. In this method the level of dispersion depends upon applied shear stress, and thermodynamic interaction between filler and polymer matrix, so optimization of these parameters, along with residence time, processing equipment and processing conditions, is required. Biopolymers that are not appropriate for exfoliate adsorption and in-situ polymerization are processed by this method (Fawaz & Mittal, 2015; Saini, Bajpai, & Jain, 2018; Shchipunov, 2012).

1.8.3 Exfoliation Adsorption

This method utilizes water or other chemical solvents such as toluene, chloroform or acetate for the process. First, the silicate layer is mixed with a solvent, which makes it swellable and dispersed in the solvent. Then this swellable silicate is mixed with polymer solution, which causes thepolymer chains to intercalate and displace the solvent within the silicate layers. Heating this complex at high temperature evaporates the solvent and eventually a multilayered structure is formed by trapping the polymer between the silicate layers. A schematic diagram of this process is shown in Figure 1.2. This method is widely used for the synthesis of

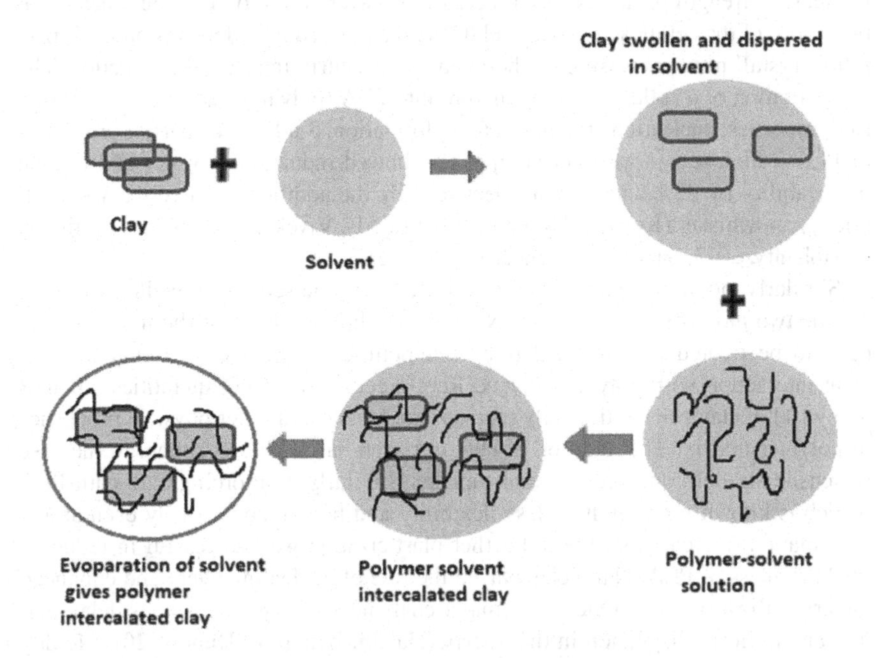

FIGURE 1.2 Schematic diagram of exfoliation adsorption process.

intercalated NCs from water-soluble polymers with low or no polarity such as polyethylene oxide and polyvinyl alcohol. This method is hazardous to the environment due to its use of toxic chemical solvents (Camargo et al., 2009; Fawaz & Mittal, 2015).

1.9 COMPARATIVE STUDY OF THE PROPERTIES OF NPCS

Polymers in general suffer from low mechanical strength and low modulus; to alleviate such limitations, fillers are reinforced in them. The addition of bio fillers into a biopolymer matrix can considerably influence the properties of resulting NCs, revealing improved electrical, thermal, mechanical, rheological, catalytic and optical properties, fire retardancy, a reduction in gas permeability and an increase in heat distortion temperature. Among many biofillers, clay and nanocellulose are the most exploited bio-origin nanofillers. Cellulose is the most abundant renewable polymer available in the biosphere. Nanocellulose is stabilized by hydrogen bonds with high levels of crystallinity, which makes cellulose nanofibers an ideal reinforcing material in polymeric materials. Reinforcement of nanocellulose in biopolymer enhanced its water resistance. The mechanical properties of cellulose NCs can be significantly affected for two reasons: first, the noncompatibility of the polar fiber and the nonpolar matrix, which results in poor fiber/matrix interfacial; second, the poor dispersion of fibers in the matrices that causes agglomeration of cellulose nanofibers, especially with thermoplastic matrix. To overcome these limitations many surface treatments have been done (Thakur & Thakur, 2015).

Cellulose nanofibers are as strong as steel; with just 5 to 10% of nanofiber loading, the tensile strength of the NCs can increase by two to three times. The mechanical properties of the cellulose nanocrystal (CNC)-reinforced NCs depends on the aspect ratio, crystallinity, processing method and CNC–matrix interfacial interaction. The incorporation of a cellulose nanostructure into PLA (polylactic acid) improved barrier properties, nucleation effects and foam formation. Study of the barrier properties of PLA–cellulose nanowhisker composites showed reduction in water and oxygen permeability by 82% and 90%, respectively, by the addition of merely 3 wt% cellulose nanowhisker (Kvien & Oksman, 2007; Reddy, Vivekanandhan, Misra, Bhatia, & Mohanty, 2013; Sanchez-Garcia & Lagaron, 2010).

Similarly montmorillonite (layered silicate clays) and sepiolite (needle-like clays) are the two mostly studied clays in NCs. These clays are hydrophilic in nature and need to be treated (transformed into organophilic) before incorporation for efficient interaction with polymer matrix. Reinforcement of small quantities (about 3 to 6 wt%) of clay can significantly increase the tensile and other physical properties of polymer matrix. The addition of clay in starch-based NCs is shown to increase its tensile strength and reduce water uptake. Similarly, reinforcing clay with PLA (which is known for exhibiting less flexibility and breaks down easily even at low deformation) increases stiffness. Further plasticization was successful in reducing the brittleness of PLA. Nano clay played the role of nucleating agent and enhanced the crystallization rates. One of the major challenges of clay layers is its delamination and uniform dispersion in the matrix (Habibi, Benali, & Dubois, 2015; Reddy et al., 2013).

1.10 COMPARATIVE STUDY OF THE QUALITY ASESSMENT OF NCS

Characterization of NCs is essential to understand many aspects of NCs such as their morphology, the quality of dispersion and the orientation of fillers in polymer matrix, interaction of fillers with polymers, the effect of filler surface modification on dispersion and other properties (Bokobza, 2018). Moreover, their application also depends upon their characteristics (that is, their intrinsic properties such as material properties, size, structure and nature of element present), and to study those characteristics we have many assessment techniques (Aimé & Coradin, 2017). These analytical techniques are broadly used to study the structure or morphology of NCs, their mechanical properties, thermal properties and so on. To study the structure or morphology different spectroscopic (infrared spectroscopy, Raman spectroscopy, wide-angle X-ray diffraction (WARD)) and microscopic (scanning electron microscope (SEM) and transmission electron miroscope) methods are used. Sometimes a combination of techniques is employed to obtain better information, for example, WARD is often used with SEM or with Fourier-transform infrared spectroscopy (FTIR). Similarly, for evaluation of thermal properties methods such as differential scanning colorimeter, thermogravity analysis, thermomechanical analysis, dynamic modulus analysis and rheology are used. Thermal analysis helps to study the temperature-based changes in the properties of materials (Lagashetty & Venkataraman, 2005). Many test are available to study the structural response of NCs to the applied mechanical stress: for example, rheology, which studies the way matter flows in response to an applied shearing force; and other convention tests such as compression, traction, bending and relaxation which assess nonlinear mechanical properties. We will explain a few of the structural and thermal analysis methods.

1.10.1 INFRARED SPECTROSCOPY

This technique is based on absorption spectroscopy. Here a beam of infrared (IR) light is passed through the polymer nanocomposites. If the frequency of IR light matches with the vibrational frequency of the chemical bonds of the sample, absorption occurs. Evaluation of transmitted light shows frequency-based absorption of IR light. This method is used to study the properties of polymers based on their vibrational properties. It is useful for the study of thin layer of polymers, as IR light is easily absorbed by the functional groups present in polymers. This method is quite simple to use and requires only a good contact between the sample and the surface of observation and provides structural information at molecular level. The IR spectrum contains various spectral bands corresponding to the functional groups present in any chemical substance (Bokobza, 2018; Lagashetty & Venkataraman, 2005).

1.10.2 X-RAY DIFFRACTION

This technique is useful to obtain the crystallinity of nanoparticles. This shows the amount of crystalline and amorphous regions in nanocomposites. An increase in crystallinity shows improved mechanical properties. Diffraction peaks were obtained at the region of the highest degree of crystallinity (Abdellaoui, Bouhfid, & Qaiss, 2017).

1.10.3 Wide-Angle X-Ray Diffraction (WAXD)

WAXD is an easy-to-use characterization technique for crystalline materials. It is frequently employed for the characterization of polymer nanocomposites (PNCs) and, most importantly, to understand the effect of nanofillers in PNCs. X-ray diffractograms are useful in analyzing the size of crystallite, percentage crystallinity, degree of dispersion of nanofillers in a polymer matrix, and the nucleating effect of nanofillers. When an X-ray beam falls on a crystalline substance through a slit pore, it gets diffracted and produces a diffraction pattern. Diffracted radiation undergoes constructive or destructive interference. Diffraction peaks obtained from constructive interference are collected and analyzed. In X-ray diffraction, crystalline materials display sharp maxima (peaks) at respective diffraction angles, whereas amorphous solids only show broad maximum (hump). Apart from characterization of various crystalline materials, WAXD can be used for the study of structures and the relative orientations of the individual crystals, for purity analysis of crystalline samples and determination of unit-cell dimensions, lattice mismatch and dislocation density. Even though X-ray diffraction is simple to interpret and is a non-destructive technique, there is a chance of high misinterpretation in nonhomogeneous samples; a probability of peak overlay still exists, it also requires a reference for inorganic compounds and lack of depth profile information may be possible (Bishnoi, Kumar, & Joshi, 2017).

1.10.4 Fourier-Transform Infrared

FTIR provides the information of the vibration and rotation of the chemical bonds present in the molecule and define the functional groups present in it. The absorption peaks present in the infrared spectrum correspond to the frequencies of vibrations between the bonds of the atoms making up the material. As each material has its own unique combination of atoms, different molecules present a different infrared spectrum. FTIR is a fast technique with less energy consumption and offers better sensitivity due to simultaneous observation of all frequencies. It also carries a few limitations, such as minimal elemental information; that molecules must be active in the infrared region and a requirement for relative transparency of the background solvent or matrix in the spectral region of interest.

1.10.5 Scanning Electron Microscope

A SEM determines the size, shape and surface morphology of nanoparticles by direct visualization. But it provides inadequate information about size distribution and true population average. For SEM observation, nanoparticles need to be converted into dry powder and then analyzed by fine beams of electrons. Secondary electrons emitted from sample surfaces define the characteristics of the nanoparticles. This technique has certain limitations, such as electron beams often damaging the polymer of nanoparticles, being time-consuming and expensive, and often needing complementary information with regard to sizing distribution (Bhatia, 2016).

1.10.6 Transmission Electron Microscope

The TEM is a powerful tool that provides diffraction, spectroscopic and imaging information at the nanometer level. It helps in studying the dispersion of nanoparticles within a polymer matrix. It allows a qualitative understanding of the size, shape, defective structures and the crystal, the internal structure such as pore size, wall thickness and pore order, and spatial distribution of the various phases through direct visualization. For TEM observation the sample needs to be in ultra-thin form, which makes it time-consuming and complex (Bhatia, 2016).

1.10.7 Zeta Potential

Zeta potential is an index to assess the interaction between colloidal particles and used to evaluate the stability of the nanoparticles in colloidal suspension. Zeta potential indicates the overall charge a particle acquires in a specific medium, which determines the interaction of nanoparticles with its environment or its agglomeration. The stability of the colloidal system is indicated by a minimum of $\pm$30mV of zeta potential (Baliah, Muthulakshmi, Sheeba, & Priyatharsini, 2018). Too large (both positive and negative) a zeta potential indicates repulsion between the particles, which helps to maintain dispersion stability, whereas a low value shows strong attractive forces between the particles. Both pH and ionic strength of solution have a strong influence in measurement of zeta potentials and need to be mentioned.

1.10.8 Differential Thermal Analysis

In this method the material that needs to be studied is heated with an inert reference material in the same oven. Both the materials undergo identical thermal cycles that are under the same heating or cooling conditions. The difference in temperature between the required and the reference material is recorded during the thermal cycles and plotted with respect to time or temperature. The plotting of temperature versus obtained temperature differences is called the DTA curve or thermogram. The area under the thermogram peak is a measure of enthalpy change. As the required material absorbs or emits heat due to a change in state, its temperature changes with respect to the reference material; this difference in temperature is reported. The DTA curve provides transformation-based information such as glass transition temperature (the temperature at which material is converted from ductile to brittle state), sublimation temperature (transition from solid to direct gas phase), melting and crystallization (Corcione & Frigione, 2012).

1.10.9 Thermo Gravimetric Analysis

TGA measures the change in weight of substance with the change in temperature in a controlled environment. Similar to DTA, here change in weight is plotted against temperature. TGA measurement is environment-dependent, which can be inert or oxidative. TGA is employed to determine the thermal stability, purity of the sample, humidity and kinetic parameter. Many factors affect the TGA thermal curve, as sample size and heating rate can raise the sample decomposition temperature; similarly, particle size or gas-flow rate may change the progress of reaction (Corcione & Frigione, 2012).

1.11 COMPARATIVE STUDY OF THE DRAWBACKS OF PNCS

The major drawback of PNCs is the incompatibility of the hydrophilic fibers with the hydrophobic polymer matrix that gives uneven dispersion of fibers within the matrix, which leads to poor mechanical properties. The required mechanical properties are achieved by improving the affinity and adhesion between polymer matrix and fibers through coupling agents or compatibilizers. Another limitation is the search for a biopolymer with good mechanical properties to be used as matrix. The large size of nanoparticles (not more than 500 nm) may reduce the ductility, which needs to be taken care of by adding suitable plasticizers. To attain the desirable properties, it is important that filler particles must be separated into the right shape, orientation and layer structure. The wrong shape and layer structure may raise the cost and workload of machinery, whereas orientation has an influence on the tensile properties of NCs (Inamuddin, 2017).

1.12 CONCLUSION

The use of biopolymers and natural fibers offers an environmentally friendly option for synthesis of nanocomposites. Moreover, the drawbacks associated with biopolymers, such as poor mechanical strength, poor barrier properties and thermal stability can be overcome by incorporation of nanosized bio-based reinforcement. Further, the surface treatment of fillers enhances the mechanical properties of nanocomposites by improving the dispersion of filler over matrix and the reduction of agglomeration. Improvement in the mechanical, thermal, antibacterial and antimicrobial properties will widen the application of green polymer nanocomposites in the biomedical field.

REFERENCES

Abdel-Raouf, M. E., El-Saeed, S. M., Zaki, E. G., & Al-Sabagh, A. M. 2018. Green chemistry approach for preparation of hydrogels for agriculture applications through modification of natural polymers and investigating their swelling properties. *Egyptian Journal of Petroleum*, **27**(4), 1345–1355. doi:10.1016/j.ejpe.2018.09.002

Abdellaoui, H., Bouhfid, R., & Qaiss, K. 2017. Preparation of bionanocomposites and bionanomaterials from agricultural wastes. In M. Jawaid, S. Boufi, & A. H. P. S. Khalil (Eds.), *Cellulose-Reinforced Nanofibre Composites* (1st ed.), (pp. 341–372). Amsterdam, the Netherlands: Elsevier.

Agrawal, A., Kaushik, N., & Biswas, S. 2014. Derivatives and applications of lignin – An insight. *The Scitech Journal*, **1**(7), 30–36

Aimé, C., & Coradin, T. (Eds.). 2017. *Bionanocomposites Integrating Biological Processes for Bioinspired Nanotechnologies* (1st ed.). Hoboken, NJ: John Wiley & Sons.

Alananbeh, K. M., Refaee, W. J. Al, & Qodah, Z. A. 2017. Antifungal effect of silver nanoparticles on selected fungi isolated from raw and waste water. *Indian Journal of Pharmaceutical Sciences*, **79**(4), 559–567. doi:10.4172/pharmaceutical-sciences.1000263

Arora, B., Bhatia, R., & Attri, P. 2018. Bionanocomposites: Green materials for a sustainable future. In C. M. Hussain & A. K. Mishra (Eds.), *New Polymer Nanocomposites for Environmental Remediation* (pp. 699–712). Amsterdam, the Netherlands: Elsevier.

Baliah, N. T., Muthulakshmi, P., Sheeba, P. C., & Priyatharsini, S. L. 2018. Green synthesis and characterization of nanocomposites. *International Research Journal of Engineering and Technology* , **5**(12), 179–186.

Balint, R., Cassidy, N. J., & Cartmell, S. H. 2014. Conductive polymers: Towards a smart biomaterial for tissue engineering. *Acta Biomaterialia*, **10**(6), 2341–2353. doi:10.1016/j.actbio.2014.02.015

Behera, S. S., & Ray, R. C. 2016. Konjac glucomannan, a promising polysaccharide of *Amorphophallus konjac* K. Koch in health care. *International Journal of Biological Macromolecules*, **92**, 942–956. doi:10.1016/j.ijbiomac.2016.07.098

Bhatia, S. 2016. *Natural Polymer Drug Delivery Systems Nanoparticles, Plants and Algae*. Cham, Switzerland: Springer.

Bhawani, S. A., Bhat, A. H., Ahmad, F. B., & Ibrahim, M. N. M. 2018. Green polymer nanocomposites and their environmental applications. In M. Jawaid & M. M. Khan (Eds.), *Polymer-Based Nanocomposites for Energy and Environmental Applications* (pp. 617–633). Amsterdam, the Netherlands: Elsevier.

Bishnoi, A., Kumar, S., & Joshi, N. 2017. Wide-Angle X-ray Diffraction (WXRD): Technique for characterization of nanomaterials and polymer nanocomposites. In S. Thomas, R. Thomas, A. K. Zachariah, & R. K. Mishra (Eds.), *Microscopy Methods in Nanomaterials Characterization* (pp. 313–337). Amsterdam, the Netherlands: Elsevier.

Bokobza, L. 2018. Spectroscopic techniques for the characterization of polymer nanocomposites: A review. *Polymers*, **10**(7), 21. doi:10.3390/polym10010007

Camargo, H. P. C., Satyanarayana, K. G., & Wypych, F. 2009. Nanocomposites: Synthesis, structure, properties and new application opportunities. *Material Research*, **12**(1), 1–39.

Chaiyasat, A., Jearanai, S., Moonmangmee, S., Moonmangmee, D., Christopher, L. P., Alam, M. N., & Chaiyasat, P. 2018. Novel green hydrogel material using bacterial cellulose. *Oriental Journal of Chemistry*, **34**(4), 1735–1740. doi:10.13005/ojc/340404

Chen, J., Yu, M., Guo, B., Ma, P. X., & Yin, Z. 2018. Conductive nanofibrous composite scaffolds based on in-situ formed polyaniline nanoparticle and polylactide for bone regeneration. *Journal of Colloid and Interface Science*, **514**, 517–527. doi:10.1016/j.jcis.2017.12.062

Costa, A. F. S., Almeida, F. C. G., Vinhas, G. M., & Sarubbo, L. A. 2017. Production of bacterial cellulose by *Gluconacetobacter hansenii* using corn steep liquor as nutrient sources. *Frontiers in Microbiology*, **8**(Oct), 1–12. doi:10.3389/fmicb.2017.02027

Corcione, C. E., & Frigione, M. 2012. Characterization of nanocomposites by thermal analysis. *Materials*, **5**(12), 2960–2980. doi:10.3390/ma5122960

Dantas de Oliveira, A., & Augusto Gonçalves Beatrice, C. 2019. Polymer nanocomposites with different types of nanofiller. In S. Sivasankaran (Ed.), *Nanocomposites: Recent Evolutions* (pp. 103–128). IntechOpen.

Elieh-Ali-Komi, D., & Hamblin, M. R. 2016. Chitin and chitosan: Production and application of versatile biomedical nanomaterials. *International Journal of Advanced Research*, **4**(3), 411–427. https://www.ncbi.nlm.nih.gov/pmc/articles/PMC5094803/

Fawaz, J., & Mittal, V. 2015. Synthesis of polymer nanocomposites. In V. Mittal (Ed.), *Synthesis Techniques for Polymer Nanocomposites*, **13**, (pp. 549–553). Weinheim, Germany: Wiley-VCH.

Fu, S., Sun, Z., Huang, P., Li, Y., & Hu, N. 2019. Some basic aspects of polymer nanocomposites : A critical review. *Nano Materials Science*, **1**(1), 2–30. doi:10.1016/j.nanoms.2019.02.006

Gao, S., Shi, Y., Zhang, S., Jiang, K., Yang, S., Li, Z., & Takayama-Muromachi, E. 2008. Biopolymer-assisted green synthesis of iron oxide nanoparticles and their magnetic properties. *Journal of Physical Chemistry C*, **112**(28), 10398–10401. doi:10.1021/jp802500a

Ghanbarzadeh, B., & Almasi, H. 2018. Biodegradable polymers babak. In R. Chamy & F. Rosenkranz (Eds.), *Long-Haul Travel Motivation by International Tourist to Penang*, I, (p. 13). IntechOpen.

Gold, K., Slay, B., Knackstedt, M., & Gaharwar, A. K. 2018. Antimicrobial activity of metal and metal-oxide based nanoparticles. *Advanced Therapeutics*, **1**(3), 1700033. doi:10.1002/adtp.201700033

Hu, L., Du, M., & Zhang, J. 2017. Hemicellulose-based hydrogels present status and application prospects: A brief review. *Open Journal of Forestry*, **8**(1), 15–28. doi:10.4236/ojf.2018.81002

Habibi, Y., Benali, S., & Dubois, P. 2015. In situ polymerization of bionanocomposites. In K. Oksman, A. P. Mathew, A. Bismarck, O. Rojas, & M. Sain (Eds.), *Handbook of Green Materials* (pp. 69–88). Singapore: World Scientific.

He, Z., & Alexandridis, P. 2017. Micellization thermodynamics of Pluronic P123 (EO20PO70EO20) amphiphilic block copolymer in aqueous Ethylammonium nitrate (EAN) solutions. *Polymers*, **10**(1), 18. doi:10.3390/polym10010032

Inamuddin (Ed.). 2017. *Green Polymer Composites Technology: Properties and Applications*. Boca Raton, FL: CRC Press.

Islam, S., Bhuiyan, M. A. R., & Islam, M. N. 2017. Chitin and chitosan: Structure, properties and applications in biomedical engineering. *Journal of Polymers and the Environment*, **25**(3), 854–866. doi:10.1007/s10924-016-0865-5

Jacob John, M., & Thomas, S. 2012. *Natural Polymers: An Overview*, **1**(16), 1–7. doi:10.1039/9781849735193-00001

Jawed, M., & Khan, M. M. (Eds.). 2018. *Polymer Based Nanocomposites for Energy and Environment Applications*. Duxford, UK: Woodhead Publishing.

Kaur, G., Adhikari, R., Cass, P., Bown, M., & Gunatillake, P. 2015. Electrically conductive polymers and composites for biomedical applications. *RSC Advances*, **5**(47), 37553–37567. doi:10.1039/c5ra01851j

Kaushik, K., Sharma, R. B., & Agarwal, S. 2016. Natural polymers and their applications. *International Journal of Pharmaceutical Sciences Review and Research*, **37**(2), 30–36.

Kenry, K., & Liu, B. 2018. Recent advances in biodegradable conducting polymers and their biomedical applications. *Biomacromolecules*, **19**(6), 1783–1803. doi:10.1021/acs.biomac.8b00275

Kvien, I., & Oksman, K. 2007. Orientation of cellulose nanowhiskers in polyvinyl alcohol. *Applied Physics*, **87**, 641–3. doi:10.1007/s00339-007-3882-3

Lagashetty, A., & Venkataraman, A. 2005. Polymer nanocomposites. *Resonance*, **10**(7), 49–57.

Liu, J., Boo, W., Clearfield, A., Sue, H., Liu, J., Boo, W., Sue, H. 2006. Intercalation and exfoliation : A review on morphology of polymer nanocomposites reinforced by inorganic layer structures intercalation and exfoliation : A review on morphology of polymer nanocomposites reinforced by inorganic layer structures. *Materials and Manufacturing Process*, **21**(2), 143–151. doi:10.1081/AMP-200068646

Lombardo, D., Kiselev, M. A., & Caccamo, M. T. 2019. Smart nanoparticles for drug delivery application: Development of versatile nanocarrier platforms in biotechnology and nanomedicine. *Journal of Nanomaterials*, February, 1–26. doi:10.1155/2019/3702518

Lone, I. H., & Akter, A. 2018. Overview of bionanocomposites. In S. Ahmed & S. Kanchi (Eds.), *Handbook of Bionanocomposites: Green and Sustainable Materials* (p. 307). Singapore: Pan Standford Publishing.

Mahendra, R., & Clemens, P. 2013. *Green Biosynthesis of Nanoparticles Mechanism and Applications,* (Vol. 3). CAB International.

Makarov, V. V, Love, A. J., Sinitsyna, O. V, Makarova, S. S., Yaminsky, I. V, Taliansky, M. E., & Kalinina, N. O. 2014. "Green" nanotechnologies: Synthesis of metal nanoparticles using plants. *Acta Naturae*, **6**(1), 35–44.

Menon, P. M., Selvakumar, R., Kumar, P. S., & Ramakrishna, S. 2017. Extraction and modification of cellulose nanofibers derived from biomass for environmental application. *RSC Advances*, **7**(68), 42750–42773. doi:10.1039/C7RA06713E

Mensink, M. A., Frijlink, H. W., van der Voort Maartschalk, K., & Hinrichs, W. L. J. 2015. Inulin, a flexible oligosaccharide I: Review of its physicochemical characteristics. *Carbohydrate Polymers*, **130**, 405–419. doi:10.1016/j.carbpol.2015.05.026

Muchlisyam, J. S., & Harahap, U. 2015. Hemicellulose: Isolation and its application in pharmacy. In V. K. Thumar & M. J. Thakur (Eds.), *Handbook of Sustainable Polymers*, (pp. 305–340). New York: Jenny Stanford Publishing.

Muthivhi, R., Parani, S., May, B., & Oluwafemi, O. S. 2018. Green synthesis of gelatin-noble metal polymer nanocomposites for sensing of Hg2+ ions in aqueous media. *Nano-Structures and Nano-Objects*, **13**, 132–138. doi:10.1016/j.nanoso.2017.12.008

Nadaroglu, H., Alayli, A., & Ince, S. 2017. Synthesis of nanoparticles by green synthesis method. *International Journal of Innovative Research and Reviews*, **1**(August 2018), 2–6.

Olatunji, O. (Ed.) 2016. *Natural Polymers: Industrial Techniques and Applications.* Cham, Switzerland: Springer International Publishing.

Pandey, S., Goswami, G. K., & Nanda, K. K. 2013. Green synthesis of polysaccharide/gold nanoparticle nanocomposite: An efficient ammonia sensor. *Carbohydrate Polymers*, **94**(1), 229–234. doi:10.1016/j.carbpol.2013.01.009

Pillai, C. K. S. 2010. Challenges for natural monomers and polymers: Novel design strategies and engineering to develop advanced polymers. *Designed Monomers and Polymers*, **13**(2), 87–121. doi:10.1163/138577210X12634696333190

Prozorova, G. F., Pozdnyakov, A. S., Kuznetsova, N. P., Korzhova, S. A., Emel'yanov, A. I., Ermakova, T. G., …, & Sosedova, L. M. 2014. Green synthesis of water-soluble non-toxic polymeric nanocomposites containing silver nanoparticles. *International Journal of Nanomedicine*, **9**(1), 1883–1889. doi:10.2147/IJN.S57865

Ralph, J. 1998. Lignin structure: Recent developments. *US Dairy Forage Research Center*, **608**, 1–16.

Reddy, M. M., Vivekanandhan, S., Misra, M., Bhatia, S. K., & Mohanty, A. K. 2013. Biobased plasticks and bionanocomposites : Current status and future opportunities. *Progress in Polymer Science*, **38**(10–11), 1653–1689. doi:10.1016/j.progpolymsci.2013.05.006

Rhein-Knudsen, N., Ale, M. T., & Meyer, A. S. 2015. Seaweed hydrocolloid production: An update on enzyme assisted extraction and modification technologies. *Marine Drugs*, **13**(6), 3340–3359. doi:10.3390/md13063340

Rinaudo, M. 2006. Chitin and chitosan: Properties and applications. *Progress in Polymer Science*, **31**(7), 603–632. doi:10.1016/j.progpolymsci.2006.06.001

Ruocco, N., Costantini, S., Guariniello, S., & Costantini, M. 2016. Polysaccharides from the marine environment with pharmacological, cosmeceutical and nutraceutical potential. *Molecules*, **21**(5), 1–16. doi:10.3390/molecules21050551

Saini, R. K., Bajpai, A. K., & Jain, E. 2018. Fundamentals of bionanocomposites. In N. G. Shimpi (Ed.), *Biodegradable and Biocompatible Polymer Composites* (pp. 351–377). Oxford, UK: Woodhead Publishing.

Sanchez-Garcia, M., & Lagaron, J. 2010. On the use of plant cellulose nanowhiskers to enhance the barrier properties of polylactic acid. *Cellulose*, **17**, 987–1004. doi:10.1007/s10570-010-9430-x

Shchipunov, Y. 2012. Bionanocomposites: Green sustainable materials for the near future. *Pure and Applied Chemistry*, **84**(12), 2579–2607. doi:10.1351/pac-con-12-05-04

Shekar, H. S. S., & Ramachandra, M. 2018. Green composites: A review. *Materials Today: Proceedings*, **5**(1), 2518–2526. doi:10.1016/j.matpr.2017.11.034

Shen, X., Shamshina, J. L., Berton, P., Gurau, G., & Rogers, R. D. 2015. Hydrogels based on cellulose and chitin: Fabrication, properties, and applications. *Green Chemistry*, **18**(1), 53–75. doi:10.1039/c5gc02396c

Singh, J., Dutta, T., Kim, K. H., Rawat, M., Samddar, P., & Kumar, P. 2018. "Green" synthesis of metals and their oxide nanoparticles: Applications for environmental remediation. *Journal of Nanobiotechnology*, **16**(1), 1–24. doi:10.1186/s12951-018-0408-4

Song, F., Li, X., Wang, Q., Liao, L., & Zhang, C. 2015. Nanocomposite hydrogels and their applications in drug delivery and tissue engineering. *Journal of Biomedical Nanotechnology*, **11**(1), 40–52. doi:10.1166/jbn.2015.1962

Thakur, V. K., & Thakur, M. K. (Eds.). 2015. *Eco-Friendly Polymer Nanocomposites: Processing and Properties*. New York: Springer.

Vishakha, K., Kishor, B., & Sudha, R. 2012. Natural polymers – A comprehensive review. *International Journal of Research in Pharmaceutical and Biomedical Sciences*, **3**(4), 1597–1613.

Vishwanatha, T., Keshavamurthy, M., Mallappa, M., Murugendrappa, M. V, Nadaf, Y. F., Siddalingeshwara, K. G., & Dhulappa, A. 2018. Biosynthesis, characterization and antibacterial activity of silver nanoparticles from *Aspergillus awamori*. *Journal of Applied Biology & Biotechnology*, **6**(5), 12–16. doi:10.7324/jabb.2018.60502

Yiu, H. H. P., Niu, H. J., Biermans, E., Van Tendeloo, G., & Rosseinsky, M. J. 2010. Designed multifunctional nanocomposites for biomedical applications. *Advanced Functional Materials*, **20**(10), 1599–1609. doi:10.1002/adfm.200902117

2 Synthesis of Green Polymeric Nanocomposites Using Electrospinning

Sweta Naik, K. Chandrasekhar and Satya Eswari Jujjavarapu

CONTENTS

2.1 INTRODUCTION

Many studies have reported on polymeric nanofibers that are prepared by using the electrospinning (ES) process. In the electrospinning technique, a high electric field is applied to induce a polymeric solution from the capillary tip toward a grounded collector (Xue et al. 2019). After applying a voltage to the polymeric solution, it further ejects a jet of the solution which can be collected in the grounded collector. Micro-sized fibers are well-known materials for various purposes for reasons of cost-effectiveness, easy processability, and good filtration efficiency. Based on the application of electrospun material in the filtration method, the method is divided into two categories: liquid and air filtration. An electrospinning apparatus, with an example, along with the process, is shown in Figure 2.1. In this chapter, we discuss the preparation of green polymeric nanocomposites via electrospinning techniques followed by their characterization and application in different fields (Zucchelli et al. 2011).

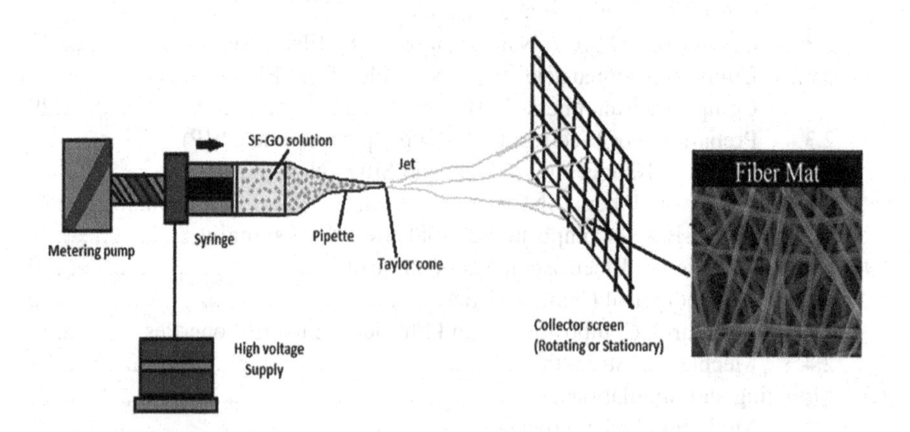

FIGURE 2.1 Principle of electrospinning technology.

2.2 PRINCIPLE OF ELECTROSPINNING

ES is a technology used to fabricate nanofibers with a nanoscopic to submicron diameter range. In ES-technology an electrical body force acts on a component of the charged fluid. Recently, ES-technology has been developed as a specialized method for the synthesis of submicron electrospun nanofibers (diameters of between 100 nm and 1 μm), with high surface areas (Karakaş 2015). Because of their high surface area, high porosity, small pore size, and unique fibers they have been recommended as excellent materials for application in filtration techniques. For the transmission of particulate pollutants, water and air are significant transportation media. In air filtration, the contaminants are the most complex particulate mixture, and usually, most contaminants are smaller than a diameter of 1,000 μm. Frequently, the biological and chemical aerosols range from 1 to10 μm (Teo, Inai, and Ramakrishna 2011). Therefore, electrospun membranes will be beneficial for use in air filtration.

2.2.1 Effect of Process Parameters on Electrospun Nanofibers

It has been demonstrated that morphological characteristics of electrospun nanofibers, such as its diameter and uniformity, are reliant on various processing parameters. These are mainly classified into three groups: (a) properties of solution, (b) processing, and (c) ambient conditions. These parameters can affect the morphological characteristics of electrospun nanofibers (Huan et al. 2015).

2.2.2 Properties of Solution Parameters

Properties of solution parameters such as electrical conductivity, viscosity, molecular weight, and concentration of the solution, surface tension, and elasticity have a substantial impact on morphological features of electrospun nanofibers (Cadafalch Gazquez et al. 2017).

2.2.3 Viscosity

The viscosity ranges are different for different nanofiber solutions, which are spinnable. The solution's viscosity is one of the most critical parameters that influence the diameter of the nanofiber. A higher viscosity of solution results in larger fiber diameter. As solution viscosity increases, the width of the bead likewise increases, and the average distance between the beads on the fibers becomes larger. In short, we can say that from more viscous solutions, beads, and beads on the fibers, can be formed (Pillay et al. 2013).

2.2.4 Concentration of the Solution

In the process of electrospinning, a minimum solution concentration is needed for nanofiber (NF) fabrication. Due to an increase in the solution's concentration, a mixture of beads and fibers are obtained. There is a change in the shape of the beads from a spherical to a spindle-form-like structure with the rise in solution concentration. The diameter of nanofibers also increases with an increase in concentration of the solution due to high viscosity resistance. However, the viscoelastic force, which generally resists frequent changes in fiber shape, may lead to uniform fiber formation at a higher concentration of the solution. Nevertheless, it is not possible to do electrospinning when the concentration or the viscosity of the solution becomes too high because of difficulty in liquid jet formation (Pillay et al. 2013).

2.2.5 Molecular Weight (MW)

MW also plays a major role in electrospinning. It also has substantial impact on the electrical and rheological properties of the nanofiber, that is, surface tension (ST), viscosity, dielectric strength, and conductivity. It was observed that a low MW solution leads to the formation of beads instead of nanofibers, and a high molecular-weight solution tends to form NFs with larger diameters (Huan et al. 2015).

2.2.6 Surface Tension (ST)

The ST of the nanofiber solution can be explained by force acting at correct positions on any line of unit length on the liquid surface. Nanofibers can be obtained without forming beads by reducing the ST of the nanofiber solution. The surface tension seems to be the function of solvent compositions but is insignificantly reliant on the concentration of the solution. Different nanofiber solutions may give different STs. Nevertheless, a solvent with lower ST may not always be required or appropriate for ES. Usually, ST defines the upper limits and lower limits of the ES window by keeping remaining variables constant. Nanofibers, beads, and the formation of droplets driven from the ST of the nanofiber solution and the lower ST of the electrospinning solution helps the ES technique to be performed at a lower electric field (Zheng et al. 2014).

2.2.7 Solution Conductivity

When electrical conductivity increases, a decrease in the diameter of the electrospun NFs is observed. Formation of beads can be seen with a drop in conductivity of the solution, which may lead to inadequate jet elongation. This is due to the electric force of the formation of uniform NF. Beads may also form due to low conductivity of the polymeric solution, and inadequate jet elongation occurs due to the electrostatic force applied to the polymeric solution to fabricate a uniform size and shape of silk NFs. The electrical conductivity of the polymeric solution is directly proportional to the diameter of the NFs. High conductivity polymer solution gives small diameter NFs. This indicates that the diameter size of the NFs decreases with a decrease in electrical conductivity of the polymer solution (Uyar and Besenbacher 2008).

2.2.8 Voltage Applied

In the ES process, it is usually assumed that due to ionic conduction of charge the electric current in the polymeric solution is small enoughto be considered negligible. For the charge transport, the flow of polymeric solution from the tip of the Taylor cone to the collector is the only mechanism. Therefore, the current in ES techniques increases with an increase in the mass flow rate of the polymeric solution from the tip to the collector when all the parameters, such as the flow rate of polymeric solution from the jet, dielectric constant, and conductivity, remain constant. Applied voltage and electrostatic force are directly proportional to each other. Electrostatic repulsive force will increase with an increase in applied voltage, which leads to the formation of small diameter NFs. The polymeric solution should be quickly removed from the jet when the Taylor cone ejects the jet; otherwise, it will lead to an increase in the diameter of the NFs (Angammana 2011).

2.2.9 Feed Rate

Feed rate also plays an important role in the fabrication of NFs by ES technology. By changing the feed rate of the solution, the morphological structure can be changed. The delivery rate of polymeric solution to the capillary tip increases when the flow rate exceeds the critical value. The delivery rate increases and is exceeded at a point at which the solution was supposed to be removed from the capillary tip by applied electrical force (Zong et al. 2002).

2.3 FABRICATION OF ELECTROSPUN NANOCOMPOSITES (NC)

2.3.1 Polyethylene Oxide, Poly(L-lactide) -Based Formed Oleic Acid-Coated Magnetite Nanocomposites by ES

Savva et al. prepared a series of micro-fibrous nanocomposite membranes using electrospinning techniques based on poly(L-lactide) (PLLA), polyethylene oxide (PEO)-formed chloroform oleic acid solution. In these solutions polymer content was constant and there was variation only in the inorganic content (OA–Fe_3O_4), which is shown in Figure 2.2. In terms of morphology (scanning electron microscopy [SEM]), composition (X-ray diffraction [XRD] spectroscopy), thermal properties (thermogravimetric analysis [TGA]), and evaluation of their magnetic characteristics (vibrational sample magnetometry) at ambient temperature, the prepared composite membranes were characterized. Prepared nanocomposite membranes on drug delivery were investigated. In drug delivery applications, prepared nanocomposite membranes were investigated by UV–Vis spectrophotometry on the release profile of paracetamol in aqueous media (Savva et al. 2012).

2.3.2 Gelatin-Based Silver Nanocomposites by Electrospinning

Marwa et al. synthesized gelatin NFs with incorporated silver nanoparticles by mixing gelatin solution with a silver nitrate solution, as represented in Figure 2.3. The combination of water and acetic acid (30:70 v/v) has been used as a common solution for gelatin and silver nitrate solution; acetic acid acts as a gelatin solvent, and silver nitrate solution reduces sugar. Silver nanoparticles were stabilized through the interaction mechanism of the O_2 atom and gelatin carbonyl group. The gelatin solution's

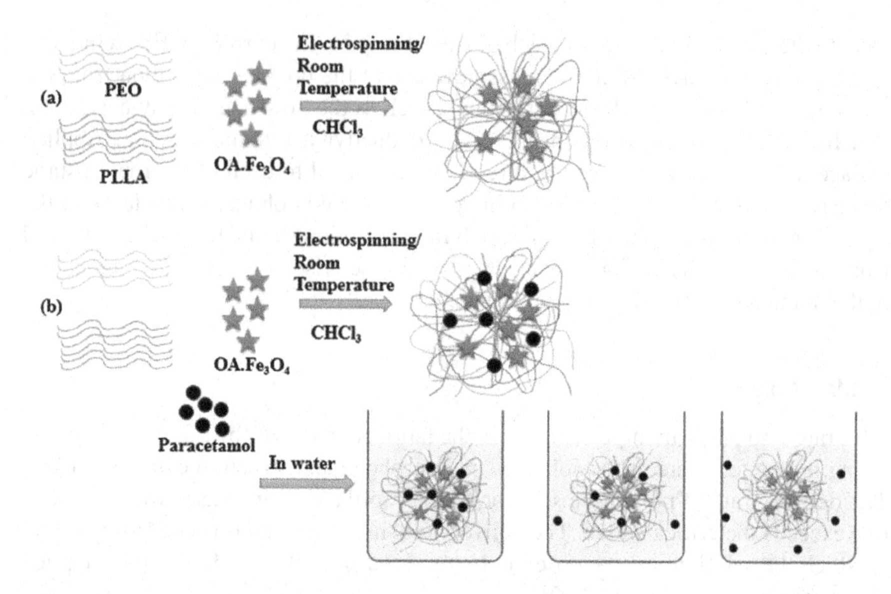

FIGURE 2.2 (a) Nanocomposite fibrous membrane fabrication via electrospinning at ambient conditions starting from a solution of PEO/PLLA/OA–Fe_3O_4 in chloroform. (b) Fabrication of drug-loaded PEO/PLLA/OA–Fe_3O_4 membranes via electrospinning followed by drug release studies in aqueous media performed by UV-Vis.

conductivity and viscosity increases with increased solution concentration. The conductivity of the gelatin solution is less than gelatin-based silver nitrate solution. The obtained nanofiber was characterized by XRD, SEM, TEM, and antimicrobial study. The results obtained from XRD and TEM study confirmed that silver nanoparticles were uniformly distributed over the nanofiber at a diameter range of 2 to 10 nm (Oraby et al. 2013).

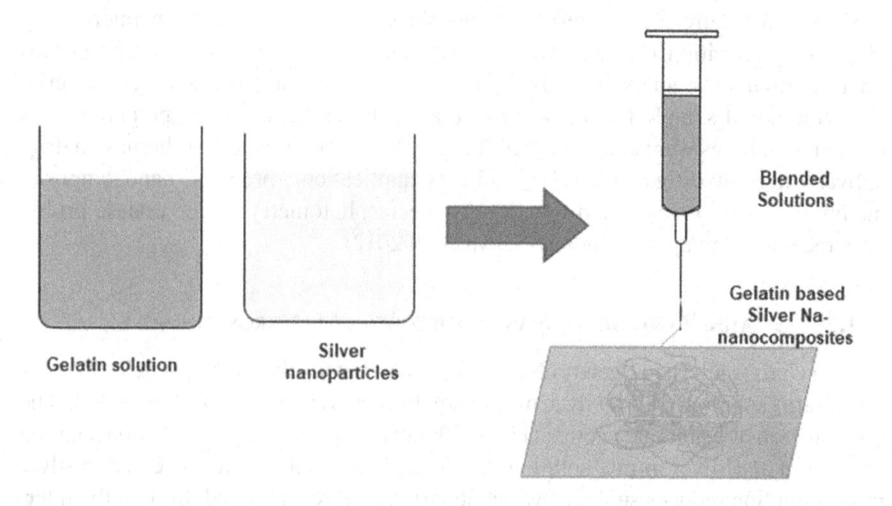

FIGURE 2.3 Gelatin-based silver nitrate nanocomposites by electrospinning.

2.3.3 Chitosan Nanoparticles Self-Assembled from Electrospun Composite Nanofibers (NF)

Due to their nanometer confinement effect and the formation of their components into a composite material, electrospun nanofibers can be good templates for manipulating molecular self-assembly (Anisha et al. 2013). Deng-Guang Yu et al. prepared hydrophilic polyvinylpyrrolidone (PVP) and chitosan (CS) composites of one-dimensional nanofiber using an elevated electrospinning process, that is, EES. Observations of SEM and TEM showed that the average size of the fibers is 77 ± 10 nm. The fibers are homogeneous with the internal structure. X-ray differential scanning (XRD), calorimetry IR spectra results revealed that the NF composites are amorphous in nature. To maintain structural homogeneity, second-order hydrogen bonding and electrostatic interactions in these amorphous nanofiber composites played a key role. An in-situ process of water addition to NFs leads to spontaneous self-assembly of CS nanoparticles up to 10 nm in diameter. A new process has been given by Guang Yu et al. (2012) which detailed combining "top-down" ES with a molecular self-assembly "bottom-up" approach. This new process has been studied for in-situ synthesis of polymeric nanoparticles.

2.3.4 Preparation of Electrospun Polyvinylpyrrolidone (PVP), Cellulose Nanocrystal (CNC), and Silver Nanoparticle Composite Fibers (SNCF)

Siwei Huang et al. used a solvent N,N′-dimethyl formamide (DMF) solvent to prepare PVP/CNC/silver nanoparticle composite fibers by ES. Compared to pure PVP solution, the composite electrospun suspensions showed better conductivity and rheological properties.

The average diameter of PVP-based electrospun nanofibers decreases with the increase in the amount of silver nanoparticles and CNCs. CNC addition leads to a reduction of the thermal stability of electrospun composite fibers. The CNCs aid the raising of the tensile strength of the composite, while the break elongation reduces it. Some studies have shown involvement of the composite nanofibers along with silver nanoparticles on enhanced antimicrobial activity on *Escherichia coli* gram-negative bacteria and *Staphylococcus aureus* gram-positive bacteria. The increased strength and antimicrobial performance of PVP/CNC/silver electrospun composite fibers makes the material more suitable for use in the biomedical area (Huang et al. 2016).

2.3.5 Starch-Based Composite Scaffolds by Electrospinning

Starch-based materials nowadays are widely used for wound-healing applications due to their properties. Starch is widely available, biocompatible, biodegradable, cost-effective, and has wound-healing properties. Vijaya Sadashiv Waghmare et al. have fabricated nanofibrous scaffolds by using ES techniques for application on the healing of wounds. The fabricated nanofibers were characterized by using field emission scanning electron microscopy (FE-SEM) and it was found that the average diameter of the nanofibers was in the range of 110 to 300 nm. The fabricated nanofibrous scaffold material mechanical strength was 0.5 to 0.8 MPa, attuned by using glutaraldehyde as a crosslinking agent, and polyvinyl alcohol as a plasticizer, to impart the scaffold materials with enough strength for tissue engineering of skin. The scaffold materials were further characterized by Attenuated Total Reflectance-Fourier Transform Infrared Spectroscopy (ATR-FTIR), differential scanning microscopy (DSC) and TGA to confirm the absence of negative interactions between the polymers. Cytotoxicity assays were performed using L929 mouse fibroblast cells, and the results of this assay showed the capability of scaffolds to stimulate cellular proliferation, without showing any lethal effect on the cells. Therefore, they concluded that fabricated nanofibrous scaffolds have the potential for wound healing applications (Waghmare et al. 2018).

2.4 CHARACTERIZATION OF ELECTROSPUN NANOCOMPOSITES

Based on the fabrication method and the properties of the material that is used for the nanocomposite preparation, the electrospun nanofibers can be characterized using various techniques. Generally, the morphological and structural properties of the electrospun nanofibers can be investigated by using microscopic methods. By using spectroscopic techniques, the compositional characterization of the electrospun nanofibers can be examined. By using different advanced tools, the thermal and mechanical properties can be studied.

2.4.1 MORPHOLOGICAL CHARACTERISTICS

Except for the composition of the NF material, various morphological properties also need to be characterized, such as morphology, diameter, orientation of NFs, etc. For the investigation of nanometre size dimensions, high-resolution imaging quality is required. Thus, structural and morphological properties of electrospun NFs can be characterized using high resolution imaging microscopic analysis techniques instead of an optical microscope such as SEM, FE- SEM, TEM, which will provide significant information on NF diameter and pore size. Additionally, the surface topography of NFs is measured by using scanning tunneling microscope (STM) and atomic force microscope (AFM) methods. For the compositional study of NFs, SEM and TEM are generally used via energy dispersive X-ray-EDX spectral scanning (Ballengee and Pintauro 2011).

2.4.2 STRUCTURAL, COMPOSITIONAL, AND PHYSICOCHEMICAL PROPERTIES

To investigate the structural, compositional, and molecular structure details of the electrospun nanofiber some spectroscopic techniques are used, such as X-ray photoelectron spectroscopes (XPS), Fourier-transform infra-red (FTIR), and nuclear magnetic resonance (NMR). These spectroscopic analysis techniques also determine the interaction between the intermolecular structure of electrospun NFs. For the study of constituent configuration in NFs, certain characterization techniques are used, such as wide-angle X-ray diffraction, angle X-ray scattering, optical birefringence, and differential scanning calorimetry (DSC). The pore size and exact surface area distribution of the NFs are investigated by the Brunauer–Emmett–Teller (BET) technique. Simultaneous differential analysis (SDTA) and TGA are used to study the thermodynamic and thermoelectric properties of nanofibers (Aruchamy, Mahto, and Nataraj 2018).

2.4.3 Mechanical Strength

Examples of mechanical strengths are tensile strength, compressive strength, and breaking modulus. These properties of the NFs can be measured by using a universal testing machine. Apart from these properties, some other mechanical properties are also present, such as the strength of individual fiber and bundle fiber, hardness, and elasticity of the NFs, and for the study of these properties, AFM is used. In recent studies, AFM has been extensively used to understand how to illustrate nanomaterials, the mechanical properties of nanofibers. Aruchamy et al. (2018) have explored various types of instructions that have been formulated for analyzing the mechanical strength of NFs and NF bundles.

2.5 MODELING AND SIMULATION

2.5.1 Modeling the ES Process

The process of electrospinning is a challenge associated with fluid dynamics. In order to control the NFs properties and geometry and perform mass production, a quantitative understanding is required of the process of electrospinning for the transformation of fluid solution into solid fibers. For this transformation, four to five orders smaller in diameter of the capillary tube of a millimeter diameter is used. A complete understanding of the electrospinning process, such as conversion of polymeric solution to nanofiber, is necessary. When electrostatic forces are applied to a polymeric solution the polymeric solution is electrified and forms a jet in the capillary tip and moves toward the ground collection for the fabrication of NFs. The process is made up of three stages (Fong 2001):

(a) Jet initiation and straight-line extension of the jet
(b) Increase of whipping instability and more stretching of the jet that may or may not go with by jet branching and splitting
(c) Jet solidification into nanofibers.

2.5.2 Jet Initiation

By using applied electrostatic forces on the polymeric solution, a fine jet will be drawn from the polymeric solution that can be taken from conducting tubes, which is shown by the Taylor cone phenomenon. As the conductive tube's potential increases, the fluid meniscus that was originally planar becomes almost conical. Taylor conferred and explained the thread formation from the viscous drops of an electrical field because of maximum instability. He explained the existence of the viscous fluid equilibrium in an electric field with the shape of a semi-vertical–angled cone, $\pi = 49.3°$. A fluid jet is developed when a semi-vertical cone angle reaches $\pi = 49.3°$, which is called the Taylor cone. The angle of the Taylor cone was individually verified by Larrondo and Manley. They detected that the angle of the semi-vertical cone was about 50° shortly before the formation of the jet (Larrondo and St. John Manley 1981). It is noted that Yarin, Koombhonges, and Reneker's recent publication claimed that the angle of the Taylor cone is 33.5° rather than 49.3°. There are two main issues that are associated with each other; one is the electrostatic force applied to the polymeric solution and the second is the initiation of the jet.

Taylor also showed that the critical voltage V (expressed in kilovolts (kV)) at which the extreme jet fluid instability improves is assumed by MacDiarmid (Yarin, Koombhongse, and Reneker 2001b)

$$VC^2 = 4H^2L^2 \ln 2LR - 1.50.117\pi R\gamma$$

where H is the distance between the collecting electrode from the capillary tip electrode, L is the capillary tube length, R is the radius of the tube, and γ is the surface tension of the fluid. The flow beyond the spinneret is mostly elongation in spinning. Hendricks et al. calculated the minimum spraying potential of a hemispherical conducting drop suspended in air as

$$V = 30020\pi r\gamma \left(\text{“r” is jet radius}\right).$$

When the medium which is surrounded is not air but is filled with nonconductive liquid which is immiscible with the spinning fluid, the drop distortion will be greater at any applied electric field. Thus, reduction in minimum spinning voltage observed. From this discussion it was concluded that a lower voltage is required when electrospinning technology is encapsulated in a vacuum.

2.5.3 Jet Thinning (JT)

The conception of jet thinning is not yet clear. What is clear is that fluid instability occurs during this stage. From the old point of view, the radial charge repulsion leads to the splitting of the primary jet into multiple streams in a course known as “splaying/spreading”. In this process, acceleration of the fluid jet which is electrified occurs and the jet becomes thin along its trajectory. Hence the final fiber size appears to be detected primarily by the number of developing supplementary jets (Reneker and Yarin 2008).

However, current studies have shown that a non-axisymmetric or whipping instability, which causes the jet to bend and stretch at very high frequencies, plays an important role in jet diameter reduction from mm to nm. Shin et al. used an asymptotic expansion technique for jet radius equations to investigate the stability of an ES-PEO jet and it was assumed that it was small (Shin et al. 2001). There may be a possibility of three types of instabilities which were obtained after resolving the equations. The first is the classical Rayleigh instability, which is axisymmetric to the center of line of the jet. The second is, again, an axisymmetric instability (Reneker et al. 2000). The third is a non-axisymmetric instability which can be called whipping instability, mostly due to the force of bending. If all of the other parameters remain unchanged, the strength of the electric field will be proportional to the level of instability. Whipping instability was observed under to the highest field and under the lowest field, Rayleigh instability was observed. Shin et al. also detected and showed an experimental phenomenon known as the inverse cone phenomenon. The inverse cone phenomenon is where the 1^0 jet is supposed to be divided into many jets, which is actually due to bending instability. The reverse cone is not due to splitting at higher resolution and with less exposure time for the electronic camera, but as a result of small lateral fluctuations in the center line of the jet. By theoretical and experimental work, some analogous phenomena were recognized later. The non-splitting phenomenon was found by both research groups using polyethylene oxide solutions with moderate concentration (2% and 6%). Nevertheless, it should be possible to branch and/or split from the primary jet as the jet fluid driven by the electrical forces is unstable during its trajectory toward the collection screen. In fact, a number of research groups have re-realized the phenomena of branching and splitting by other updated experimental set-ups. Yarin et al. recognized that several polymeric solutions such as HEMA, PVDF, polystyrene-PS, and poly(ether imide) (PE) were dividing and branching from the 1^0 jet with concentrations of weight exceeding 10% theoretically. Yarin et al. investigated the excitation condition of an electrically charged jet's axisymmetric and non-axisymmetric disturbances.

The linearized issue was investigated with respect to approximation of the surface frozen-charge. In the longitudinal electrical field, the stability control of the affecting electrified jet is a possible indication of their solutions. By altering the properties of the liquid and electrical intensity, the axisymmetric mode instability that typically leads the jet to decline, drops, which can be significantly decreased and increments of no axisymmetric means m = 1 and 2, where m is called the azimuthal wave number. An increase in wave number is observed with no axis.

Regarding the diameter of the fluid jet, Baumgarten noted that the spinning drop altered from roughly hemispheric to conical as the viscosity (μ) of the polymer solvent increased (Baumgarten 1971). Baumgarten obtained an expression using an equipotential line estimate calculation to calculate the radius (r) of a spherical drop (jet) (Baumgarten 1971):

$$r_0^3 = 4\varepsilon m_0' k\pi\rho\left(\text{gram/sec}\right)$$

at the moment where r is to be calculated, k is a dimensionless parameter associated with the electric currents, electric conductivity (σ units amp/volt cm), and density (ρ - g/cm). Spivak et al. communicated a steady state electrospinning electrohydrodynamic model in a single jet, taking into account inertial, viscous, hydrostatic, electrical, and surface-tension forces. The average physical quantity over the jet cross-section was derived from a 1D differential equation for the radius of jet as a function of the distance from the jet tip to the collection plate (Spivak, Dzenis, and Reneker 2000).

2.5.4 Jet Solidification

Yarin et al. derived a quasi-1D equation to explain the fluid jet's mass reduction and volume change due to evaporation and solidification of the jet with the assumption that 1^0 jets do not include a branch or split. They also calculated that the dry characterization cross-sectional radius was 1.31×10 that of an initial weight concentration and other parameters of the process, even though the rate of solidification changed with polymer concentration. However, other issues, such as electrostatic field rate variation, gap distance, porous dimension control, and distribution throughout solidification, were not addressed clearly (Yarin, Koombhongse, and Reneker 2001a).

2.6 APPLICATIONS OF ELECTROSPUN NANOCOMPOSITES

Electrospun nanofibers in the form of non-woven fibrous mats are generally classified as high surface area materials; for example, they carry a surface-to-weight or volume ratio. Due to its unique features, electrospinning technology has been extensively used for the fabrication of nanofibers with numerous applications, such as in tissue engineering, drug delivery, wastewater treatment, air pollution control, food packaging, etc. This technique has gone through outstanding modifications since its reinvention in the 1990s, initially a single-needle spinning method in lab-scale, then to mass production with a multi-needle spinning method and industrial-scale production of nanofibers using an advanced bubble-spinning technique. Depending upon the use and applications of NFs, the ES process can be modified to obtain special types of electrospun NFs. In the form of wound dressing pads, filtration membranes, and air purifiers the process can produce items from simple randomly systematized nanofiber webs to highly oriented and packed commercial nanofiber mats (Aruchamy et al. 2018). Different applications of electrospun nanocomposites are illustrated in Figure 2.4.

2.6.1 Wound Dressing

Recent advances in ES technology possess greater potentials for making multidimensional NF, which is massively involved in applications of wound healing. ES technology permits control of the diameter and pore size of the NFs according to the applications. These nanofibrous-based scaffolds also provide a 3D network for many diagnostic studies. Initially, for scaffold preparation, PLGA was used and for nanocomposite preparation carbon nanotubes were included with the polymeric solution. The prepared scaffold was tested for three-dimensional tissue engineering applications and the cytotoxicity of the prepared scaffold was investigated by using NR6-mouse fibroblast cells. The old drug delivery system is affected by the restrictive response of therapeutic agents applied for wound healing. The electrospun nanofibrous scaffold allows a lower response time and more control over the release rate of therapeutic agents for wound healing. Based on the different applications of ES technology, various materials are used for the preparation of NFs, such as natural and artificial polymers, proteins, nanoparticles, composites, etc. There are many parameters involved in ES technology to control the size, shape, diameter,

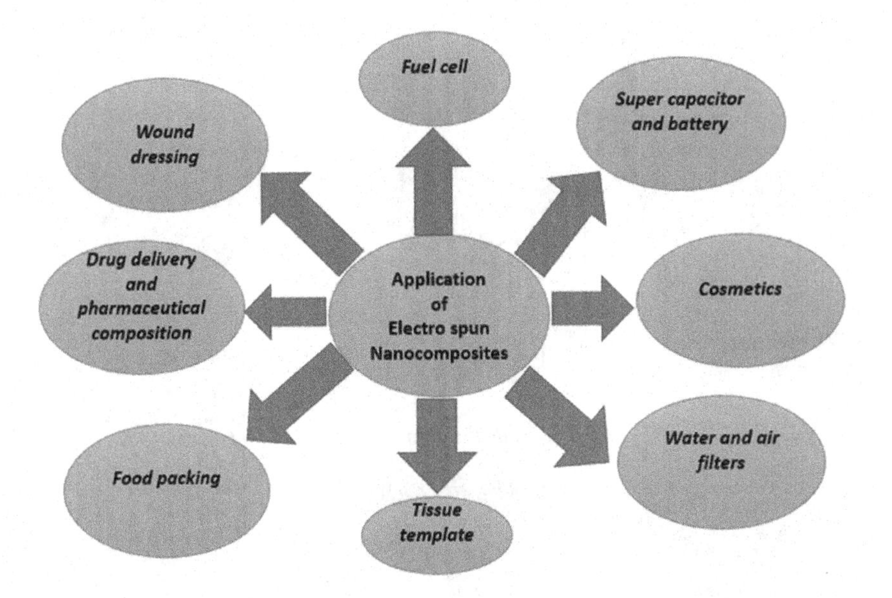

FIGURE 2.4 Application of electrospun nanocomposites.

and morphology of NFs. These characteristics of the NFs can be controlled by changing the conductivity and viscosity of the polymeric solution also (Memic et al. 2019). On the other hand, nanofiber entanglement depends upon the polymer MW, the concentration of the solution, and the therapeutic agent during the electrospinning process. To obtain a stabilized liquid jet, that results in uniform fibers and it is necessary to design a polymer/therapeutic agent solution with ample chain entanglement. Biopolymers such as gelatin, collagen, chitosan, alginate, and polysaccharide-based dope solutions are used for the therapeutic drug charging and are claimed to be the most positive dope polymers so far. NFs containing therapeutic agents for anti-inflammatory drugs, antioxidants, antimicrobials, enzymes, anesthetics, and growth-factors are numerous products that are formed for the dressing of wounds (Liu et al. 2019). Carrier polymers and therapeutic agents play a crucial role in designing stable nanofiber wound dressing patches. Emulsion ES or coaxial ES have been successfully used to prepare polymer nanofiber with hydrophilic therapeutic moods, although, as a post-modification step, hydrophobic moieties are comprised of electrospun nanofibers. Figure 2.5 illustrates how electrospun fibers accelerate wound healing.

2.6.2 Food Packing

Nowadays the need for healthy and natural food has increased (K. Chandrasekhar, Amulya, and Venkata Mohan 2015; Chandrasekhar et al. 2018; Kuppam Chandrasekhar, Lee, and Lee 2015; Kuppam et al. 2017; Chandrasekhar and Venkata Mohan 2014a; Chandrasekhar and Venkata Mohan 2014b; Kumar et al. 2018; Reddy, Devi, et al. 2011; Sivagurunathan, Kuppam, et al. 2018; Mohan and Chandrasekhar 2011b; Venkata Mohan et al. 2019). A recent development in food

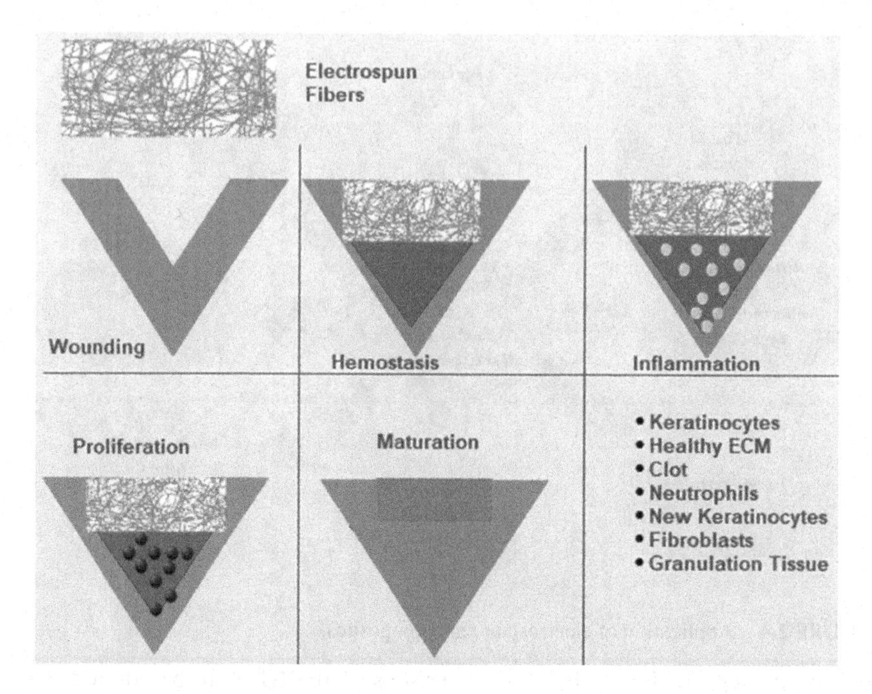

FIGURE 2.5 Electrospun fiber accelerates wound healing.

packaging and the materials of food processing has mainly targeted protective flavors and the quality of packaged food. An increase in concern for the environment has led to the development of improved and advanced packaging materials to yield less food-packaging waste and spoiled food waste in our landfills (Torres-Giner 2011). A special food-packaging technique should be developed to improve the value of food and avert food-borne diseases by discharging active biocidic elements to increase food shelf life. Recent advances in active food-wrapping ideas are gaining more attention in response to consumer demand and market changes. Innovation in the development of active packaging plans involves a combination of antimicrobial and antioxidant agents in a material that is bio-degradable and has tremendous liquid and gas barrier qualities. In this way, packaging of food with material that contains anti-microbial properties is gaining more interest in the scenario of food-borne pathogens associated with health problems. Consequently, packing material designed with lesser preservative loads is being implemented to increase the quality of food and shelf-life. Until now, various biopolymers have been used as active food-wrapping materials by altering composition with polysaccharides, proteins, and lipids for application in food wrapping. The recent advances of the materials in food-packaging applications have focused on the improvement of the materials' shelf-life activity and the convenience of their use in food processing. The materials prepared by ES technology are most extensively used and more promising in food-packaging applications. Due to the small pore size of the electrospun material, it acts as a physical barrier for the entry of microbes into food, which can cause contamination (Dai 2016).

Nanofibers based on polymers can be extended to host active feature impregnation, which is beneficial for food protection. Active proxies such as nutraceuticals can also be encapsulated in and on the surfaces of nanofibers and as functional foods to act as a bioactive packaging for food ingredients, in addition to providing nutritional value. Vitamins and anti-oxidants such as carotene and 3-fatty acids have been capably merged with NFs using the electrospinning technique. In addition, the bioactivity of vitamins and anti-oxidants on the quality of food in a wrapped form was tested and was observed to be activated when the package film is opened.

2.6.3 Water and Air Filters

The spectrum of applications of nanofibrous filtration membranes has become wide in recent years due to their pore size and filtration capability. Pressure-driven membrane-based filtration processes can be classified into ultrafiltration (UF), microfiltration (MF), nanofiltration (NF), and two osmotic gradient-driven processes, forward osmosis (FO) and reverse osmosis (RO). The improvements in electrospinning technology offer a feasible way to strategize and maximize the scale-up of electrospun nanofibrous membranes useful in filtration applications. Among all the filtration processes, UF and MF processes have better potential to use electrospun nanofibers as filtration media due to their ability to decrease the resistance to water flow (Doyle et al. 2013). The high yield water flux in nanofiltration approaches create water channels in the obstacle layer, in that way inducing anti-fouling properties in the membrane by reducing solute fouling and scaling as a result of the reduction of resistance to water flow. In both circumstances, the tailor-prepared nanofibrous membrane was used as an affinity-membrane according to the type of contaminant used for treatment. The preparation method using conventional filtration membranes usually follows the technique of phase-inversion to bring permeability to the resultant membranes. Conventional phase reversal techniques, in which solvent and non-solvent exchange create pores in the polymer matrix and thermally induce porosity, determine pore size and membrane distribution. Polymer nanofiber mats made with electrospinning technology are employed as inventive membrane constituents for water treatment owing to their interrelated fiber grid that induces specific porosity to the fibrous mat. Parameters such as fiber mats' light weight, high surface area and inherent fiber porosity make them attractive for filtration applications (Thavasi, Singh, and Ramakrishna 2008).

2.6.4 Super-Capacitor and Battery

Nowadays the need for sustainable energy and eco-friendly resources for the transformation of energy and storage application is growing due to climate change, misuse of fossil-fuel-based energy resources, and environmental degradation triggering severe difficulties to the life (Kadier, Jiang, Lai, Rai, Chandrasekhar, Mohamed, and Mohd. Sahaid Kalil 2018; Chandrasekhar 2019; Chandrasekhar and Ahn 2017; Chandrasekhar and Venkata Mohan 2012; Deval et al. 2017; Enamala et al. 2018; Kadier, Chandrasekhar, and Kalil 2017; Kadier et al. 2018; Kadier, Kalil, et al. 2016; Kadier, Kalil, et al. 2018; Kadier, Simayi, et al. 2016; Kadier et al. 2015; Kakarla et al.

2017; Kiran Kumar et al. 2012; Kumar et al. 2017; Kumar et al. 2018; Pandit, Sarode, and Chandrasekhar 2018; Pandit, Sarode, Sargunaraj, et al. 2018; Patel, Pandit, and Chandrasekhar 2017; Sivagurunathan, Sivagurunathan, et al. 2018; Cadafalch Gazquez et al. 2017; Reddy, Chandrasekhar, and Mohan 2011; Sivagurunathan, Kuppam, et al. 2018; Venkata Mohan et al. 2019; Mohan and Chandrasekhar 2011b; Mohan et al. 2013). Energy production, energy transformation and storage devices such as batteries, supercapacitors, solar cells, and fuel cells have been extensively used for a few decades. Especially, electrospinning of conducting polymers such as polyacetylene (PA), polyacrylonitrile (PAN), and polyaniline (PANI) liquids mixed by diverse types of polymers, and complicated with various inorganic precursors have been efficiently employed for attaining electrospun NFs (Liu et al. 2018). These NFs were further stimulated, graphitized, and carbonized at a high temperature to obtain increased surface area, high porosity, and highly conductive electrode materials which are beneficial and aimed at energy use. Still, nowadays, new advances are required to increase the activity of, for example, power density, durability, conversion efficacy, and harvest efficiency. Nowadays electrospun nanofibers have been extensively employed for sustainable energy use in the form of films, membranes, mats, and webs. Over the last few decades it is well known from the literature that electrospun nanofiber materials offer progressive and improved properties as compared to traditional old electrode materials made via several methods, including sol–gel templating, chemical vapor deposition, solid-state methods, the polymer precursor method, self-assembly, and electrospinning technology. The latest advances in designing approaches indicate that electrospun materials offer unique and advanced properties among different existing electrode materials (Levitt et al. 2019). Electrospun materials can be employed as electrodes, diffusion layers, and separators in energy applications, energy transformation applications, and storage use such as fuel cells, supercapacitors, and batteries (Chee et al. 2017).

2.6.5 Tissue Template

The major objectives in the field of tissue engineering and biomaterials to heal damaged organs or tissues in a human or animal body are to design the best scaffold or synthetic matrices that can mimic the structural properties and biological purposes of the normal extracellular matrix. The smaller diameter of electrospun nanofiber facilitates the easy attachment of human cells to the cells around the nanofibers (Laurencin et al. 1999). In this manner, the electrospun nanofibrous supports can offer an ideal pattern for cells to seed, proliferate, and grow. Effective rejuvenation of natural organs and tissues demands the improvement of nanofibrous assemblies through nanofiber constructions advantageous for cell proliferation and cell deposition. In the field of tissue engineering, researchers are getting more interested in the formation of biocompatible and reproducible 3D frameworks for cellular growth in synthetic matrix complexes for tissue repair and replacement methods.

In recent times, researchers have started working on the preparation of frameworks with biopolymer and artificial biopolymer nanofibers (Fertala, Han, and Ko 2001). It is observed that converting biopolymers into nanofibers and networks that can mimic the natural structure of the biological tissue or organs will eventually increase the efficacy of these nanofibrous materials since large-diameter fibers do not mimic the morphological characteristics of natural fibrils (Lei Huang et al. 2000).

2.6.6 Drug Delivery and Pharmaceutical Composition

One of the most pressing concerns in the field of medicine is how to deliver drugs to patients in the most physiologically appropriate way. Generally, the drug is better absorbed by a human being when the drug and the coating material encapsulating the drug are small. Drug delivery with polymeric nanofibers is generally based upon the principle of the dissolution rate of particulate drugs, which increases with increase in surface area of both the drug and the carrier if necessary. Kenawy et al. examined the distribution of tetracycline hydrochloride based upon the nanofibrous transport media of poly(lactic acid), poly(ethylene-co-vinyl acetate), and their mixture (Kenawy et al. 2002). E. Zussman et al. invented poly(lactic acid)-based bioabsorbable nanofibrous membranes, which are targeted at the prevention of surgery-induced adhesions and which have also been employed for drug loading. The membrane was also used to load an antibiotic drug, Mefoxin, in another study utilizing bioabsorbable nanofiber membranes of poly (lactic acid) aimed at preventing surgical-induced adhesions. This nanofiber membrane's preliminary efficiency compared to the bulk film has been demonstrated (Zussman, Yarin, and Weihs 2002).

Ignatious et al. demonstrated electrospun polymer nanofibers for pharmacological alignments that can be intended to offer immediate, rapid, modified, or late suspension and pulsative sustained-release characteristics (Ignatious et al. 2010) since the drug and carrier resources for the electrospinning of nanofibers are able to mix. The drug's likely modes in the subsequent nanostructured yields are (1) as a particulate drug attached to the carrier surface in the form of nanofibers; (2) both carrier and drug in the form of nanofibers; consequently, two types of NF are woven together; (3) a mixture of medicine and transporter ingredients combined into a single type of fiber comprising both components; and (4) the carrier material is electrospun into a tubular form in which the drug particulates are encapsulated. The third and fourth modes may be preferred. However, medical transport, actual delivery after production, and efficiency in NF processes are yet to be determined and will need further research in the future.

2.6.7 Cosmetics

Recent skin-care coatings such as lotions, ointments, or state-of-the-art creams may comprise a powder or fluid sprays, which may be expected more than fibrous materials to transfer into delicate parts of the body such as eyes or the nose when the skin coating is applied to the face. Nowadays electrospun bio-polymeric materials are used in cosmetic products for skin care such as skin-healing products, skin-cleansing products, medicines, and other therapeutic products using different additives (Kaul et al. 2018). An electrospun skin cover with microscropic pores and excellent surface area can enable far better application. It also increases the transfer rate of additives into the skin for the most complete potential of the preservative. An electrospun skin mask can be applied softly without pain as well as directly to the 3D topography of the skin to offer curing and careful treatment of the skin (Huang et al. 2003).

2.6.8 Fuel Cells

Fuel cells function as a secondary battery and consist of electrodes (anode and cathode) and an electrolyte (conductive solution). They do not require charging; fuel cells transform potential chemical energy into electrical energy promptly at point of usage. Fuel cells require a continuous source of compounds as a substrate, which is called a fuel and which the cell obtains from outside (Sood et al. 2016). As long as fuel is supplied to the cells, fuel cells can generate electrical power for an indefinite period. Fuel cells have several advantages over batteries (Kadier et al. 2018; Enamala et al. 2018; Pandit et al. 2017; Chang 2019; Saratale et al. 2017; Mohan and Chandrasekhar 2011a; Mohan and Chandrasekhar 2011b). Similar to a battery, a fuel cell also comprises two solid electrode materials (anode and cathode) which have been either submerged in the electrolyte or sandwiched around an electrolyte solution. Fuel is fed into the cell in the form of hydrogen, ethanol, and methanol through a gas diffusion layer which passes to the anode. Microbes present in the anode decompose or oxidize the organic component, as a result of which numerous electrons and protons will generate. The electrons produced are separated from the anodic solution over and done with an external circuit and protons travel to the cathode through the proton exchange membrane (PEM) where they are oxidized with O_2 and electrons to generate H_2O and heat.

However, fuel cells also suffer from numerous disadvantages, such as the high price of fuel, electrode stability, deterioration of electrolytes, durability, and safety. Apart from fuel cost, other parameters are also directly associated with electrode and electrolyte materials employed in the cell (Zhang 2010). Separators or electrolyte/polyelectrolyte also play an essential part in a fuel cell system. In this system, separators that are also intended to act as extremely proton-conductive membranes play an added role in maintaining neutrality with electrodes.

Similarly, PVA-based separators have been employed as PEM to enable unrestricted power generation in secondary batteries. Here, the electrospun membrane can be used as a PEM which has high efficiency for transferring protons from anode to cathode.

2.7 CONCLUSION AND FUTURE TRENDS

Electrospinning is now considered old knowledge, having been apparent in literature for more than six decades. However, while it is not yet fully developed, it is the best feasible method of continuous NF fabrication. In this chapter, a comprehensive and state-of-the-art review of this method was conducted along with a review of the applications of the polymer NFs it produces. In the future, extensive research and development are needed in all these areas. Some additional remarks are given as follows for each of them. Interesting nanofiber properties include their mechanical behavior and features such as bio-compatibility. No systematic information has been reported on the relationship between the mechanical properties of the electrospun nanofibers and electrospinning parameters such as elongational viscosity or zero-shear, electric field strength, electrical conductivity, dielectric permeability, relaxation time, volatility, surface tension, and solvent quality, as well as the molecular weight of the polymers, their concentration and their polydispersity. In the electrospinning procedure there are many challenges, and several essential queries remain open. Nafion nanofibers have satisfied several uses in different product forms (Shi et al. 2015). Electrospinning of biopolymers is becoming increasingly important as the world is moving to sustainable material sources. Even though electrospinning has taken a big leap in extensive production with the development of bubble spinning and the multi-nozzle spinneret, the production rate still needs to be increased to meet commercial requirements. On the other hand, the challenge of obtaining clog-free electrospun nanofiber is the biggest difficulty in processing biopolymers such as chitosan, polycaprolactone, etc., even with the enhanced conductivity of modified solutions (Niu and Lin 2012). In applications such as skin substitutes and wound dressings, determining the barrier property of electrospun nanomaterials is an important concern. The lack of standard methods to assess microbial barrier property boundaries has hindered the development of electron nanofibers as bioengineering and biomaterial products. Conversely, it is a tedious task to control pore packing for membrane filtration applications in nanofiber webs. Studies have developed methods for a solvent replacement to define opening sizes. However, several products have already been placed on the market as air filters; barriers have been surmounted to remove bacteria from wastewater. Functionalized nanomaterials are also becoming more important for water remediation as a result of the absence of effective polymer membrane-based processes as film-shaped nanofiber mats offer alternatives for hazardous contaminants.

REFERENCES

Anisha, B. S., R. Biswas, K. P. Chennazhi, and R. Jayakumar. 2013. "Chitosan–Hyaluronic Acid/Nano Silver Composite Sponges for Drug Resistant Bacteria Infected Diabetic Wounds." *International Journal of Biological Macromolecules* **62**:310–20.

Angammana, Chitral Jayasanka. 2011. A Study of the Effects of Solution and Process Parameters on the Electrospinning Process and Nanofiber Morphology. *UWSpace.* http://hdl.handle.net/10012/6214

Aruchamy, Kanakaraj, Ashesh Mahto, and S. K. Nataraj. 2018. "Electrospun Nanofibers, Nanocomposites and Characterization of Art: Insight on Establishing Fibers as Product." *Nano-Structures & Nano-Objects* 16:45–58.

Ballengee, J. B. and P. N. Pintauro. 2011. "Morphological Control of Electrospun Nafion Nanofiber Mats." *Journal of the Electrochemical Society* 158(5):B568.

Baumgarten, Peter K. 1971. "Electrostatic Spinning of Acrylic Microfibers." *Journal of Colloid and Interface Science* 36(1):71–9.

Cadafalch Gazquez, Gerard, Vera Smulders, Sjoerd A. Veldhuis, Paul Wieringa, Lorenzo Moroni, Bernard A. Boukamp, and Johan E. Ten Elshof. 2017. "Influence of Solution Properties and Process Parameters on the Formation and Morphology of YSZ and NiO Ceramic Nanofibers by Electrospinning." *Nanomaterials* 7(1):16.

Chandrasekhar, K. 2019. "Effective and Nonprecious Cathode Catalysts for Oxygen Reduction Reaction in Microbial Fuel Cells." Pp. 485–501 in *Microbial Electrochemical Technology.* Amsterdam: Elsevier.

Chandrasekhar, K. and Young-Ho Ahn. 2017. "Effectiveness of Piggery Waste Treatment Using Microbial Fuel Cells Coupled with Elutriated-Phased Acid Fermentation." *Bioresource Technology* 244(Pt 1):650–57.

Chandrasekhar, K., K. Amulya, and S. Venkata Mohan. 2015. "Solid Phase Bio-Electrofermentation of Food Waste to Harvest Value-Added Products Associated with Waste Remediation." *Waste Management* 45:57–65.

Chandrasekhar, K. and S. Venkata Mohan. 2012. "Bio-Electrochemical Remediation of Real Field Petroleum Sludge as an Electron Donor with Simultaneous Power Generation Facilitates Biotransformation of PAH: Effect of Substrate Concentration." *Bioresource Technology* 110:517–25.

Chandrasekhar, K. and S. Venkata Mohan. 2014a. "Bio-Electrohydrolysis as a Pretreatment Strategy to Catabolize Complex Food Waste in Closed Circuitry: Function of Electron Flux to Enhance Acidogenic Biohydrogen Production." *International Journal of Hydrogen Energy* 39(22):11411–22.

Chandrasekhar, K. and S. Venkata Mohan. 2014b. "Induced Catabolic Bio-Electrohydrolysis of Complex Food Waste by Regulating External Resistance for Enhancing Acidogenic Biohydrogen Production." *Bioresource Technology* 165:372–82.

Chandrasekhar, Kuppam, Abudukeremu Kadier, Gopalakrishnan Kumar, Rosa Anna Nastro, and Velpuri Jeevitha. 2018. "Challenges in Microbial Fuel Cell and Future Scope." Pp. 483–99 in *Microbial Fuel Cell*. Cham, Switzerland: Springer.

Chandrasekhar, Kuppam, Yong-Jik Lee, and Dong-Woo Lee. 2015. "Biohydrogen Production: Strategies to Improve Process Efficiency through Microbial Routes." *International Journal of Molecular Sciences* 16(12):8266–93.

Chang, Young-Cheol. 2019. *Microbial Biodegradation of Xenobiotic Compounds*. Boca Raton, FL: CRC Press.

Chee, W. K., H. N. Lim, Z. Zainal, I. Harrison, N. M. Huang, Y. Andou, K. F. Chong, and A. Pandikumar. 2017. "Electrospun Nanofiber Membranes as Ultrathin Flexible Supercapacitors." *RSC Advances* 7(20):12033–40.

Dai, Minhui. 2016. *ScholarWorks@UMass Amherst Functionalized Electrospun Nanofibers for Food Science Applications*. http://scholarworks.umass.edu

Deval, Animesh S., Harita A. Parikh, Abudukeremu Kadier, K. Chandrasekhar, Ashok M. Bhagwat, and Anil K. Dikshit. 2017. "Sequential Microbial Activities Mediated Bioelectricity Production from Distillery Wastewater Using Bio-Electrochemical System with Simultaneous Waste Remediation." *International Journal of Hydrogen Energy* 42(2):1130–41.

Doyle, James J., Santosh Choudhari, Seeram Ramakrishna, and Ramesh P. Babu. 2013. "Electrospun Nanomaterials: Biotechnology, Food, Water, Environment, and Energy." *Conference Papers in Materials Science* 2013:1–14.

Enamala, Manoj Kumar, Swapnika Enamala, Murthy Chavali, Jagadish Donepudi, Rajasri Yadavalli, Bhulakshmi Kolapalli, Tirumala Vasu Aradhyula, Jeevitha Velpuri, and Chandrasekhar Kuppam. 2018. "Production of Biofuels from Microalgae: A Review on Cultivation, Harvesting, Lipid Extraction, and Numerous Applications of Microalgae." *Renewable and Sustainable Energy Reviews* 94:49–68.

Fertala, Andrzej, Wendy B. Han, and Frank K. Ko. 2001. "Mapping Critical Sites in Collagen II for Rational Design of Gene-Engineered Proteins for Cell-Supporting Materials." *Journal of Biomedical Materials Research* 57(1):48–58.

Fong, H. 2001. "Electrospinning and the Formation of Nanofibers." Pp. 225–46 in *Structure Formation in Polymeric Fibers*, edited by D. R. Salem. Cincinnati, OH: Hansen.

Guang Yu, Deng-, Gareth R. Williams, Jun-he Yang, Xia Wang, Wei Qian1, and Ying Li. 2012. "Chitosan Nanoparticles Self-Assembled from Electrospun Composite Nanofibers." *Journal of Textile Science & Engineering* 2(1):1–5.

Huan, Siqi, Guoxiang Liu, Guangping Han, Wanli Cheng, Zongying Fu, Qinglin Wu, and Qingwen Wang. 2015. "Effect of Experimental Parameters on Morphological, Mechanical and Hydrophobic Properties of Electrospun Polystyrene Fibers." *Materials* 8(5):2718–34.

Huang, Siwei, Ling Zhou, Mei-Chun Li, Qinglin Wu, Yoichi Kojima, and Dingguo Zhou. 2016. "Preparation and Properties of Electrospun Poly (Vinyl Pyrrolidone)/Cellulose Nanocrystal/Silver Nanoparticle Composite Fibers." *Materials* 9(7):523.

Huang, Zheng-Ming, Y. Z. Zhang, M. Kotaki, and S. Ramakrishna. 2003. "A Review on Polymer Nanofibers by Electrospinning and Their Applications in Nanocomposites." *Composites Science and Technology* 63(15):2223–53.

Ignatious, Francis, Linghong Sun, Chao-Pin Lee, and John Baldoni. 2010. "Electrospun Nanofibers in Oral Drug Delivery." *Pharmaceutical Research* 27(4):576–88.

Kadier, A., M. S. Kalil, P. Abdeshahian, K. Chandrasekhar, A. Mohamed, N. F. Azman, W. Logroño, Y. Simayi, and A. A. Hamid. 2016. "Recent Advances and Emerging Challenges in Microbial Electrolysis Cells (MECs) for Microbial Production of Hydrogen and Value-Added Chemicals." *Renewable and Sustainable Energy Reviews* **61**:501–25.

Kadier, A., K. Chandrasekhar, and M. S. Kalil. 2017. "Selection of the best barrier solutions for Liquid Displacement Gas Collecting Metre to Prevent Gas Solubility in Microbial Electrolysis Cells." *International Journal of Renewable Energy Technology* **8**(2):93–103.

Kadier, Abudukeremu, Yong Jiang, Bin Lai, Pankaj Kumar Rai, Kuppam Chandrasekhar, Azah Mohamed, and Mohd Sahaid Kalil. 2018. "Biohydrogen Production in Microbial Electrolysis Cells from Renewable Resources." Pp. 331–56 in *Bioenergy and Biofuels*, edited by O. Konur. Boca Raton, FL: CRC Press.

Kadier, A., M. S. Kalil, K. Chandrasekhar, G. Mohanakrishna, G. D. Saratale, R. G. Saratale, G. Kumar, A. Pugazhendhi, and P. Sivagurunatham. 2018. "Surpassing the Current Limitations of High Purity H_2 Production in Microbial Electrolysis Cell (MECs): Strategies for Inhibiting Growth of Methanogens." *Bioelectrochemistry* **119**:211–19.

Kadier, Abudukeremu, Mohd Sahaid Kalil, Peyman Abdeshahian, K. Chandrasekhar, Azah Mohamed, Nadia Farhana Azman, Washington Logroño, Yibadatihan Simayi, and Aidil Abdul Hamid. 2016. "Recent Advances and Emerging Challenges in Microbial Electrolysis Cells (MECs) for Microbial Production of Hydrogen and Value-Added Chemicals." *Renewable and Sustainable Energy Reviews* 61:501–25.

Kadier, Abudukeremu, Mohd Sahaid Kalil, Kuppam Chandrasekhar, Gunda Mohanakrishna, Ganesh Dattatraya Saratale, Rijuta Ganesh Saratale, Gopalakrishnan Kumar, Arivalagan Pugazhendhi, and Periyasamy Sivagurunathan. 2018. "Surpassing the Current Limitations of High Purity H2 Production in Microbial Electrolysis Cell (MECs): Strategies for Inhibiting Growth of Methanogens." *Bioelectrochemistry* 119:211–19.

Kadier, Abudukeremu, Yibadatihan Simayi, Peyman Abdeshahian, Nadia Farhana Azman, K. Chandrasekhar, and Mohd Sahaid Kalil. 2016. "A Comprehensive Review of Microbial Electrolysis Cells (MEC) Reactor Designs and Configurations for Sustainable Hydrogen Gas Production." *Alexandria Engineering Journal* 55(1):427–43.

Kadier, Abudukeremu, Yibadatihan Simayi, K. Chandrasekhar, Manal Ismail, and Mohd Sahaid Kalil. 2015. "Hydrogen Gas Production with an Electroformed Ni Mesh Cathode Catalysts in a Single-Chamber Microbial Electrolysis Cell (MEC)." *International Journal of Hydrogen Energy* 40(41):14095–103.

Kakarla, Ramesh, Chandrasekhar Kuppam, Soumya Pandit, Abudukeremu Kadier, and Jeevitha Velpuri. 2017. "Algae—The Potential Future Fuel: Challenges and Prospects." Pp. 239–51 in *Microbial Applications, Vol. 1*, edited by V. C. Kalia and P. Kumar. Cham, Switzerland: Springer.

Karakaş, Hale. 2015. *Electrospinning of Nanofibers and Their Applications*. Istanbul, Turkey: Istanbul Technical University.

Kaul, Shreya, Neha Gulati, Deepali Verma, Siddhartha Mukherjee, and Upendra Nagaich. 2018. "Role of Nanotechnology in Cosmeceuticals: A Review of Recent Advances." *Journal of Pharmaceutics* 2018:3420204.

Kenawy, El-Refaie, Gary L. Bowlin, Kevin Mansfield, John Layman, David G. Simpson, Elliot H. Sanders, and Gary E. Wnek. 2002. "Release of Tetracycline Hydrochloride from Electrospun Poly(Ethylene-Co-Vinylacetate), Poly(Lactic Acid), and a Blend." *Journal of Controlled Release* 81(1–2):57–64.

Kiran Kumar, A., M. Venkateswar Reddy, K. Chandrasekhar, S. Srikanth, and S. Venkata Mohan. 2012. "Endocrine Disruptive Estrogens Role in Electron Transfer: Bio-Electrochemical Remediation with Microbial Mediated Electrogenesis." *Bioresource Technology* 104:547–56.

Kumar, Gopalakrishnan, Periyasamy Sivagurunathan, Arivalagan Pugazhendhi, Ngoc Bao Dung Thi, Guangyin Zhen, Kuppam Chandrasekhar, and Abudukeremu Kadier. 2017. "A Comprehensive Overview on Light Independent Fermentative Hydrogen Production from Wastewater Feedstock and Possible Integrative Options." *Energy Conversion and Management* 141:390–402.

Kumar, Prasun, Kuppam Chandrasekhar, Archana Kumari, Ezhaveni Sathiyamoorthi, Beom Kim, Prasun Kumar, Kuppam Chandrasekhar, Archana Kumari, Ezhaveni Sathiyamoorthi, and Beom Soo Kim. 2018. "Electro-Fermentation in Aid of Bioenergy and Biopolymers." *Energies* 11(2):343.

Kuppam, Chandrasekhar, Soumya Pandit, Abudukeremu Kadier, Chakradhar Dasagrandhi, and Jeevitha Velpuri. 2017. "Biohydrogen Production: Integrated Approaches to Improve the Process Efficiency." Pp. 189–210 in *Microbial Applications, Vol. 1*, edited by V. C. Kalia and P. Kumar. Cham, Switzerland: Springer.

Larrondo, L. and R. St. John Manley. 1981. "Electrostatic Fiber Spinning from Polymer Melts. I. Experimental Observations on Fiber Formation and Properties." *Journal of Polymer Science: Polymer Physics Edition* 19(6):909–20.

Laurencin, C. T., A. M. A. Ambrosio, M. D. Borden, and J. A. Cooper. 1999. "Tissue Engineering: Orthopedic Applications." *Annual Review of Biomedical Engineering* 1(1):19–46.

Lei Huang, R. Andrew McMillan, Robert P. Apkarian, Benham Pourdeyhimi, Vincent P. Conticello, and Elliot L. Chaikof. 2000. "Generation of Synthetic Elastin-Mimetic Small Diameter Fibers and Fiber Networks." *Macromolecules* 33(8):2989–97.

Levitt, Ariana S., Mohamed Alhabeb, Christine B. Hatter, Asia Sarycheva, Genevieve Dion, and Yury Gogotsi. 2019. "Electrospun MXene/Carbon Nanofibers as Supercapacitor Electrodes." *Journal of Materials Chemistry A* 7(1):269–77.

Liu, Xinhua, Max Naylor Marlow, Samuel J. Cooper, Bowen Song, Xiaolong Chen, Nigel P. Brandon, and Billy Wu. 2018. "Flexible All-Fiber Electrospun Supercapacitor." *Journal of Power Sources* 384:264–69.

Liu, Yan, Shiya Zhou, Yanlin Gao, and Yinglei Zhai. 2019. "Electrospun Nanofibers as a Wound Dressing for Treating Diabetic Foot Ulcer." *Asian Journal of Pharmaceutical Sciences* 14(2):130–43.

Memic, Adnan, Tuerdimaimaiti Abudula, Halimatu S. Mohammed, Kasturi Joshi Navare, Thibault Colombani, and Sidi A. Bencherif. 2019. "Latest Progress in Electrospun Nanofibers for Wound Healing Applications." *ACS Applied Bio Materials* 2(3):952–69.

Mohan, S. Venkata and K. Chandrasekhar. 2011a. "Self-Induced Bio-Potential and Graphite Electron Accepting Conditions Enhances Petroleum Sludge Degradation in Bio-Electrochemical System with Simultaneous Power Generation." *Bioresource Technology* 102(20):9532–41.

Mohan, S. Venkata and K. Chandrasekhar. 2011b. "Solid Phase Microbial Fuel Cell (SMFC) for Harnessing Bioelectricity from Composite Food Waste Fermentation: Influence of Electrode Assembly and Buffering Capacity." *Bioresource Technology* 102(14):7077–85.

Mohan, S. Venkata, K. Chandrasekhar, P. Chiranjeevi, and P. Suresh Babu. 2013. "Biohydrogen Production from Wastewater." Pp. 223–57 in *Biohydrogen*, edited by A. Pandey, J.-S. Chang, P. Hallenbeck, and C. Larroche. Amsterdam: Elsevier.

Niu, Haitao and Tong Lin. 2012. "Fiber Generators in Needleless Electrospinning." *Journal of Nanomaterials* 2012:1–13.

Oraby, Marwa A., Ahmed I. Waley, Ahmed I. El-Dewany, Ebtesam A. Saad, and Bothaina M. Abd El-Hady. 2013. "Electrospinning of Gelatin Functionalized with Silver Nanoparticles for Nanofiber Fabrication." *Modeling and Numerical Simulation of Material Science* 3:95–105.

Pandit, Soumya, Kuppam Chandrasekhar, Ramesh Kakarla, Abudukeremu Kadier, and Velpuri Jeevitha. 2017. "Basic Principles of Microbial Fuel Cell: Technical Challenges and Economic Feasibility." Pp. 165–88 in *Microbial Applications, Vol. 1*, edited by V. C. Kalia and P. Kumar. Cham, Switzerland: Springer.

Pandit, Soumya, Shruti Sarode, and Kuppam Chandrasekhar. 2018. "Fundamentals of Bacterial Biofilm: Present State of Art." Pp. 43–60 in *Quorum Sensing and Its Biotechnological Applications*, edited by V. C. Kalia. Singapore: Springer.

Pandit, Soumya, Shruti Sarode, Franklin Sargunaraj, and Kuppam Chandrasekhar. 2018. "Bacterial-Mediated Biofouling: Fundamentals and Control Techniques." Pp. 263–84 in *Biotechnological Applications of Quorum Sensing Inhibitors*, edited by V. K. Kalia. Singapore: Springer.

Patel, Vinay, Soumya Pandit, and Kuppam Chandrasekhar. 2017. "Basics of Methanogenesis in Anaerobic Digester." Pp. 291–314 in *Microbial Applications, Vol. 2*, edited by V. C. Kalia. Cham, Switzerland: Springer.

Pillay, Viness, Clare Dott, Yahya E. Choonara, Charu Tyagi, Lomas Tomar, Pradeep Kumar, Lisa C. du Toit, and Valence M. K. Ndesendo. 2013. "A Review of the Effect of Processing Variables on the Fabrication of Electrospun Nanofibers for Drug Delivery Applications." *Journal of Nanomaterials* 2013:1–22.

Reddy, M. Venkateswar, K. Chandrasekhar, and S. Venkata Mohan. 2011. "Influence of Carbohydrates and Proteins Concentration on Fermentative Hydrogen Production Using Canteen Based Waste under Acidophilic Microenvironment." *Journal of Biotechnology* 155(4):387–95.

Reddy, M. Venkateswar, M. Prathima Devi, K. Chandrasekhar, R. Kannaiah Goud, and S. Venkata Mohan. 2011. "Aerobic Remediation of Petroleum Sludge through Soil Supplementation: Microbial Community Analysis." *Journal of Hazardous Materials* 197:80–87.

Reneker, Darrell H. and Alexander L. Yarin. 2008. "Electrospinning Jets and Polymer Nanofibers." *Polymer* 49(10):2387–425.

Reneker, Darrell Hyson, Alexander L. Yarin, Hao Fong, Sureeporn Koombhongse, and Darrell H. Reneker. 2000. "Bending Instability of Electrically Charged Liquid Jets of Polymer Solutions in Electrospinning." *Journal of Applied Physics* 87(9):4531–47.

Saratale, Rijuta Ganesh, Chandrasekar Kuppam, Ackmez Mudhoo, Ganesh Dattatraya Saratale, Sivagurunathan Periyasamy, Guangyin Zhen, László Koók, Péter Bakonyi, Nándor Nemestóthy, and Gopalakrishnan Kumar. 2017. "Bioelectrochemical Systems Using Microalgae – A Concise Research Update." *Chemosphere* 177:35–43.

Savva, I., D. Constantinou, L. Evaggelou, Om Marinica, A. Taculescu, and L. Vekas. 2012. "PEO/PLLA and PVP/PLLA-Based Magnetoresponsive Nanocomposite Membranes: Fabrication via Electrospinning, Characterization and Evaluation in Drug Delivery." *Procedia Engineering* 44:1052–3.

Shi, Xiaomin, Weiping Zhou, Delong Ma, Qian Ma, Denzel Bridges, Ying Ma, and Anming Hu. 2015. "Electrospinning of Nanofibers and Their Applications for Energy Devices." *Journal of Nanomaterials* 2015:1–20.

Shin, Y. M., M. M. Hohman, M. P. Brenner, and G. C. Rutledge. 2001. "Experimental Characterization of Electrospinning: The Electrically Forced Jet and Instabilities." *Polymer* 42(25):09955–67.

Sivagurunathan, Periyasamy, Chandrasekhar Kuppam, Ackmez Mudhoo, Ganesh D. Saratale, Abudukeremu Kadier, Guangyin Zhen, Lucile Chatellard, Eric Trably, and Gopalakrishnan Kumar. 2018. "A Comprehensive Review on Two-Stage Integrative

Schemes for the Valorization of Dark Fermentative Effluents." *Critical Reviews in Biotechnology* 38(6):868–82.

Sivagurunathan, Periyasamy, Periyasamy Sivagurunathan, Abudukeremu Kadier, Ackmez Mudhoo, Gopalakrishnan Kumar, Kuppam Chandrasekhar, Takuro Kobayashi, and Kaiqin Xu. 2018. "Nanomaterials for Biohydrogen Production." Pp. 217–37 in *Nanomaterials: Biomedical, Environmental, and Engineering Applications*. Hoboken, NJ: John Wiley & Sons.

Sood, Rakhi, Sara Cavaliere, Deborah J. Jones, and Jacques Rozière. 2016. "Electrospun Nanofiber Composite Polymer Electrolyte Fuel Cell and Electrolysis Membranes." *Nano Energy* 26:729–45.

Spivak, A. F., Y. A. Dzenis, and D. H. Reneker. 2000. "A Model of Steady State Jet in the Electrospinning Process." *Mechanics Research Communications* 27(1):37–42.

Teo, Wee-Eong, Ryuji Inai, and Seeram Ramakrishna. 2011. "Technological Advances in Electrospinning of Nanofibers." *Science and Technology of Advanced Materials* 12(1):013002.

Thavasi, V., G. Singh, and S. Ramakrishna. 2008. "Electrospun Nanofibers in Energy and Environmental Applications." *Energy & Environmental Science* 1(2):205.

Torres-Giner, S. 2011. "Electrospun Nanofibers for Food Packaging Applications." Pp. 108–25 in *Multifunctional and Nanoreinforced Polymers for Food Packaging*, edited by J.-M. Lagarón. Oxford, UK: Woodhead Publishing.

Uyar, Tamer and Flemming Besenbacher. 2008. "Electrospinning of Uniform Polystyrene Fibers: The Effect of Solvent Conductivity." *Polymer* 49(24):5336–43.

Venkata Mohan, S., P. Chiranjeevi, K. Chandrasekhar, P. Suresh Babu, and Omprakash Sarkar. 2019. "Acidogenic Biohydrogen Production from Wastewater." Pp. 279–320 in *Biohydrogen, 2nd edition*, edited by A. Pandey, V. K. Mohan, J.-S. Chang, P. C. Hallenbeck, and C. Larroche. Amsterdam: Elsevier.

Waghmare, Vijaya Sadashiv, Pallavi Ravindra Wadke, Sathish Dyawanapelly, Aparna Deshpande, Ratnesh Jain, and Prajakta Dandekar. 2018. "Starch Based Nanofibrous Scaffolds for Wound Healing Applications." *Bioactive Materials* 3(3):255–66.

Xue, Jiajia, Tong Wu, Yunqian Dai, and Younan Xia. 2019. "Electrospinning and Electrospun Nanofibers: Methods, Materials, and Applications." *Chemical Reviews* 119(8):5298–5415.

Yarin, A. L., S. Koombhongse, and D. H. Reneker. 2001a. "Bending Instability in Electrospinning of Nanofibers." *Journal of Applied Physics* **89**(5):3018–26.

Yarin, A. L., S. Koombhongse, and D. H. Reneker. 2001b. "Taylor Cone and Jetting from Liquid Droplets in Electrospinning of Nanofibers." *Journal of Applied Physics* 90(9):4836–46.

Zhang, Wenjing 2010. "Electrospinning Pt/C Catalysts into a Nanofiber Fuel Cell Cathode." *The Electrochemical Society Interface* 19(4):51.

Zheng, Jian-Yi, Ming-Feng Zhuang, Zhao-Jie Yu, Gao-Feng Zheng, Yang Zhao, Han Wang, and Dao-Heng Sun. 2014. "The Effect of Surfactants on the Diameter and Morphology of Electrospun Ultrafine Nanofiber." *Journal of Nanomaterials* 2014:1–9.

Zong, Xinhua, Kwangsok Kim, Dufei Fang, Shaofeng Ran, Benjamin S. Hsiao, and Benjamin Chu. 2002. "Structure and Process Relationship of Electrospun Bioabsorbable Nanofiber Membranes." *Polymer* 43(16):4403–12.

Zucchelli, Andrea, Maria Letizia Focarete, Chiara Gualandi, and Seeram Ramakrishna. 2011. "Electrospun Nanofibers for Enhancing Structural Performance of Composite Materials." *Polymers for Advanced Technologies* 22(3):339–49.

Zussman, E., A. Yarin, and D. Weihs. 2002. "A Micro-Aerodynamic Decelerator Based on Permeable Surfaces of Nanofiber Mats." *Experiments in Fluids* 33(2):315–20.

3 Synthesis and Characterization of Nanocomposites of Animal Origin

Sweta Naik, Anita Tirkey, and Satya Eswari Jujjavarapu

CONTENTS

3.1 INTRODUCTION

Bionanocomposites have recently gained the tremendous interest of scientists and researchers as this area offers exciting platforms and it will provide an interface between biology, nanotechnology, and material science. Currently, bionanocomposites are very useful advantageous applications in polymer science, nanotechnology, tissue engineering, biomaterial, etc. Generally, the existing bionanocomposites are made from biopolymers, including proteins, polypeptides, polysaccharides, polynucleic acids, and aliphatic polyesters; while fillers include metal nanoparticles, hydroxyapatite, and clays. Bionanocomposites are also knowns as "biocomposites," "nanobiocomposites," "bioplastics," or "biohybrids," and "green composites" (Hasan et al., 2018). Bionanocomposites show various properties, such as water solubility, thermal stability, biodegradability, and biocompatibility (Arora, Bhatia, & Attri, 2018). These properties regulate the preparation techniques, the function of the composites, and the applicability of the material in different fields. Bionanocomposites are comprised of nanoparticles in a one-dimensional range (1–100nm) and constituted by biological material or biopolymers (Mishra, Ha, Verma, & Tiwari, 2018). Nanocomposites and composites are distinct terms. Generally, petroleum-based polymers are used as an organic constituent in "nanocomposites" and "composites" are composed of nanoscale size inorganic additives (Shchipunov, 2012).

Furthermore, one significant distinction between bionanocomposites and biocomposites is that biocomposites are consituted of biopolymers but they don't contain nanoscale-sized inorganic additives. These biopolymers are soluble in polar solvents (e.g. water), while in organic solvents petroleum-derived polymers are more soluble (Mohan, Oluwafemi, Kalarikkal, Thomas, & Songca, 2016). Even though there are some similarities with nanocomposites, there are also some variations, including biodegradability, biocompatibility, properties, and preparation techniques (Jamróz et al., 2019). Surprisingly, the shape of nanoparticles plays an important role in the formation of bionanocomposites and in the determination of the properties they exhibit (Müller et al., 2017). On the basis of geometry, the nanoparticulates can be divided into different categories, such as nanofibers, nanorods, nanotubes, nanoparticles, and nanoplatelets (Abdulkareem Ghassan, Mijan, & Hin Taufiq-Yap, 2019). Fabrication of nanocomposites is itself one of the major challenges. The pre-polymerization or post-polymerization preparation of nanocomposites has several disadvantages. Nanoparticulates enter the aquatic ecosystems through different paths, including industrial release or wastewater disposal, etc. (Müller et al., 2017), through various chemical and physical processes in the environment. Higher organisms can ingest these nanoparticles directly (Hussain & Mishra, 2018). In addition, both aquatic and terrestrial organisms can accumulate nanoparticles. Due to all these factors, a new type of material known as a nanocomposite was formed (Hlongwane, Sekoai, Meyyappan, & Moothi, 2019). Bionanocomposites are known as one of the most cost-effective materials with better performance in comparison with petroleum-based materials, as they consist of a significant amount of biobased material and maintain a positive balance between economics, technology, and ecology (Mhd Haniffa, Ching, Abdullah, Poh, & Chuah, 2016).

Therefore, these materials can be considered to be green. The terminology "green" is not a label but it is a new category of materials, products, and technologies (Irimia-Vladu, 2014). Generally, green refers to materials, products, and technologies that are less toxic or harmful to the environment, as well as to human health, than traditional equivalents. Biodegradation involves the decomposition of material in the natural environment in such a way that the materials convert to other compounds in terms of their structural properties, chemical composition, and mechanical properties (Mierzwa-Hersztek, Gondek, & Kopeć, 2019). These biodegradable materials prove to be non-toxic and less harmful to the environment or appear to be beneficial to the environment in some way. Composite materials that tend to be biodegradable and renewable are regarded as "green," compostable, or disposable without harming the environment (Lambert & Wagner, 2017). However, the main problem is the use of "green" materials to produce green composites. Naturally obtained polymers from different sources include chitosan, alginate, silk fibroin, etc. (Gao, Shih, Pan, Chueh, & Chen, 2018). Certain polymers from synthetic sources are also classified as biopolymers (such as polyvinyl alcohol, aliphatic and aromatic polyesters, etc.) and can be degraded in the environment, though they can not be renewed (Imre & Pukánszky, 2013). Thus, they do not fully follow the definition of biodegradability and renewability. In this chapter, the fabrication methods, characterization techniques, and applications of a large number of polymers of animal origin that have been exploited as green nanocomposites in their are discussed.

3.2 ECO-FRIENDLY BIO FABRICATION OF POLYMERIC NANOCOMPOSITES

Bionanocomposites belong to an evolving class of bio-inorganic nanostructured hybrid materials arising from the assembly of polymeric species of biological origin to inorganic substrates through nanometric-scale interactions (Hood, Mari, & Muñoz-Espí, 2014). Bionanocomposites naturally produced by living organisms show a significant hierarchical arrangement of organic and inorganic constituents from nanoscale to the macroscopic scale (Jeevanandam, Barhoum, Chan, Dufresne, & Danquah, 2018). Researchers have developed different types of biomimetic materials based upon the physical and chemical characteristics of bionanocomposites (F. Zhao et al., 2015). For example, hydroxyapatite-based synthetic materials and biopolymers such as gelatine and collagen are typical biomimetic materials that have been widely researched, particularly because they are of interest in tissue engineering (Saveleva et al., 2019). Recently, silicates from the mineral family of clay, and other inorganic solids such as layered perovskites, carbonaceous solids, and layered double hydroxides, have been explored in terms of the assembly of various biopolymers.

3.2.1 Collagen/Gelatine-Based Nanocomposites

Gelatine is a structural protein used in various applications. The rheological characteristics of this protein have rendered it an irreplaceable material in techniques such as photographic paper coating, where it provided, and still offers, a clear matrix for precipitating silver halide or bromide crystals for precipitation, or food processing, where there is a well-known "melt-in-the-mouth effect" for which no equivalent has yet been found. However, the potential of gelatine as a biocompatible, biodegradable, and extremely flexible polymer is presently showing itself through state-of-the-art techniques such as wound dressings, arterial implants, bone scaffolds, or regulated drug delivery systems. Changing the mechanical and functional characteristics of gelatine is crucial in many of these applications (Ramos et al., 2016). Use of inorganic or hybrid nanoparticles is now one of the most significant approaches for altering the mechanical and functional characteristics of biopolymers while preserving their biocompatibility (Ali et al., 2016). Zhuang et al. studied the effects of collagen and collagen hydrolysate from jellyfish on mice skin. First, they denatured the collagen protein, from which a portion of the polymer's structure is inherited. Collagen also acts at certain temperatures as a high-weight linear polymer with a random spatial arrangement. Even though the conversion between these two states is commonly recognized, the collagen protein has critical implications for the preparation and implementation of gelatine-based composites, as its mechanical characteristics, for example, differ dramatically depending on the protein-conforming state. However, this protein's complexity doesn't adhere to its conformational transition between triple helix and random coil since, as previously mentioned, gelatin to some extent preserves the primary structure of collagen. This is defined by a repeat unit (Gly-Xaa-Yaa) with an elevated content of glycine (Gly) and where the positions of Xaa and Yaa are usually occupied by residues of hydroxyproline and proline, together with the minor components inserted in the phase of denaturation of collagen such as arginine or glutamic acid.

Therefore, the following elements must be taken into account in a thorough view of gelatine and derived products:

1. Below transition temperature, gelatine can be observed as a physical gel exhibiting collagen-like triple helix conformation.
2. Above transition temperature, gelatine has a linear random polymer conformation.
3. Due to its anionic and cationic residues in the polymer backbone, it exhibits polyampholytic behavior.

These are the key components that allow some degree of insight into the properties of gelatine, the continuous phase in gelatin-based nanocomposites.

While structural bionanocomposites based on gelatine, referred to in the previous section, are more prevalent, these hybrid materials' functional features are similarly staggering. This section looks at the most appropriate and exceptional applications of gelatine-based functional nanocomposites, some of which will be analyzed in detail in the following parts. Darder et al. presented findings revealing the regulated variability that can be brought into the gelatine gelling temperature by various clay minerals such as laponite, cloisite, or sepiolite. As noted earlier, the gelling temperature in gelatine-based nanocomposites is critical to the processability and applicability of bio-nanocomposites under certain conditions. Another critical feature, also associated with the structural strengthening of gelatine-based nanocomposites, is the biocompatibility of prepared scaffolds for bone regeneration therapies (Darder, Colilla, & Ruiz-Hitzky, 2003). Numerous attempts have been made in this context to develop approaches that could improve biocompatibility, such as the introduction of bone morphogenetic proteins. Bone-tissue engineering applications, combined with biocompatibility, often involve a sensitive equilibrium between biodegradability and osteoconduction. The chemical crosslinking of the gelatine matrix with cytotoxic compounds such as glutaraldehyde or formaldehyde generally achieves such an equilibrium. Recent progress has shown the use of chemical cross-linkers with low toxicity such as oligomeric proanthocyanidins. Chen et al. or Sisson et al. can substitute elevated toxicity counterparts efficiently, creating a sustainable cell development environment. As the products currently on the pharmaceutical market such as Gelfoam R or Floseal R Haemostatic Matrix show, gelatine, and the resulting nanocomposites, are also widely used in hemostatic applications. However, in various processes that use gelatine-based products, such as endoscopic sinus surgery, cardiovascular surgery, or laparoscopic surgery, further inquiry is under way to improve hemostasis. Other biomedical applications of gelatine-based composites include drug delivery and gene therapy from vitamins to peptides engaged in DNA transfection owing to the possibility of functionalizing these materials with the required bioactive element (Chen & Yao, 2011).

To sum up, a broad variety of apps are being constructed for functional nanocomposites based on gelatine. The wide range of inorganic solids offering interesting functionalities that can be combined into gelatine, or the option of greasing the required functional groups into gelatin-based hybrids, gives a wide range of methods that will assist us to create advanced functional materials from such a prevalent polymer.

3.2.2 Silk Fibroin-Based Nanocomposites

Silk fibroin (SF) is a natural polymeric material found abundantly in the natural environment. It can be extracted from Bombyx mori silkworms. The chemical structure of silk fibroin comprises of L-chain, H-chain, and P25 glycoprotein. The H-chain comprises various amino acid sequences, but around 94% of the sequence of amino acids is made up of glycine-alanine-glycine-alanine-glycine-serine (GAGAGS) repeats. This sequence assembles together to form a strong crystalline antiparallel β-sheet structure. Engineered silk fibroin material can be prepared by adjusting the β-sheet content and crystallinity index in its structure (Melke, Midha, Ghosh, Ito, & Hofmann, 2015). Due to this assembly, SF offers high mechanical strength. Along with the side chain, SF is also composed of some amino acids such as tyrosine, glutamic, aspartic, threonine, and serine. These amino acids offer reactive sites for functional and chemical modification. Silk fibroin materials can be prepared in various forms such as hydrogel, nanofiber, scaffold, SF beads, mesospheres, etc (Joseph & Raj, 2012). Because of its unique features, and biological properties such as biodegradability, biocompatibility, and low *in vivo* inflammatory response, silk fibroin is an excellent candidate for biomaterial for various biological applications and can be used as a template in a number of *in vivo* and *in vitro* studies. Silk fibroin has recently been used in various biomedical areas involving the delivery of drugs, tissue engineering, and implantable devices. Specifically, due to their exceptional mechanical strength, degummed SF fibers have been studied for ligament and tendon repair (Kundu, Rajkhowa, Kundu, & Wang, 2013).

3.2.2.1 Preparation Methods of SF-Nanoparticle Composites

Generally, there are three main methods that are used to introduce nanoparticles into the SF matrix, that is, directly mixing, in-situ synthesis, and silkworm feeding method. In the first method, the nanoparticles are directly introduced into the SF solution and, by physical interaction, the macromolecules present in the nanoparticles will bind with SF molecules to form nanocomposite in different forms such as a hydrogel, scaffold, nanofiber, film, etc. In the second method, SF materials are directly mixed with the nanoparticles' precursor solution, as SF works as a template for the growth and in-situ nucleation of the nanoparticles. In the third method, silkworms are fed with nanoparticles to obtain naturally spun SF-nanoparticle-based composite fiber (Zhao, Li, & Xie, 2015).

3.2.2.2 Direct Mixing

This method can be easily manipulated. One of the major problems with this method is an aggregation of nanoparticles due to their high surface area. Therefore, for uniform distribution of nanoparticles in the SF solution generally ultrasonication and mechanical stirring methods are used in this process. After sonication, the nanoparticles mixed with the SF solution are able to retain their homogeneity and stability even after a month (Xu, Shi, Yang, & Zhu, 2019). The direct mixing method for preparation of silfibroin-nanoparticle composite is illustrated in Figure 3.1.

Wang et al. prepared a SF-GO composite film. The graphene oxide (GO) was dispersed by an ultrasonication method and then added dropwise into the SF solution to get a composite mixture. The obtained mixture was used for SF-GO composite film preparation by a solvent evaporation method. The resultant film was characterized by scanning electron micrograph (SEM), showing that the graphene oxide sheet was embedded properly and evenly distributed in the SF matrix without forming aggregation. This results clearly indicated that both SF and GO were compatible with each other. The crystal structure formation in the SF-GO composite is due to the addition of graphene. This crystal structure forms due to strong hydrogen

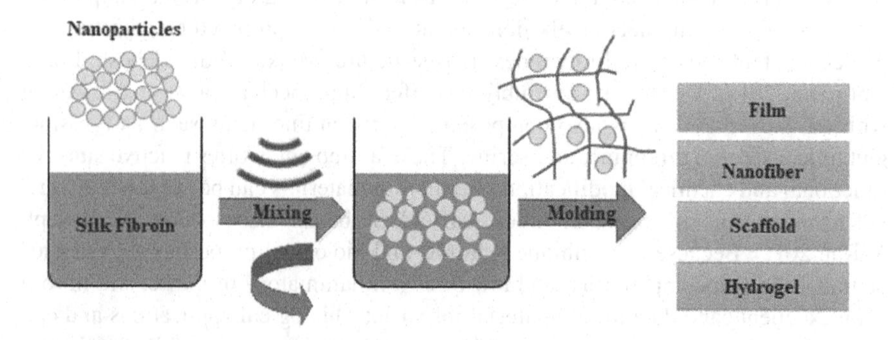

FIGURE 3.1 Silk fibroin-nanoparticle composite preparation by direct mixing method.

bonds/intermolecular forces between the polar group of SF chains and the functional group of GO sheets. The mechanical strength, thermal stability, and enzymatic degradation resistivity of the composite film have significantly improved with the addition of a low concentration of GO (Wang et al., 2014).

Feng et al. fabricated silk fibroin and cellulose nanocomposite films with high mechanical strength for tissue engineering applications. Silk fibroin was extracted from Bombyx mori silk and cellulose nanofibrils are extracted from microcrystalline cellulose by using an aqueous lithium bromide solution for the fabrication of nanocomposites. Cellulose nanofibrils also possess excellent biocompatibility and high mechanical strength. Both materials, silk fibroin and cellulose nanofibrils, were separately dissolved in aqueous lithium bromide solution and then mixed together to form nanofilm or nanocomposites as shown in Figure 3.2. As a result, the obtained silk fibroin/cellulose nanocomposite films show improved mechanical strength as compared to individual silk fibroin and cellulose nanofibrils. The solution used to dissolve the nanofibrils, that is, aqueous lithium bromide solution, is also reported as a novel solvent for dissolution by Feng et al. This nanocomposite also promotes cell viability and good biodegradability, which can make this nanocomposite an excellent potential candidate for application in tissue engineering (Feng et al., 2017).

The direct mixing approach provides a convenient and simple method for the fabrication of SF-nanoparticle composites. A low concentration of nanoparticles is good for biomedical applications and a high concentration may lead to an increase in toxicity and reduce the function.

3.2.2.3 In-Situ Synthesis

In this method, silk fibroin acts as a template for nanoparticle synthesis, its growth and in-situ nucleation. The intrinsic nature of the nanocomposite can be achieved by functional coating and grafting groups. This process is illustrated in Figure 3.3,

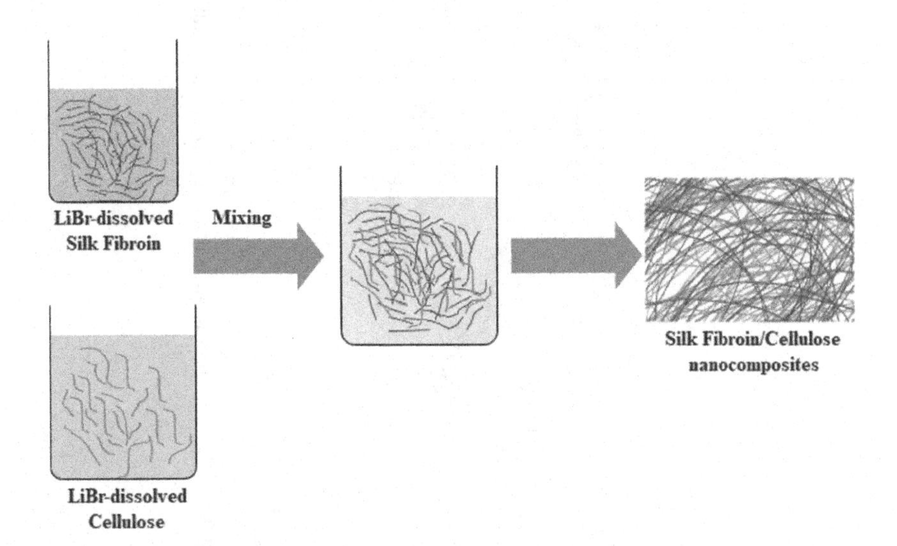

FIGURE 3.2 Silk fibroin/cellulose nanocomposite films using direct mixing method.

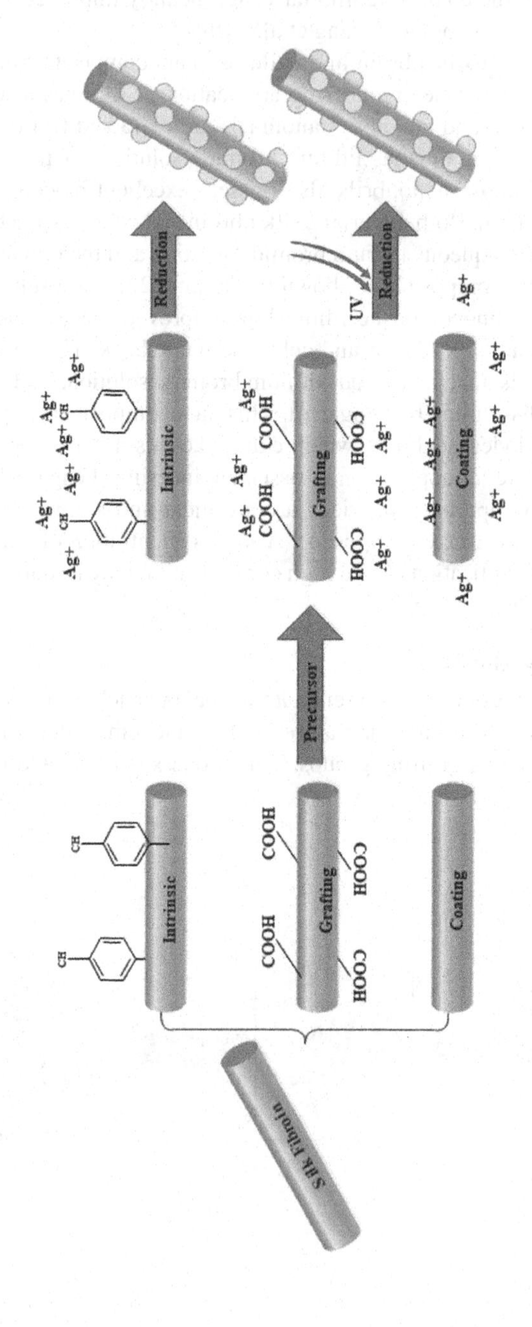

FIGURE 3.3 Silk fibroin-nanoparticle composite preparation by in-situ synthesis method.

where Ag+ is used as an example of a precursor in the in-situ synthesis method (Xu et al., 2019).

3.2.2.4 Silkworm Feeding Method

In this method, nanomaterials are given in the diet of silkworms to get naturally spun fiber. The nanoparticles are absorbed by the posterior silk glands of the silkworm and enter into the fibroin and produc SF fiber-containing nanoparticles (see Figure 3.4). All three silk nanoparticles composite preparation methods work very well, each with their own advantages and disadvantages in terms of the properties and applications of the nanocomposite. In the direct mixing method, the nanoparticles integrated pderfectly into the SF. It is a much easier method for the fabrication of silk-based nanocomposites compared to other methods (Xu et al., 2019). Using the direct mixing method, various composites can be prepared, such as hydrogels, scaffolds, and nanofibers for application in tissue engineering. In the in-situ synthesis method, SF acts as a template for nanoparticle synthesis, cellular growth, and mediate nucleation. The silkworm feeding method is a technique used for the preparation of nanoparticle-doped silk fiber. However, the nanofiber synthesized from this method is limited to certain applications due to its nature. To resolve this problem, scientists are now choosing other techniques according to their needs and to demand. In Table 3.1 various forms of SF-NP composites prepared by using these three methods are summarized.

3.2.3 Alginate-Based Nanocomposites

Alginate is a naturally occurring water-soluble anionic polymer obtained from brown seaweed. It is composed of primarily (1–4)-linked units of β-d-mannuronic acid and α-l-guluronic acid. Alginate is the only polysaccharide that includes carboxyl groups naturally in each residue and has different functionalities, as shown in Figure 3.5.

Alginate is known as an excellent biomaterial, which has been used in various applications in nanotechnology, biomedical science, food additives, pharmaceutical additives, enzyme carriers, and tissue engineering due to its promising characteristics, such as biodegradability, cytotoxicity, biocompatibility, and ease of gelation. In particular, alginate hydrogels nowadays have become more attractive in the fields of wound healing, drug delivery, and tissue engineering applications. This alginate hydrogel retains very good structural similarity with respect to the extracellular

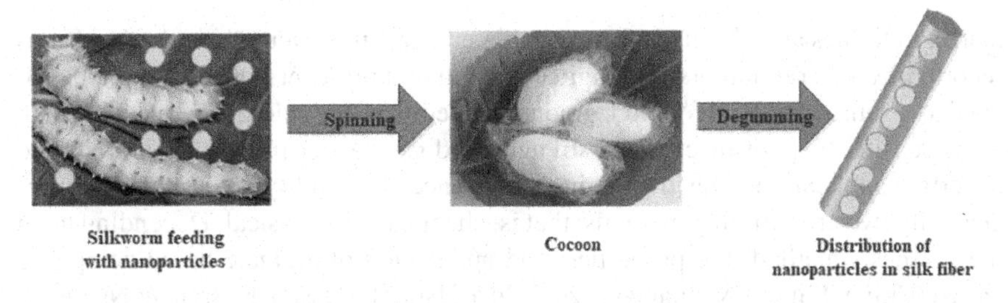

FIGURE 3.4 Synthesis of silk-fibroin nanoparticle composite by silkworm feeding methods.

TABLE 3.1
SF-NPs Nanocomposite and Their Applications

Form of Silk Fibroin Material	Types of Nanoparticles	Application	References
Film	GO (Graphene Oxide)	Area of biomedical aspects of engineering	Izyan Syazana Mohd Yusoff, Uzir Wahit, Jaafar, & Wong (2018); Zhao et al. (2018); Rodríguez-Lozano et al. (2014)
Fibers and films	CdTe QDs	Bio-imaging	Sohail Haroone et al. (2018)
Hydro-gel	LAP	Engineered bone-tissue aspects	Su, Jiang, Chen, Dong, & Shao (2016)
Scaffolds	TiO_2	Bone-tissue engineering	Johari, Madaah Hosseini, & Samadikuchaksaraei (2018)
Fiber	Ag	Antibacterial	Liu, Yin, Yi, & Duan (2016)
Nanofiber	Au (gold)	Tissue engineering	Cohen-Karni et al. (2012)
Fiber	AgCl	Antibacterial	Xie, Xu, Hu, He, & Zhu (2017)
Nanofiber	MWCNT	Animal tissue-engineering aspects	Kang & Jin (2007)
Membranes	GO (Graphene Oxide)	Nanosensing aspects in biological ways	Wang et al. (2016)
Membranes	Silica	Bio-sensor	Kharlampieva et al. (2010)
Scaffolds	HAP	Engineered bone tissue aspects	Ye, Yu, Deng, She, & Huang (2017); L. Liu, Liu, Kong, Cai, & Yao (2011)
Films	Silica	Engineered bone tissue aspects	Mieszawska et al. (2010)
Hydrogel	Au (gold)	Infection-treatment	Kojic et al. (2012)
Nanofibrils	MoS_2	Tumor-treatment	Li, Yang, Yao, Shao, & Chen (2017)
Film	SWCNT	Neuroregeneration	Mottaghitalab et al. (2013)
Scaffold	Au (gold)	Neuroregeneration	Cohen-Karni et al. (2012)
Membrane	Au (gold)	Bio-sensor	Vargas-Bernal, Rodrguez-Miranda, & Herrera-Prez (2012)
Hydrogel	CdTe QDs	Bioimaging	Zheng et al. (2015)
Nanosphere	Nanodiamond	Bioimaging	Khalid et al. (2015)
Nanoparticle	Fe_3O_4	Cancer treatment	Zhang, Ma, Cao, Wang, & Zhu (2014)
Nanoparticle	Fe_3O_4	Drug delivery	Zhao, Li, Xie, & Bing (2015)

matrix in tissues and also it can be manipulated to act in various critical roles. It was reported by several authors that alginate also has significant potential to be developed as a source material for biodegradation or edible films, although, due to its high water sensitivity, poor mechanical strength, and gas barrier properties, it is limited to certain applications, specifically in the presence of humidity and water. There are generally two crosslinking methods, that is, chemical and physical. Depending upon the synthesis method, the properties, and application of alginate-based composite varied (Ching, Bansal, & Bhandari, 2017; Idris, Ismail, Hassan, Misran, & Ngomsik, 2012). For the synthesis of alginate-based composite, there are four methods

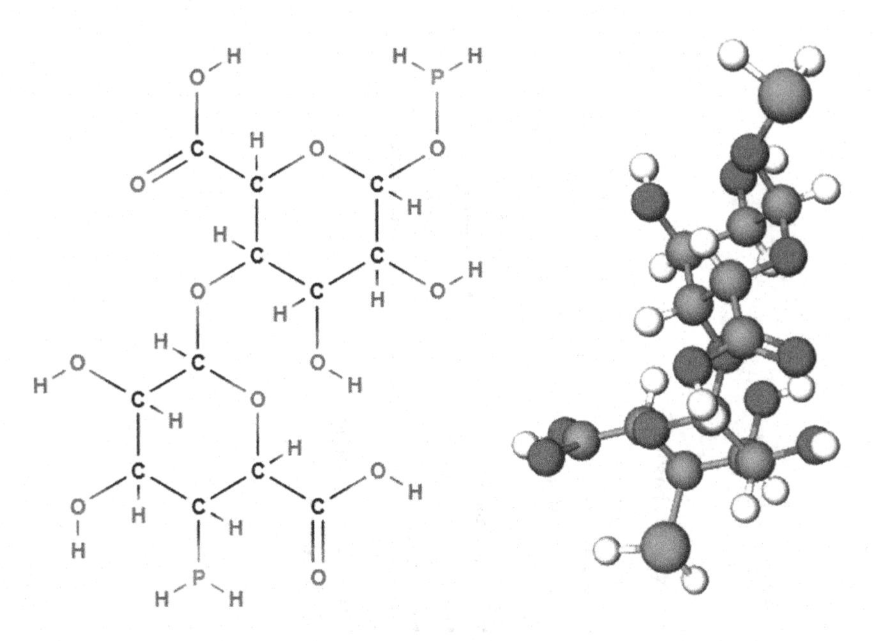

FIGURE 3.5 Structural representation of alginate.

available, namely electrostatic complexation, emulsification, ionic crosslinking, and self-assembly (Akhtar, Hanif, & Ranjha, 2016; Mane, Ponrathnam, & Chavan, 2015; Paques, van der Linden, van Rijn, & Sagis, 2014). Physically crosslinked composite can be prepared by protein interaction, hydrophobized polysaccharides, stereocomplex formation, crystallization, and ionic interaction. Nanocomposite prepared by the chemically crosslinked method is synthesized by chain-growth polymerization. Alginate composite prepared by mixing the nanoparticles with the alginate solution before the gelation process starts as illustrated in Figure 3.6. Selection of materials is one of the important factors that enhances the mechanical and physical properties of the nanocomposite.

First, the alginate beads provide a stable polymeric matrix for the small particle absorbents. These absorbents are so small that it is very difficult to separate them from the aqueous solution. These types of absorbents are usually carbon-based materials including graphene oxide, biochar, activated carbon, and carbon nanotubes (Mohammadi, Khani, Gupta, Amereh, & Agarwal, 2011; Gupta, Khamparia, Tyagi, Jaspal, & Malviya, 2015). Second, nanocomposites prepared from magnetic nanoparticles and alginate show excellent absorption capacity and are also less toxic to the environment. Nowadays, magnetic nanoparticle-based nanocomposites are widely used for wastewater treatment and water quality analysis (Han, Zeng, Li, & Chang, 2013; Theron, Walker, & Cloete, 2008). Alginate- and nanoparticle-conjugated nanocomposites provide an improved absorption capacity to the material. Third, alginate offers itself as a carrier for microbes for the optimization of microbial technology for agricultural and environmental applications (Cohen, 2001; Covarrubias, de-Bashan, Moreno, & Bashan, 2012). Alginate microorganism composites provide

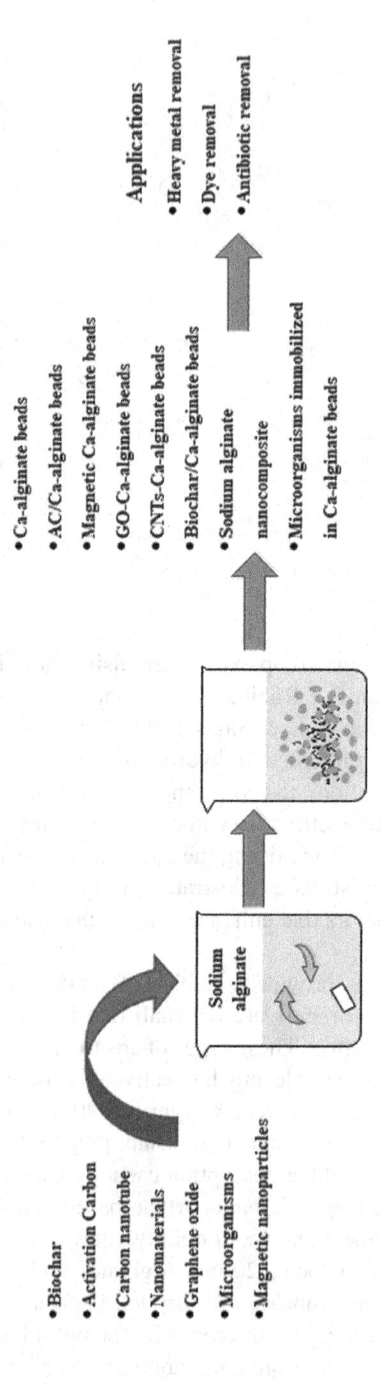

FIGURE 3.6 Fabrication of different types of alginate based nanocomposite.

various advantages compared to other conventional systems, including more resistance to toxicants, a high metabolic activity, and high biomass.

Furthermore, these types of nanocomposites can be used many times without decrease in activity of the microorganisms immobilized on the nanocomposite (Junter & Jouenne, 2004; Cai, Chen, Ren, Cai, & Zhang, 2011; T. An, Zhou, Li, Fu, & Sheng, 2008). That is why alginate-based microbial technology has gained lots of attention for wastewater treatment (T. An et al., 2008). Abdollahi et. al testified to the use of cellulose nanoparticles prepared from alginate solutions by sulfuric acid hydrolysis. Nanocomposite film was prepared by adding a different concentration of cellulose nanoparticles (such as 1, 3, 5, and 10% (w/w)) to an alginate solution followed by five minutes' sonication of the final mixture solutions. After sonication, the obtained solution mixture was stirred for one hour and at room temperature was homogenized at 1000 rpm for five minutes. The mixed solution was degassed by using a vacuum pump for the removal of bubbles and the nanocomposite film was fabricated by casting the mixed solution. The impact of nanocrystal loading on alginate nanocomposites was analyzed in terms of optical, physicomechanical, and microstructural properties (Abdollahi, Alboofetileh, Behrooz, Rezaei, & Miraki, 2013). The obtained results indicated that the water vapor permeability and water solubility of the nanocomposites decreased by 17% and 40% and the number of cellulose nanoparticles increased to 10%. The high crystallinity index of cellulose nanoparticles and the tendency of a strong hydrogen bond with the alginate matrix reduced the wt.% nanocrystal from 18.03 to 22.4 MPa. Further rises in nanocrystal content, however, resulted in negative results owing to partial nanocrystal agglomeration. These findings have shown that cellulose nanoparticles have excellent potential in food packaging applications as a reinforcing filler to overcome the constraints of alginate film. Further rises in nanocrystal content, however, resulted in negative results owing to partial nanocrystal agglomeration. Abdollahi et al. (2013) also demonstrated that cellulose nanoparticles are more appropriate than inorganic nanoclays for creating fully renewable and environmentally friendly alginate nanocomposites.

Huq et al. (2012) prepared alginate-based cellulose nanocrystals by using a solution casting method. The content of nanocrystals in the alginate matrix was varied from 1 to 8 %. From that study, it was detected that a 5 wt.% nanocrystal containing film exhibited the highest tensile strength, which was increased by 37%. Addition of nanocrystals (5 wt.%) decreased nanocomposite water vapor permeability by 31%. In a Fourier transform infrared spectroscopy (FTIR) study, it was observed that hydrogen bonding interaction increases between alginate and nanocrystalline cellulose (NCC). The XRD studies confirmed, in the presence of the NCC, the formation of extra crystalline peaks in alginate film, indicating the powerful interaction between the NCC and the matrix of alginates. In the same study, the physiochemical and thermal properties of alginate-based films were found to be impacted significantly by a small number of cellulose nanoparticles (3 to 5%) (Huq et al., 2012).

3.2.4 Chitosan-Based Nanocomposites

Chitosan is a naturally occurring polymer and it is a derivative of chitin. Chitin is found in the exoskeleton of insects, crustaceans, and in a few varieties of fungi.

Commercially, chitin can be extracted from the shells of lobsters, shrimp waste, crabs, and krill. Chitin can be converted to chitosan through enzymatic or chemical deacetylation in the presence of an alkali. In a few varieties of fungi this conversion may occur naturally. After cellulose, it is the second most abundant cationic polysaccharide in nature. It is the only cationic biopolymer whose structure is very similar to cellulose (Elieh-Ali-Komi & Hamblin, 2016). The structure of chitin and chitosan is illustrated in Figure 3.7.

Nowadays, chitosan has gained more scientific attention in the fields of biomaterials, tissue engineering, and drug delivery systems due to its tremendous biological properties. Chitosan shows excellent biological properties such as biodegradability, biocompatibility, anticarcinogenic, fungistatic, anticholesteremic, bacteriostatic, and hemostatic, etc. Depending upon their application and function, it can be used in various forms (Saikia, Gogoi, & Maji, 2015). Various forms of chitosan nanocomposite are illustrated in Figure 3.8 along with their preparation methods. For the selection of appropriate preparation methods, various factors need to be taken into account, such as kinetic profile, type of drug delivery system, particle size, bioactive agents, thermal and chemical stability, etc.

The chitosan-based nanocomposites have potential applications in various areas such as environmental protection, textiles, cosmetics, agriculture, biotechnology, medicine, food industries, etc. This type of nanocomposite exhibits excellent improvement in its properties at low concentrations of nanofillers compared to others with a high percentage of nanofillers. If the percentage of chitosan is high in the nanocomposite then the fabricated product will show good biocompatibility, improved mechanical strength, high transparency, good thermal stability, and bioactivity (Cheaburu-Yilmaz, Yilmaz, & Vasile, 2015). Nanofillers are classified into three groups based on their aspect ratio and geometry: (i) nanoplate, (ii) nanoparticles (spherical-shaped), and (iii) nanofiber as shown in Figure 3.9.

Recently, the scientific world and researchers have been focusing on various chemotherapy agents used in drug delivery systems. These agents are generally used to target drug delivery sites by simultaneously reducing the systemic distribution. Chitosan nanocomposites are gaining tremendous attention in various diagnostic applications due to their easy formulation, and are being introduced as imaging agents and therapeutic agents, which is one major challenge. Due to their unique properties, chitosan nanocomposites are drawing more consideration in cancer imaging and diagnostics applications (Patra et al., 2018).

Venkatesan et al. fabricated hydroxyapatite-chitosan nanocomposite along with celecoxib as a drug for colon cancer. Some of the chitosan nanocomposite along with its application is illustrated in Table 3.2.

3.2.5 Albumin-Based Nanocomposites

Albumin is a natural polymer. It is one of the important components of human blood. Albumin has been widely used in various formulations as a clinical excipient as listed in Table 3.3. Due to its successful applicability in various clinical and medicinal applications, researchers have recently begun to focus on albumin for its development and employment in nanomedicine. Because albumin is found throughout the body, in blood, albumin, in the form of Megatope and Janatope, has been used

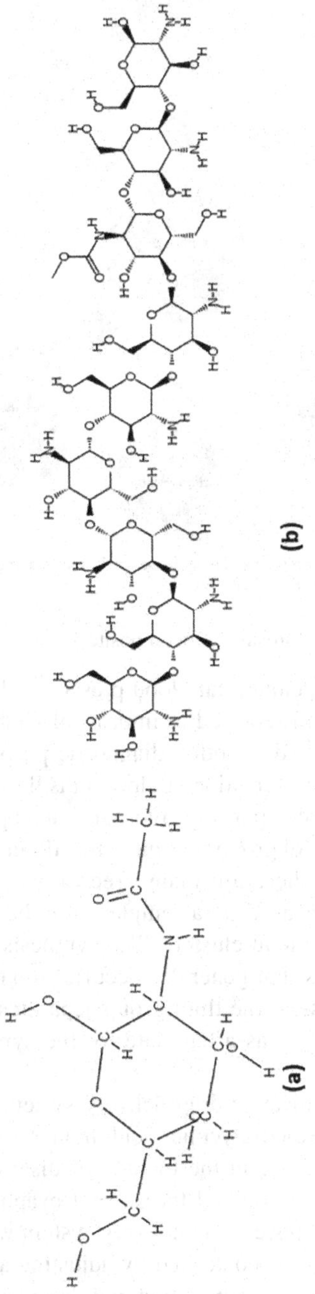

(a)

(b)

FIGURE 3.7 Structure of chitin and chitosan.

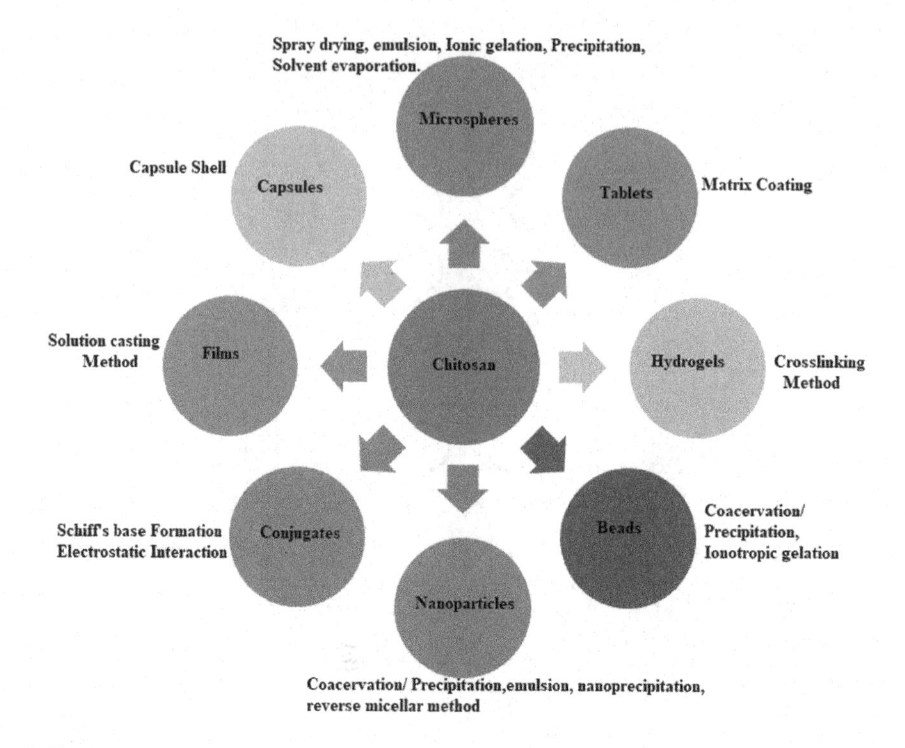

FIGURE 3.8 Various forms of chitosan nanocomposite.

to detect the volume of plasma and total blood present in the human body. Further investigation of albumin is still required to improve the fabrication method, on its application in drug delivery, and for other diagnostic purposes (Zhao, Jia, Shi, & Huang, 2019). A structural representation of albumin is illustrated in Figure 3.10.

There are several scientific reports on inorganic nanoparticle synthesis using thiols. Cysteine-34 is a free thiol group present in the albumin structure. According to several scientific reports, there are many free thiols present in the albumin. Therefore, albumin has been used as a template for the synthesis of inorganic nanocrystals and fluorescent metal clusters. The synthesis process is the same as the biomineralization process that generally occurrs in natural organisms (An & Zhang, 2017). In both processes, the thiol group separates and interacts with the metal ions and offers a scaffold as a template for the synthesis of nanoparticles (see Figure 3.11).

Misak et al. created an innovative drug delivery system that incorporates magnetite nanoparticles, poly(lactic-co-glycolic) acid, human serum albumin, and therapeutic agents for prospective use in the therapy of diseases such as skin cancer and rheumatoid arthritis. They modified the solvent evaporation and an oil-in-oil emulsion method to prepare a novel drug delivery system with a diameter range of 0.5 to 2 μm. The diameter can be controlled by adjusting albumin viscosity in the discontinuous stage of the solvent evaporation and the oil-in-oil emulsion method. From the drug delivery system, it was observed that mostly the release of albumin

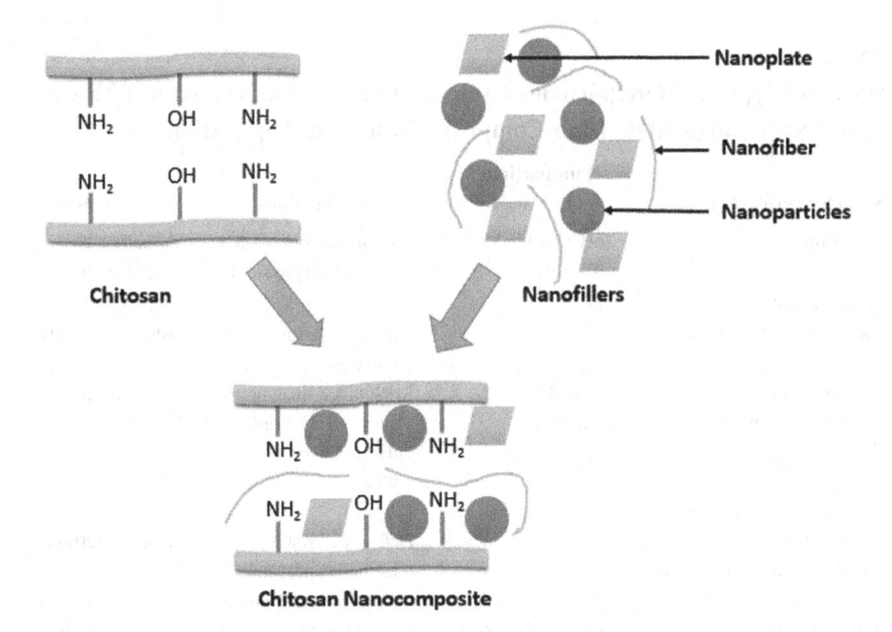

FIGURE 3.9 Diagrammatical representation of chitosan nanocomposite containing different types of nanofillers.

and drug depends upon the albumin concentration in the drug delivery system, which is very closely related to the occlusion-mesopore drug model. Cell viability studies showed that cell viability increases with an increase in the concentration of albumin in the drug delivery system, potentially owing to enhanced system biocompatibility as compared to the low content of albumin in the drug delivery system. From the overall studies, it was concluded that the novel system could become a viable option as a drug delivery system for the treatment of certain diseases such as breast cancer, skin cancer, and rheumatoid arthritis (Misak et al., 2014).

Shadpour Mallakpour et al. investigated the physicochemical characteristics of polyvinyl chloride (PVC) and zinc oxide nanocomposite. First, zinc oxide nanoparticles were modified by using bovine serum albumin as a biocompatible substance and organo-modified through an ultrasound irradiation method. These modified materials are cost effective,eco friendly, and easy to use. Nanocomposite (NC) films were produced inside the PVC by loading different ZnO / BSA NP ratios (3, 6 and 9 wt.%). The morphology and physical properties of the zinc-oxide-based bovine serum albumin nanoparticles and nanocomposite film structure were studied via FTIR, X-ray diffraction (XRD), thermogravimetric analysis (TGA), field emission scanning electron microscopy (FESEM), and transmission electron microscopy (TEM). Thermal stability improved based on their results from TGA. Also, the findings of the contact angle study showed that the hydrophilic behaviors of NCs increased with an increase in the percentage of zinc-oxide-based bovine serum albumin nanoparticles in PVC (Mallakpour & Darvishzadeh, 2018).

TABLE 3.2
Different Types of Nanoparticles Used in Chitosan Matrix to Form Chitosan-Based Nanocomposites, Their Composite Types and Application

Nanocomposite Types	Nanoparticulates Used	Applications	References
Thin Film	MCM-41 or MCM-41-APS	Metformin controlled drug delivery system	Shariatinia & Zahraee (2017)
Surface modification by using magnetic nanoparticles	Fe_3O_4	Cellular imaging and controlled drug delivery	Arum, Oh, Kang, Ahn, & Oh (2015)
N-naphthyl-O-dimethylmaleoyl-drug loaded magnetic nanoparticles (NChitosan-DMNPs)	Fe and Mn magnetic nanocrystals (MNCs)	MR-guided imaging for cancer therapy and pH-sensitive drug release	Lim et al. (2013)
Covalent functionalized graphene oxide with chitosan	Graphene oxide	Delivery of water insoluble anti-cancer drug CPT and Genes	Bao et al. (2011)
Hydrogel beads	Zinc oxide – Zno	Controlled drug delivery system	Liu, Jiao, & Zhang (2007)
ZnO-chitosan mediated gel	Zinc oxide	Wound care application	Sudheesh Kumar et al. (2012)
Ferrous oxide nanoparticles (NP) decorated with chitosan and functionalized with polyethylene glycolated methotrexate and cyanine dye	Ferrous oxide	Fluorescence and magnetic resonance imaging as self-targeted therapeutic delivery device for cancer treatment	Lin et al. (2015)
Magnetite/chitosan mediated core/shell nanocomposite	Fe_3O_4	Cancer therapy via stimuli-sensitive nanomedicine	Arias, Reddy, & Couvreur (2012)
Magnetic hydrogel	Fe_3O_4	Intravesical drug delivery	Ragab & Rohani (2016)

3.2.6 Hyaluronic Acid-Based Composites

Hyaluronic acid is a nonsulfated glycosaminoglycan, which is found in epithelial, neural, and connective tissues (see Figure 3.12). By crosslinking hyaluronic acid (HA) and sodium alginate (SAL), Yong-Hao Chen et al. prepared a new hydrogel nanocomposite with enhanced structural morphology and mechanical characteristics. Fourier-transform infrared spectroscopy verified the presence of covalent bonds or amide bonds in the hydrogel that they intended to form by crosslinking. From the morphological study of the hydrogel nanocomposite, using SEM, they observed that the pore size of the nanocomposite is greater than 100 μm. Analysis of the texture profile stated that the hardness of the hydrogels increased with a rise in the

TABLE 3.3
Different Molecular Types of Albumin Formulations and Their Applications

Molecular Type Albumin	Clinical Application	References
Peptide-HSA conjugate	Diabetes mellitus, Type 2	Bukrinski et al. (2017)
Paclitaxel-HSA-bound nanoparticles	Metastatic breast cancer, non-small-cell lung cancer, Adenocarcinoma of the pancreas	Cucinotto et al. (2013)
HAS labeled with 99mTc	Breast SPECT scan for node identification	Wang, Chen, Li, & Chuang (2011)
Perflutren Protein-type A Microsphere Injectable Suspension	Contrast agent for ultrasound imaging	"Optison (Perflutren Protein-Type A Microspheres): Side Effects, Interactions, Warning, Dosage & Uses" (2019)
Albumin iodinated I-125 serum	Estimation of total blood volume and plasma volume	(Dworkin, 2007)
Albumin iodinated I-131 serum	Determination of total blood volume and plasma volumes, cardiac output, cardiac blood and pulmonary volumes time of blood circulation, heart and large vessel delineation.	("Iodinated I 131 Albumin Drug Information, Professional," 1994)
Human-Albumin	Repair and maintenance of circulating blood volume	(Iijima, Brandstrup, Rodhe, Andrijauskas, & Svensen, 2013)
Human-Albumin	Emphysema	(Pérez-Rial et al., 2014)
Human-Albumin	Cancers & virus infections	(Liao, 2006)
Human-Albumin	Multiple-sclerosis	(LeVine, 2016)
Human-Albumin	Pulmonary embolism	(Folsom, Lutsey, Heckbert, & Cushman, 2010)

concentration of the polymer, but decreased with a rise in the molar ratio of HA/SAL. With the increased concentration of the polymer and carbodiimide (EDC)/HA molar ratio, the swelling capacity was decreased and increased by a rise in the molar ratio of HA/SAL. Hyaluronidase strength was negatively associated with the hydrogel percentage of HA and strongly correlated with the molar ratio of EDC/HA (Y.-H. Chen et al., 2015).

3.3 OPTIMIZATION OF THE VARIOUS PHYSICOCHEMICAL PARAMETERS

Indeed, nanocomposite materials have distinctive features compared to the other materials present in the bulk solution, and these give them useful features; ironically, they can also give them distinctive toxicity processes. Toxicity was

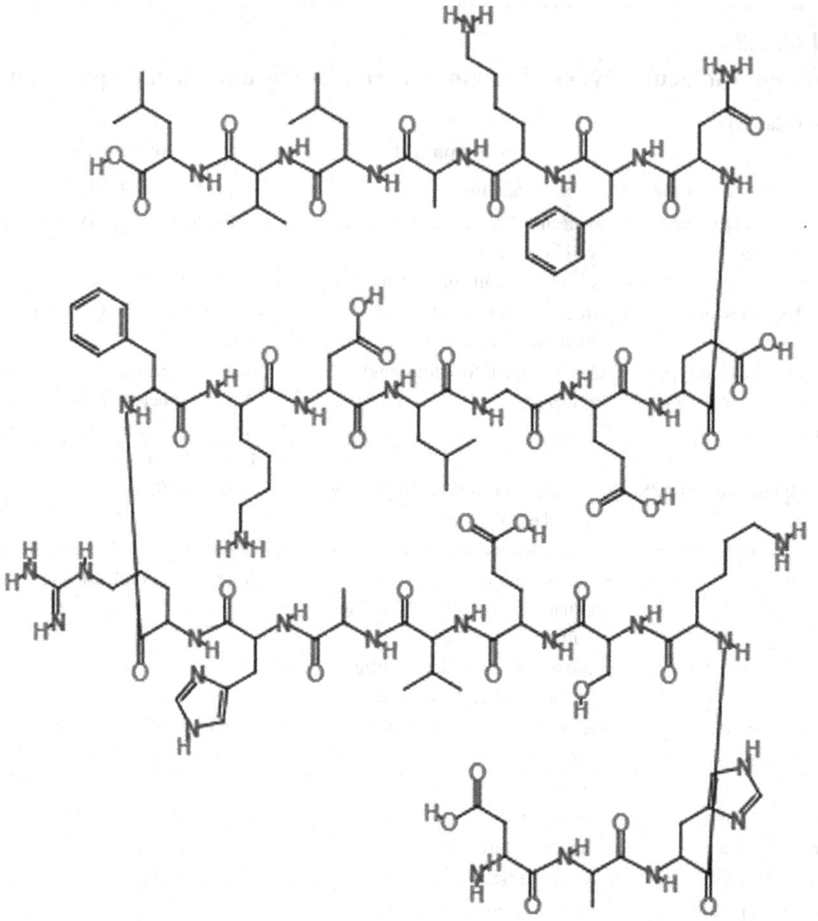

FIGURE 3.10 Structure of albumin.

generally believed to depend on the surface area and size of nanomaterials, their structure, forms, and so on, as will be discussed in the following sections (Gatoo et al., 2014).

3.3.1 Size and Surface Area of the Nanocomposites

It is known that the size and surface area of particles plays a significant part in the material interaction of a biological system. The reactivity of a material can be hugely increased by the continuous decrease of its particle size, which is due to the increase in the surface area of nanoparticles. The system's rejection of particles is also dependent on the size and surface area. Additionally, various biological mechanisms occurring in the body, such as endocytosis and cell absorption, are dependent on particle size (Nel, Xia, Mädler, & Li, 2006; Aillon, Xie, El-Gendy, Berkland, & Forrest, 2009). Various scientists have studied the cytotoxicity of various nanoparticles

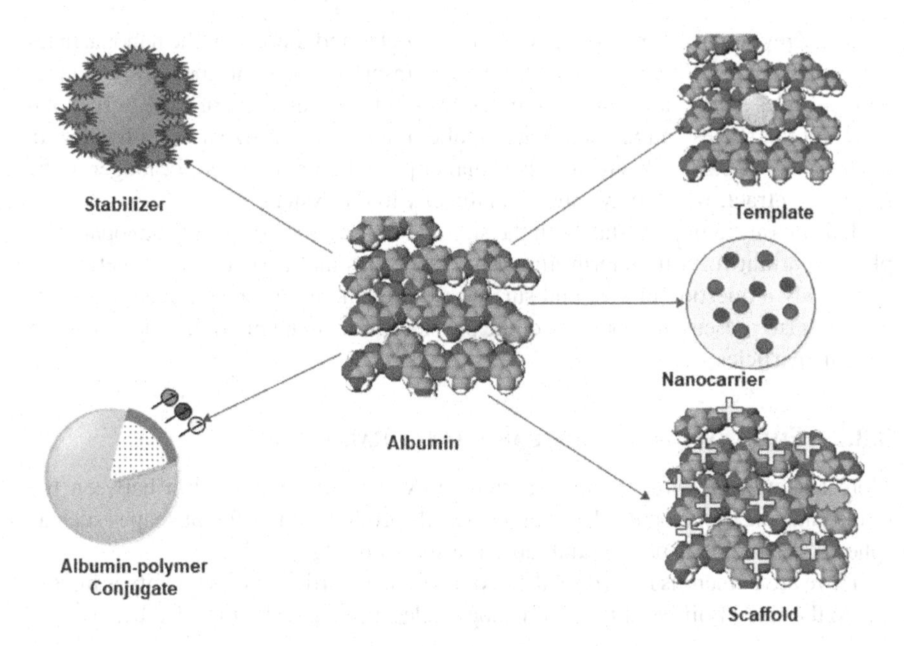

FIGURE 3.11 Albumin-based nanocomposites and their application.

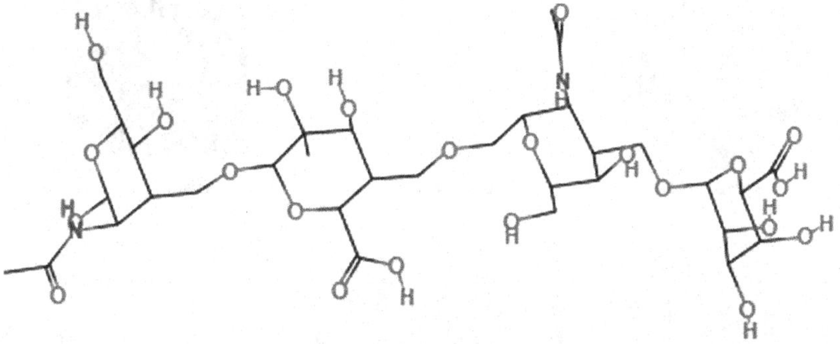

FIGURE 3.12 Structure of hyaluronic acid.

for several different cell types, sizes, growing conditions, etc. but *in vitro* studies are proving to be difficult due to the complex nature of bodily conditions. Several researchers are employing indirect methods in testing the effects *in vitro*. In particular, the toxicity of size-dependent nanoparticles can be attributed to their capacity to enter the biological ecosystems (Lovrić et al., 2005) and then alter the composition of different macromolecules (Aggarwal, Hall, McLeland, Dobrovolskaia, & McNeil, 2009), thus interfering with critical biological functions. The size of nanoparticles also controls their pharmacological behavior. Investigations show that nanoparticles with a size smaller than 50 nm quickly transverse to all the tissues and transmit toxic exposer to many tissues but the nanoparticles whose size is greater than 50 nm are directly taken by the reticuloendothelial system which blocks their pathway to other tissues. Kreyling et al. prepared Ir-192 nanoparticles and the size of the nanoparticles is around 80 nm. When this nanoparticle was inserted into a rat, they observed that nanoparticles started accumulating in the liver of the rat at a rate of 0.1% of the total amount of particles and nanoparticles smaller than 15 nm in size accumulated at an extent of 0.3 to 0.5%. This means that smaller particles are retained for longer in the respiratory tract, which may cause oral toxicity to the system.

This literature implies that both the size and the surface area of the nanoparticles play important roles in determining their properties and toxicity, but sometimes it is not only a question of size and surface area, as there are several chemical components also present that may affect the physiochemical and biological properties of the nanoparticles.

3.3.2 Effect of Particle Shape and Aspect Ratio

More research and studies are required to evaluate the relationship between the shape and size of nanoparticles. Nanomaterials produced in different shapes such as spheres, rings, fibers, planes, and tubes are shown in Figure 3.13.

There are numerous reports on the toxicity of nanoparticles. This toxicity is mainly due to the size of different types of nanoparticles, such as gold, nickel, silica, carbon

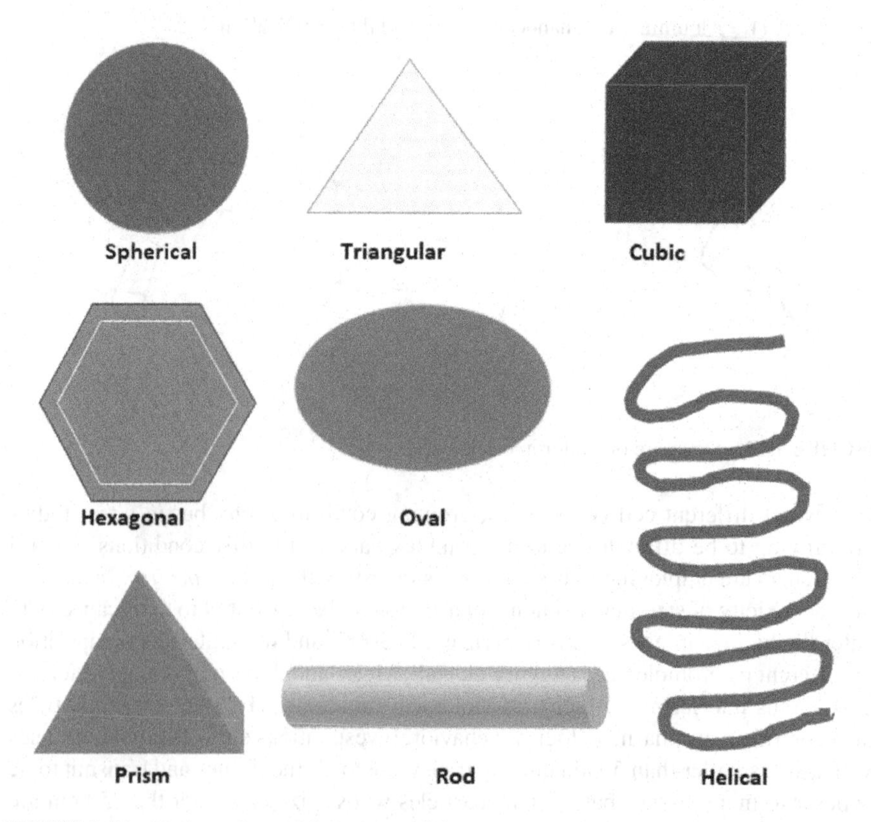

FIGURE 3.13 Various shapes of nanomaterials.

nanotubes, titanium nanoparticles, etc. In an *in vivo* study of nanoparticles, during phagocytosis or endocytosis the shape-dependent toxicity generally affects the membrane wrapping process. It was reported that phagocytosis of spherical-shaped nanomaterials is faster and easier than nanofibers or rod-shaped nanomaterials. Spherical-shaped nanomaterials are also comparatively less toxic than those in both hetero- and homogenous types of nanomaterials (Gatoo et al., 2014). The nonspherical-shaped nanomaterials are more prominent and are easily inclined to flow through the capillaries which cause other biological functions. Researchers have investigated that rod-shaped carbon nanotube materials can more easily block the K^+ ion path than spherical-shaped carbon materials. Some allotropic nanomaterials have been used as a food additive but the same material with a crystalline nature is considered to be a carcinogen. There are certain nanofiber materials present that are easily degradable but some fibers are nondegradable in nature, and uptake of that nanofiber can cause pulmonary toxicity due to longer exposure to the material in the lungs. Some of the factors involved in shape-dependent toxicity are unresolved, and may need further investigation in order to improve the safety of nanotechnology-based systems.

3.3.3 The Effect of Aggregation and Concentration

The concentration and aggregation of nanoparticles also impacts the toxicity of nanomaterials. Generally, aggregation of nanoparticles is mainly dependent on the composition, size, and surface charge of the materials. Investigations showed that some of the nanoparticles accumulated mainly in the lungs, the spleen, and the liver without causing any severe toxicity to the system. But due to the longer period of aggregation in the system, they generally induce cytotoxic effects. Aggregated or agglomerated nanomaterial shows more adverse effects than do the well-dispersed nanomaterials, which may lead to pulmonary interstitial fibrosis. Furthermore, it was observed that the toxicity of the nanomaterial decreases with an increase in concentration.

3.3.4 The Effect of Surface Charge

The surface charge of the nanoparticles also influences the toxicity of the nanoparticles. There are several aspects of nanomaterials that are regulated by the surface charge of the materials, such as transmembrane permeability, blood–brain barrier integrity, plasma protein binding, colloidal behavior, and selective adsorption of nanoparticles. Positively charged nanomaterial shows more significant cellular uptake than that of the neutral and negatively charged nanomaterials. Due to this, it may lead to platelet aggregation and can induce hemolysis, which can cause critical toxicity to the biological system.

3.3.5 The Effect of Surface Coating and Surface Roughness

The surface properties of nanoparticles plays a significant role in the toxicity of nanoparticles. The surface coating and surface roughness play a critical role as they determine the output of their interaction with the cells and biological entities. The physiochemical properties of nanoparticles such as electrical, chemical reactivity, optical, and magnetic properties can be modified and, due to this modification, surface coating of nanomaterials can affect their cytotoxic properties. In the presence of oxygen radicals, oxygen, transition metal, and ozone on the surface of nanoparticles leads to reactive oxygen species (ROS) generation and it induces inflammation. Fubini et al. have reported that cytotoxicity of silica occurs specifically only due to the presence of surface free radicals and reactive oxygen species on the surface of silica nanoparticles.

3.3.6 Effect of Solvents/Medium

The medium conditions and types of solvent used for nanocomposite preparation are two major factors that influence the dispersion of the nanoparticles and agglomeration status. They also affect the toxicity of the nanoparticles and the size of the particles. It was observed that the sizes of the titanium oxide nanoparticles and zinc oxide nanoparticles prepared using phosphate-buffered saline (PBS) were larger than the size of the nanoparticles prepared using water. Generally, nanoparticles show distinct diameter variation in the biological environment (Sager et al., 2007; Jiang, Oberdörster, & Biswas, 2009). Therefore, the toxic impacts of nanoparticles reveal variation depending on the structure of the medium in which they are suspended; when dissolved in different mediums, the same nanoparticles indicate different toxic effects (Colvin, 2003; Hou, Westerhoff, & Posner, 2013). While the dispersing mediator may enhance the physicochemical characteristics and solution properties of formulations of nanomaterials, the toxicity of nanomaterials may also be adversely affected.

3.4 PHYSICAL CHARACTERIZATIONS

Characterization is an essential and important part of all research dealing with materials. The essential elements of the physical characterization of materials are structure (such as functional group, chemical bonding, ultrastructure, and atomic coordinates), chemical composition and chemical homogeneity, and analysis and identification of impurities and defects that influence the properties of the material. Hence, characterization defines all the structural and compositional properties of a sample material which would be sufficient for replicating the material. Recent developments in the field of characterization techniques from the last few years have been remarkable, especially in the fields of material characterization and structural studies. Among the various characterization methods, the three significant methods are X-ray diffraction (XRD), infrared (IR) spectroscopy, and scanning electron micrograph (SEM). Since the beginning of this century, X-ray diffraction has played a key role in defining and characterizing solids. In determining the characteristics of a polymer, structural phenomena play a significant part (Kumirska et al., 2010). Mechanical characteristics are not only determined by the shape of materials and the movement of individual polymer molecules, but also by the conduct of bigger and more complicated structural structures. Using SEM, structural characteristics can be explored. Infrared (IR) spectroscopy is one of the strongest analytical methods that can be used to identify chemicals. This method can be used for quantitative assessment when combined with intensity measurements. For measuring the physiochemical characteristics of materials, thermal analysis can be well-defined as a function of temperature. Generally, there are two main techniques available for thermal analysis of materials, tha is, thermogravimetric analysis (TGA), and differential thermal analysis (DTA). TGA analysis automatically records the change in weight of the sample materials as a function of either time or temperature; DTA measures change in heat content by recording the difference in temperature between the sample material and the inert reference material as a function of temperature.

3.5 NANOCOMPOSITES, THEIR USES, AND APPLICATIONS

A nanocomposite is a matrix that has been added to nanoparticles to enhance a specific material property. Nanocomposite characteristics have led scientists and firms to consider using this material in several areas. The following study of applications of nanocomposites introduces you to many of the uses being investigated, including:

3.5.1 Food and Beverage Packaging

In food and beverage packaging, the nanocomposite acts as a sensor, as an antimicrobial agent, and as a barrier material. Nanocomposites act as tough barrier packaging materials in the form of nanoplates (Idumah, Zurina, Ogbu, Ndem, & Igba, 2019).

3.5.2 Tissue Engineering

In tissue engineering, polymeric nanocomposites are used for scaffold preparation. A scaffold is nothing other than engineered material to which cells are added, and the scaffold should allow the cells to grow without showing any toxic effect. The nanocomposite should have a biodegradability quality. In tissue engineering, the polymeric scaffold is used for bone, tendon, cartilage, blood vessels, etc. (Dhandayuthapani, Yoshida, Maekawa, & Kumar, 2011).

3.5.3 Drug Delivery Systems

Polymeric nanocomposites act as a template for the *in vivo* study of drug delivery systems (Liechty, Kryscio, Slaughter, & Peppas, 2010).

3.5.4 Environmental Protection and Wastewater Treatment

In this field, polymeric nanocomposite film is used as a filtration membrane for air quality detection and wastewater treatment. The polymeric nanocomposite can be used to absorb suspended pollutants present in the water (Dongre et al., 2019).

3.5.5 Application in Cosmetics

Generally, in cosmetics, polymeric nanocomposites are used in sunscreen lotion. These composites can easily be absorbed by the skin and will deflect the sun's UV rays. Sometimes they absorb the energy of UV rays and convert it into harmless energy (Haniffa et al., 2016).

3.5.6 Making Lightweight Sensors with Nanocomposites

Electricity is conducted by a polymer-nanotube nanocomposite; how well it performs relies on the spacing of the nanotube. This property enables patches of nanocomposite polymer-nanotube to act as stress sensors on the blades of wind turbines. The nanocomposite will also bend when a powerful wind blows the blades, and with changes in bending, the electrical conductance of the nanocomposite sensor will cause an alarm to sound. This alarm makes it possible to shut down the wind turbine before extreme damage occurs (Okpala, n.d.).

3.5.7 Making Flexible Batteries Using Nanocomposites

To create a conductive paper, a nanocomposite of cellular components and also nanotubes can be used. When an electrolyte soaks this conductive paper, a flexible battery is created (Jabbour et al., 2010).

3.5.8 Making Tumors More Visible and Easier to Remove

Researchers are trying to incorporate fluorescent and magnetic nanoparticles into a nanocomposite particle. The magnetic properties of the nanocomposite particles make the tumor more visible during a pre-operative MRI examination. The nanocomposite particles' fluorescent properties could assist the surgeon to better see the tumor during surgery (Wolinsky, Colson, & Grinstaff, 2012).

3.6 CONCLUSION AND FUTURE PERSPECTIVE

Nanocomposites have been widely used in various applications due to their numerous benefits, such as having powerful mechanical strength, being lightweight, chemical resistant, having outstanding barrier characteristics, etc. The nanocomposites manufactured by biodegradable polymers protected with natural fibers are greatly valued as being the eco-friendliest and, as a result, a variety of research has been dedicated to exploring the "green" nanocomposites with ecofriendly manufacturing processes (Mohammed, Ansari, Pua, Jawaid, & Islam, 2015). It is desirable to use new polymers with intrinsic eco-friendly characteristics such as biodegradability and renewability, and a sequence of natural polymeric materials such as silk fibroin, alginate, albumin, chitosan, etc. Environmentally friendly nanocomposites also face some challenges, requiring significant consideration in terms of loose fiber orientation control, bad matrix–fill bonding, and difficulties in shaping nanoscale particulates (Müller et al., 2017). Polymeric nanocomposites have unique features such as high mechanical strength and can act as a barrier material even in low concentrations of added nanoparticles. All the physical and chemical characteristics such as size, shape, roughness, surface area, concentration of nanoparticles, etc play a very important role in developing a better nanocomposite. Various preparation methods are available for the synthesis of nanocomposites but every method has its own advantages and disadvantages based on their application and the availability of resources (MacLeod et al., 2006).

REFERENCES

Abdollahi, M., Alboofetileh, M., Behrooz, R., Rezaei, M., & Miraki, R. (2013). Reducing water sensitivity of alginate bio-nanocomposite film using cellulose nanoparticles. *International Journal of Biological Macromolecules, 54*, 166–173. doi:10.1016/j.ijbiomac.2012.12.016

Abdulkareem Ghassan, A., Mijan, N.-A., & Hin Taufiq-Yap, Y. (2019). Nanomaterials: An overview of nanorods synthesis and optimization. In M. S. Ghamsari & S. Dhara (Eds.), *Nanorods: An Overview from Synthesis to Emerging Device Applications* . IntechOpen. doi:10.5772/intechopen.84550

Aggarwal, P., Hall, J. B., McLeland, C. B., Dobrovolskaia, M. A., & McNeil, S. E. (2009). Nanoparticle interaction with plasma proteins as it relates to particle biodistribution, biocompatibility and therapeutic efficacy. *Advanced Drug Delivery Reviews, 61*(6), 428–437. doi:10.1016/j.addr.2009.03.009

Aillon, K. L., Xie, Y., El-Gendy, N., Berkland, C. J., & Forrest, M. L. (2009). Effects of nanomaterial physicochemical properties on in vivo toxicity. *Advanced Drug Delivery Reviews, 61*(6), 457–466. doi:10.1016/j.addr.2009.03.010

Akhtar, M. F., Hanif, M., & Ranjha, N. M. (2016). Methods of synthesis of hydrogels … A review. *Saudi Pharmaceutical Journal, 24*(5), 554–559. doi:10.1016/j.jsps.2015.03.022

Ali, A., Zafar, H., Zia, M., Ul Haq, I., Phull, A. R., Ali, J. S., & Hussain, A. (2016). Synthesis, characterization, applications, and challenges of iron oxide nanoparticles. *Nanotechnology, Science and Applications, 9*, 49–67. doi:10.2147/NSA.S99986

An, F.-F., & Zhang, X.-H. (2017). Strategies for preparing albumin-based nanoparticles for multifunctional bioimaging and drug delivery. *Theranostics, 7*(15), 3667–3689. doi:10.7150/thno.19365

An, T., Zhou, L., Li, G., Fu, J., & Sheng, G. (2008). Recent patents on immobilized microorganism technology and its engineering application in wastewater treatment. *Recent Patents on Engineering, 2*(1), 28–35. doi:10.2174/187221208783478543

Arias, J. L., Reddy, L. H., & Couvreur, P. (2012). Fe_3O_4/chitosan nanocomposite for magnetic drug targeting to cancer. *Journal of Materials Chemistry, 22*(15), 7622. doi:10.1039/c2jm15339d

Arora, B., Bhatia, R., & Attri, P. (2018). Bionanocomposites: Green materials for a sustainable future. *New Polymer Nanocomposites for Environmental Remediation, 2018*, 699–712. doi:10.1016/B978-0-12-811033-1.00027-5

Arum, Y., Oh, Y.-O., Kang, H. W., Ahn, S.-H., & Oh, J. (2015). Chitosan-coated Fe_3O_4 magnetic nanoparticles as carrier of cisplatin for drug delivery. *Fisheries and Aquatic Sciences, 18*(1), 89–98. doi:10.5657/FAS.2015.0089

Bao, H., Pan, Y., Ping, Y., Sahoo, N. G., Wu, T., Li, L., … Gan, L. H. (2011). Chitosan-functionalized graphene oxide as a nanocarrier for drug and gene delivery. *Small, 7*(11), 1569–1578. doi:10.1002/smll.201100191

Bukrinski, J. T., Sønderby, P., Antunes, F., Andersen, B., Schmidt, E. G. W., Peters, G. H. J., & Harris, P. (2017). Glucagon-like peptide 1 conjugated to recombinant human serum albumin variants with modified neonatal Fc receptor binding properties. Impact on molecular structure and half-life. *Biochemistry, 56*(36), 4860–4870. doi:10.1021/acs.biochem.7b00492

Cai, T., Chen, L., Ren, Q., Cai, S., & Zhang, J. (2011). The biodegradation pathway of triethylamine and its biodegradation by immobilized Arthrobacter protophormiae cells. *Journal of Hazardous Materials, 186*(1), 59–66. doi:10.1016/J.JHAZMAT.2010.10.007

Cheaburu-Yilmaz, C. N., Yilmaz, O., & Vasile, C. (2015). Eco-friendly chitosan-based nanocomposites: Chemistry and applications. In V. K. Thakur & M. K. Thakur (Eds.), *Eco-Friendly Polymer Nanocomposites*. Elsevier. doi:10.1007/978-81-322-2473-0_11

Chen, Z., Mo, X., He, C., & Wang, H. (2008). Intermolecular interactions in electrospun collagen–chitosan complex nanofibers. *Carbohydrate Polymers*, **72**(3), 410–418. doi:10.1016/j.carbpol.2007.09.018

Chen, K.-Y., & Yao, C.-H. (2011). Repair of bone defects with gelatin-based composites: A review. *BioMedicine*, *1*(1), 29–32. doi:10.1016/j.biomed.2011.10.005

Chen, Y.-H., Li, J., Hao, Y.-B., Qi, J.-X., Dong, N.-G., Wu, C.-L., & Wang, Q. (2015). Preparation and characterization of composite hydrogels based on crosslinked hyaluronic acid and sodium alginate. *Journal of Applied Polymer Science*, *132*(19), 41898. doi:10.1002/app.41898

Ching, S. H., Bansal, N., & Bhandari, B. (2017). Alginate gel particles–A review of production techniques and physical properties. *Critical Reviews in Food Science and Nutrition*, *57*(6), 1133–1152. doi:10.1080/10408398.2014.965773

Cohen-Karni, T., Jeong, K. J., Tsui, J. H., Reznor, G., Mustata, M., Wanunu, M., … Kohane, D. S. (2012). Nanocomposite gold-silk nanofibers. *Nano Letters*, *12*(10), 5403–5406. doi:10.1021/nl302810c

Cohen, J. (2001). Defining Identification: A Theoretical Look at the Identification of Audiences With Media Characters. *Mass Communication & Society*, *4*(3), 245–264. Retrieved from http://citeseerx.ist.psu.edu/viewdoc/download?doi=10.1.1.469.63&rep=rep1&type=pdf

Colvin, V. L. (2003). The potential environmental impact of engineered nanomaterials. *Nature Biotechnology*, *21*(10), 1166–1170. doi:10.1038/nbt875

Covarrubias, S. A., de-Bashan, L. E., Moreno, M., & Bashan, Y. (2012). Alginate beads provide a beneficial physical barrier against native microorganisms in wastewater treated with immobilized bacteria and microalgae. *Applied Microbiology and Biotechnology*, *93*(6), 2669–2680. doi:10.1007/s00253-011-3585-8

Cucinotto, I., Fiorillo, L., Gualtieri, S., Arbitrio, M., Ciliberto, D., Staropoli, N., … Tagliaferri, P. (2013). Nanoparticle albumin bound Paclitaxel in the treatment of human cancer: Nanodelivery reaches prime-time? *Journal of Drug Delivery*, *2013*, 905091. doi:10.1155/2013/905091

Darder, M., Colilla, M., & Ruiz-Hitzky, E. (2003). Biopolymer–clay nanocomposites based on chitosan intercalated in montmorillonite. doi:10.1021/CM0343047

Dhandayuthapani, B., Yoshida, Y., Maekawa, T., & Kumar, D. S. (2011). Polymeric scaffolds in tissue engineering application: A review. *International Journal of Polymer Science*, *2011*, 1–19. doi:10.1155/2011/290602

Dongre, R. S., Sadasivuni, K. K., Deshmukh, K., Mehta, A., Basu, S., Meshram, J. S., … Karim, A. (2019). Natural polymer based composite membranes for water purification: A review. *Polymer-Plastics Technology and Materials*, *58*(12), 1–16. doi:10.1080/25740881.2018.1563116

Dworkin, H. J. (2007). Blood volume; Plasma volume; Red cell volume; I-125 RISA. *The American Journal of the Medical Sciences*, *334*, 37–40. Retrieved from www.daxor.com/wp-content/uploads/2014/10/Dworkin.pdf

Elieh-Ali-Komi, D., & Hamblin, M. R. (2016). Chitin and chitosan: Production and application of versatile biomedical nanomaterials. *International Journal of Advanced Research*, *4*(3), 411–427. Retrieved from http://www.ncbi.nlm.nih.gov/pubmed/27819009

Feng, Y., Li, X., Li, M., Ye, D., Zhang, Q., You, R., & Xu, W. (2017). Facile preparation of biocompatible silk fibroin/cellulose nanocomposite films with high mechanical performance. *ACS Sustainable Chemistry & Engineering*, *5*(7), 6227–6236. doi:10.1021/acssuschemeng.7b01161

Folsom, A. R., Lutsey, P. L., Heckbert, S. R., & Cushman, M. (2010). Serum albumin and risk of venous thromboembolism. *Thrombosis and Haemostasis, 104*(1), 100–104. doi:10.1160/TH09-12-0856

Fubini, B., & Hubbard, A. (2003). Reactive oxygen species (ROS) and reactive nitrogen species (RNS) generation by silica in inflammation and fibrosis. *Free Radical Biology and Medicine* **34**(12), 1507–1516. doi:10.1016/S0891-5849(03)00149-7

Gao, M., Shih, C.-C., Pan, S.-Y., Chueh, C.-C., & Chen, W.-C. (2018). Advances and challenges of green materials for electronics and energy storage applications: From design to end-of-life recovery. *Journal of Materials Chemistry A, 6*(42), 20546–20563. doi:10.1039/C8TA07246A

Gatoo, M. A., Naseem, S., Arfat, M. Y., Dar, A. M., Qasim, K., & Zubair, S. (2014). Physicochemical properties of nanomaterials: Implication in associated toxic manifestations. *BioMed Research International, 2014*, 498420. doi:10.1155/2014/498420

Gupta, V. K., Khamparia, S., Tyagi, I., Jaspal, D., & Malviya, A. (2015). Decolorization of mixture of dyes: A critical review. *Global Journal of Environmental Science and Management, 1*(1), 71–94. doi:10.7508/GJESM.2015.01.007

Han, Y., Zeng, Q., Li, H., & Chang, J. (2013). The calcium silicate/alginate composite: Preparation and evaluation of its behavior as bioactive injectable hydrogels. *Acta Biomaterialia, 9*(11), 9107–9117. doi:10.1016/j.actbio.2013.06.022

Haniffa, M. A. C. M., Ching, Y. C., Abdullah, L. C., Poh, S. C., & Chuah, C. H. (2016). Review of bionanocomposite coating films and their applications. *Polymers, 8*(7). doi:10.3390/polym8070246

Hasan, A., Morshed, M., Memic, A., Hassan, S., Webster, T. J., & Marei, H. E.-S. (2018). Nanoparticles in tissue engineering: Applications, challenges and prospects. *International Journal of Nanomedicine, 13*, 5637–5655. doi:10.2147/IJN.S153758

Hlongwane, G. N., Sekoai, P. T., Meyyappan, M., & Moothi, K. (2019). Simultaneous removal of pollutants from water using nanoparticles: A shift from single pollutant control to multiple pollutant control. *Science of The Total Environment, 656*, 808–833. doi:10.1016/J.SCITOTENV.2018.11.257

Hood, M. A., Mari, M., & Muñoz-Espí, R. (2014). Synthetic strategies in the preparation of polymer/inorganic hybrid nanoparticles. *Materials, 7*(5), 4057–4087. doi:10.3390/ma7054057

Hou, W.-C., Westerhoff, P., & Posner, J. D. (2013). Biological accumulation of engineered nanomaterials: A review of current knowledge. *Environmental Science: Processes Impacts, 15*(1), 103–122. doi:10.1039/C2EM30686G

Huq, T., Salmieri, S., Khan, A., Khan, R. A., Le Tien, C., Riedl, B., … Lacroix, M. (2012). Nanocrystalline cellulose (NCC) reinforced alginate based biodegradable nanocomposite film. *Carbohydrate Polymers, 90*(4), 1757–1763. doi:10.1016/j.carbpol.2012.07.065

Hussain, C. M., & Mishra, A. K. (Eds.) (2018). *New Polymer Nanocomposites for Environmental Remediation*. Elsevier.

Idris, A., Ismail, N. S. M., Hassan, N., Misran, E., & Ngomsik, A.-F. (2012). Synthesis of magnetic alginate beads based on maghemite nanoparticles for Pb(II) removal in aqueous solution. *Journal of Industrial and Engineering Chemistry, 18*(5), 1582–1589. doi:10.1016/j.jiec.2012.02.018

Idumah, C. I., Zurina, M., Ogbu, J., Ndem, J. U., & Igba, E. C. (2019). A review on innovations in polymeric nanocomposite packaging materials and electrical sensors for food and agriculture. *Composite Interfaces, 27*(1), 1–72. doi:10.1080/09276440.2019.1600972

Iijima, T., Brandstrup, B., Rodhe, P., Andrijauskas, A., & Svensen, C. H. (2013). The maintenance and monitoring of perioperative blood volume. *Perioperative Medicine (London, England), 2*(1), 9. doi:10.1186/2047-0525-2-9

Imre, B., & Pukánszky, B. (2013). Compatibilization in bio-based and biodegradable polymer blends. *European Polymer Journal, 49*(6), 1215–1233. doi:10.1016/J.EURPOLYMJ.2013.01.019

Iodinated I 131 Albumin Drug Information, Professional. (1994). Retrieved July 1, 2019 from https://www.drugs.com/mmx/iodinated-i-131-albumin.html

Irimia-Vladu, M. (2014). "Green" electronics: Biodegradable and biocompatible materials and devices for sustainable future. *Chemical Society Reviews, 43*(2), 588–610. doi:10.1039/C3CS60235D

Izyan Syazana Mohd Yusoff, N., Uzir Wahit, M., Jaafar, J., & Wong, T.-W. (2018). Characterization of graphene-silk fibroin composites film. *Materials Today: Proceedings, 5*(10), 21853–21860. doi:10.1016/J.MATPR.2018.07.042

Jabbour, L., Gerbaldi, C., Chaussy, D., Zeno, E., Bodoardo, S., & Beneventi, D. (2010). Microfibrillated cellulose–graphite nanocomposites for highly flexible paper-like Li-ion battery electrodes. *Journal of Materials Chemistry, 20*(35), 7344. doi:10.1039/c0jm01219j

Jamróz, E., Kulawik, P., Kopel, P., Jamróz, E., Kulawik, P., & Kopel, P. (2019). The effect of nanofillers on the functional properties of biopolymer-based films: A review. *Polymers, 11*(4), 675. doi:10.3390/polym11040675

Jeevanandam, J., Barhoum, A., Chan, Y. S., Dufresne, A., & Danquah, M. K. (2018). Review on nanoparticles and nanostructured materials: History, sources, toxicity and regulations. *Beilstein Journal of Nanotechnology, 9*, 1050–1074. doi:10.3762/bjnano.9.98

Jiang, J., Oberdörster, G., & Biswas, P. (2009). Characterization of size, surface charge, and agglomeration state of nanoparticle dispersions for toxicological studies. *Journal of Nanoparticle Research, 11*(1), 77–89. doi:10.1007/s11051-008-9446-4

Johari, N., Madaah Hosseini, H. R., & Samadikuchaksaraei, A. (2018). Novel fluoridated silk fibroin/ TiO2 nanocomposite scaffolds for bone tissue engineering. *Materials Science and Engineering: C, 82*, 265–276. doi:10.1016/J.MSEC.2017.09.001

Joseph, B., & Raj, S. J. (2012). Therapeutic applications and properties of silk proteins from *Bombyx mori. Frontiers in Life Science, 6*(3–4), 55–60. doi:10.1080/21553769.2012.760491

Junter, G.-A., & Jouenne, T. (2004). Immobilized viable microbial cells: From the process to the proteome… or the cart before the horse. *Biotechnology Advances, 22*(8), 633–658. doi:10.1016/j.biotechadv.2004.06.003

Kang, M., & Jin, H.-J. (2007). Electrically conducting electrospun silk membranes fabricated by adsorption of carbon nanotubes. *Colloid and Polymer Science, 285*(10), 1163–1167. doi:10.1007/s00396-007-1668-y

Khalid, A., Mitropoulos, A. N., Marelli, B., Simpson, D. A., Tran, P. A., Omenetto, F. G., & Tomljenovic-Hanic, S. (2015). Fluorescent nanodiamond silk fibroin spheres: Advanced nanoscale bioimaging tool. *ACS Biomaterials Science & Engineering, 1*(11), 1104–1113. doi:10.1021/acsbiomaterials.5b00220

Kharlampieva, E., Kozlovskaya, V., Wallet, B., Shevchenko, V. V., Naik, R. R., Vaia, R., … Tsukruk, V. V. (2010). Co-cross-linking silk matrices with silica nanostructures for robust ultrathin nanocomposites. *ACS Nano, 4*(12), 7053–7063. doi:10.1021/nn102456w

Kojic, N., Pritchard, E. M., Tao, H., Brenckle, M. A., Mondia, J. P., Panilaitis, B., … Kaplan, D. L. (2012). Focal infection treatment using laser-mediated heating of injectable silk hydrogels with gold nanoparticles. *Advanced Functional Materials, 22*(18), 3793–3798. doi:10.1002/adfm.201200382

Kreyling, W. G., Semmler, M., Erbe, F., Mayer, P., Takenaka, S., Schulz, H., Oberdörster, G., & Ziesenis, A. (2002). Translocation of ultrafine insoluble iridium particles from lung epithelium to extrapulmonary organs is size dependent but very low. *Journal of Toxicology and Environmental Health, Part A*, **65**(20), 1513–1530.

Kumirska, J., Czerwicka, M., Kaczyński, Z., Bychowska, A., Brzozowski, K., Thöming, J., … Stepnowski, P. (2010). Application of spectroscopic methods for structural analysis of chitin and chitosan. *Marine Drugs, 8*(5), 1567–1636. doi:10.3390/md8051567

Kundu, B., Rajkhowa, R., Kundu, S. C., & Wang, X. (2013). Silk fibroin biomaterials for tissue regenerations. *Advanced Drug Delivery Reviews, 65*(4), 457–470. doi:10.1016/J.ADDR.2012.09.043

Lambert, S., & Wagner, M. (2017). Environmental performance of bio-based and biodegradable plastics: The road ahead. *Chemical Society Reviews, 46*(22), 6855–6871. doi:10.1039/C7CS00149E

LeVine, S. M. (2016). Albumin and multiple sclerosis. *BMC Neurology, 16*(1), 47. doi:10.1186/s12883-016-0564-9

Li, Z., Yang, Y., Yao, J., Shao, Z., & Chen, X. (2017). A facile fabrication of silk/MoS 2 hybrids for Photothermal therapy. *Materials Science and Engineering: C, 79*, 123–129. doi:10.1016/j.msec.2017.05.010

Liao, J. B. (2006). Viruses and human cancer. *The Yale Journal of Biology and Medicine, 79*(3–4), 115–122. Retrieved from www.ncbi.nlm.nih.gov/pubmed/17940621

Liechty, W. B., Kryscio, D. R., Slaughter, B. V, & Peppas, N. A. (2010). Polymers for drug delivery systems. *Annual Review of Chemical and Biomolecular Engineering, 1*, 149–173. doi:10.1146/annurev-chembioeng-073009-100847

Lim, E.-K., Sajomsang, W., Choi, Y., Jang, E., Lee, H., Kang, B., … Huh, Y.-M. (2013). Chitosan-based intelligent theragnosis nanocomposites enable pH-sensitive drug release with MR-guided imaging for cancer therapy. *Nanoscale Research Letters, 8*(1), 467. doi:10.1186/1556-276X-8-467

Lin, J., Li, Y., Li, Y., Wu, H., Yu, F., Zhou, S., … Hou, Z. (2015). Drug/dye-loaded, multifunctional PEG–chitosan–iron oxide nanocomposites for methotraxate synergistically self-targeted cancer therapy and dual model imaging. *ACS Applied Materials & Interfaces, 7*(22), 11908–11920. doi:10.1021/acsami.5b01685

Liu, L., Liu, J., Kong, X., Cai, Y., & Yao, J. (2011). Porous composite scaffolds of hydroxyapatite/silk fibroin via two-step method. *Polymers for Advanced Technologies, 22*(6), 909–914. doi:10.1002/pat.1595

Liu, X., Yin, G., Yi, Z., & Duan, T. (2016). Silk fiber as the support and reductant for the facile synthesis of Ag–Fe3O4 nanocomposites and its antibacterial properties. *Materials, 9*(7). doi:10.3390/MA9070501

Liu, Z., Jiao, Y., & Zhang, Z. (2007). Calcium-carboxymethyl chitosan hydrogel beads for protein drug delivery system. *Journal of Applied Polymer Science, 103*(5), 3164–3168. doi:10.1002/app.24867

Lovrić, J., Bazzi, H. S., Cuie, Y., Fortin, G. R. A., Winnik, F. M., & Maysinger, D. (2005). Differences in subcellular distribution and toxicity of green and red emitting CdTe quantum dots. *Journal of Molecular Medicine, 83*(5), 377–385. doi:10.1007/s00109-004-0629-x

MacLeod, D., Dowman, J., Hammond, R., Leete, T., Inoue, K., & Abeliovich, A. (2006). The familial parkinsonism gene LRRK2 regulates neurite process morphology. *Neuron, 52*(4), 587–593. doi:10.1016/j.neuron.2006.10.008

Mallakpour, S., & Darvishzadeh, M. (2018). Nanocomposite materials based on poly(vinyl chloride) and bovine serum albumin modified ZnO through ultrasonic irradiation as a green technique: Optical, thermal, mechanical and morphological properties. *Ultrasonics Sonochemistry, 41*, 85–99. doi:10.1016/j.ultsonch.2017.09.022

Mane, S., Ponrathnam, S., & Chavan, N. (2015). Effect of chemical cross-linking on properties of polymer microbeads: A review, *3*, 473–485. doi:10.13179/canchemtrans.2015.03.04.0245

Melke, J., Midha, S., Ghosh, S., Ito, K., & Hofmann, S. (2015). Silk fibroin as biomaterial for bone tissue engineering. *Acta Biomaterialia*. doi:10.1016/j.actbio.2015.09.005

Mierzwa-Hersztek, M., Gondek, K., & Kopeć, M. (2019). Degradation of polyethylene and biocomponent-derived polymer materials: An overview. *Journal of Polymers and the Environment*, *27*(3), 600–611. doi:10.1007/s10924-019-01368-4

Mieszawska, A. J., Fourligas, N., Georgakoudi, I., Ouhib, N. M., Belton, D. J., Perry, C. C., & Kaplan, D. L. (2010). Osteoinductive silk–silica composite biomaterials for bone regeneration. *Biomaterials*, *31*(34), 8902–8910. doi:10.1016/J.BIOMATERIALS.2010.07.109

Misak, H. E., Asmatulu, R., Gopu, J. S., Man, K.-P., Zacharias, N. M., Wooley, P. H., & Yang, S.-Y. (2014). Albumin-based nanocomposite spheres for advanced drug delivery systems. *Biotechnology Journal*, *9*(1), 163–170. doi:10.1002/biot.201300150

Mishra, R. K., Ha, S. K., Verma, K., & Tiwari, S. K. (2018). Recent progress in selected bionanomaterials and their engineering applications: An overview. *Journal of Science: Advanced Materials and Devices*, *3*(3), 263–288. doi:10.1016/J.JSAMD.2018.05.003

Mohammadi, N., Khani, H., Gupta, V. K., Amereh, E., & Agarwal, S. (2011). Adsorption process of methyl orange dye onto mesoporous carbon material–kinetic and thermodynamic studies. *Journal of Colloid and Interface Science*, *362*(2), 457–462. doi:10.1016/j.jcis.2011.06.067

Mohammed, L., Ansari, M. N. M., Pua, G., Jawaid, M., & Islam, M. S. (2015). A Review on natural fiber reinforced polymer composite and its applications. *International Journal of Polymer Science*, *2015*, 1–15. doi:10.1155/2015/243947

Mohan, S., Oluwafemi, O. S., Kalarikkal, N., Thomas, S., & Songca, S. P. (2016). Biopolymers – application in nanoscience and nanotechnology. In F. K. Perveen (Ed.), *Recent Advances in Biopolymers*. InTech. doi:10.5772/62225

Mottaghitalab, F., Farokhi, M., Zaminy, A., Kokabi, M., Soleimani, M., Mirahmadi, F., … Sadeghizadeh, M. (2013). A biosynthetic nerve guide conduit based on Silk/SWNT/Fibronectin nanocomposite for peripheral nerve regeneration. *PLoS ONE*, *8*(9), e74417. doi:10.1371/journal.pone.0074417

Müller, K., Bugnicourt, E., Latorre, M., Jorda, M., Echegoyen Sanz, Y., Lagaron, J., … Schmid, M. (2017). Review on the processing and properties of polymer nanocomposites and nanocoatings and their applications in the packaging, automotive and solar energy fields. *Nanomaterials*, *7*(4), 74. doi:10.3390/nano7040074

Nel, A., Xia, T., Mädler, L., & Li, N. (2006). Toxic potential of materials at the nanolevel. *Science*, *311*(5761), 622–627. doi:10.1126/science.1114397

Optison (Perflutren Protein-Type A Microspheres): Side effects, interactions, warning, dosage & Uses. (2019). Retrieved July 1, 2019 from https://www.rxlist.com/optison-drug.htm

Okpala, C. C. (n.d.). The benefits and applications of nanocomposites. *International Journal of Advanced Engineering Technology*. Retrieved from www.technicaljournalsonline.com/ijeat/VOL V/IJAET VOL V ISSUE IV OCTBER DECEMBER 2014/Vol V Issue IV Article 3.pdf

Paques, J. P., van der Linden, E., van Rijn, C. J. M., & Sagis, L. M. C. (2014). Preparation methods of alginate nanoparticles. *Advances in Colloid and Interface Science*, *209*, 163–171. doi:10.1016/j.cis.2014.03.009

Patra, J. K., Das, G., Fraceto, L. F., Campos, E. V. R., Rodriguez-Torres, M. del P., Acosta-Torres, L. S., … Shin, H.-S. (2018). Nano based drug delivery systems: Recent developments and future prospects. *Journal of Nanobiotechnology*, *16*(1), 71. doi:10.1186/s12951-018-0392-8

Pérez-Rial, S., del Puerto-Nevado, L., Girón-Martínez, Á., Terrón-Expósito, R., Díaz-Gil, J. J., González-Mangado, N., & Peces-Barba, G. (2014). Liver growth factor treatment reverses emphysema previously established in a cigarette smoke exposure mouse model. *American Journal of Physiology-Lung Cellular and Molecular Physiology*, *307*(9), L718–L726. doi:10.1152/ajplung.00293.2013

Ragab, D., & Rohani, S. (2016). Multifunctional ionic liquid-Fe_3O_4/chitosan shell magnetic hydrogel for magnetic responsive, pH-triggered and long-acting hormone therapy. *Advances in Biotechnology & Microbiology*, *1*(1), 555554. doi:10.19080/AIBM.2016.01.555554

Ramos, M., Valdés, A., Beltrán, A., Garrigós, M., Ramos, M., Valdés, A., ... Garrigós, M. C. (2016). Gelatin-based films and coatings for food packaging applications. *Coatings*, *6*(4), 41. doi:10.3390/coatings6040041

Rodríguez-Lozano, F. J., García-Bernal, D., Aznar-Cervantes, S., Ros-Roca, M. A., Algueró, M. C., Atucha, N. M., ... Cenis, J. L. (2014). Effects of composite films of silk fibroin and graphene oxide on the proliferation, cell viability and mesenchymal phenotype of periodontal ligament stem cells. *Journal of Materials Science: Materials in Medicine*, *25*(12), 2731–2741. doi:10.1007/s10856-014-5293-2

Sager, T. M., Porter, D. W., Robinson, V. A., Lindsley, W. G., Schwegler-Berry, D. E., & Castranova, V. (2007). Improved method to disperse nanoparticles for *in vitro* and *in vivo* investigation of toxicity. *Nanotoxicology*, *1*(2), 118–129. doi:10.1080/17435390701381596

Saikia, C., Gogoi, P., & Maji, T. K. (2015). Chitosan: A promising biopolymer in drug delivery applications. *Journal of Molecular and Genetic Medicine*, *s4*(1), 1–10. doi:10.4172/1747-0862.S4-006

Saveleva, M. S., Eftekhari, K., Abalymov, A., Douglas, T. E. L., Volodkin, D., Parakhonskiy, B. V., & Skirtach, A. G. (2019). Hierarchy of hybrid materials—the place of inorganics-in-organics in it, their composition and applications. *Frontiers in Chemistry*, *7*, 179. doi:10.3389/fchem.2019.00179

Shariatinia, Z., & Zahraee, Z. (2017). Controlled release of metformin from chitosan–based nanocomposite films containing mesoporous MCM-41 nanoparticles as novel drug delivery systems. *Journal of Colloid and Interface Science*, *501*, 60–76. doi:10.1016/j.jcis.2017.04.036

Shchipunov, Y. (2012). Bionanocomposites: Green sustainable materials for the near future. *Pure and Applied Chemistry*, *84*(12), 2579–2607. doi:10.1351/PAC-CON-12-05-04

Sisson, K., Zhang, C., Farach-Carson, M. C., Chase, D. B., & Rabolt, J. F. (2009). Evaluation of cross-linking methods for electrospun gelatin on cell growth and viability. *Biomacromolecules*, **10**(7), 1675–1680. doi:10.1021/bm900036s

Sohail Haroone, M., Li, L., Ahmad, A., Huang, Y., Ma, R., Zhang, P., ... Lu, J. (2018). Luminous composite ultrathin films of CdTe quantum dots/silk fibroin co-assembled with layered doubled hydroxide: Enhanced photoluminescence and biosensor application. *Journal of Materiomics*, *4*(2), 165–171. doi:10.1016/J.JMAT.2018.05.002

Su, D., Jiang, L., Chen, X., Dong, J., & Shao, Z. (2016). Enhancing the gelation and bioactivity of injectable silk fibroin hydrogel with laponite nanoplatelets. *ACS Applied Materials & Interfaces*, *8*(15), 9619–9628. doi:10.1021/acsami.6b00891

Sudheesh Kumar, P. T., Lakshmanan, V.-K., Anilkumar, T. V., Ramya, C., Reshmi, P., Unnikrishnan, A. G., ... Jayakumar, R. (2012). Flexible and microporous chitosan hydrogel/nano ZnO composite bandages for wound dressing: In vitro and in vivo evaluation. *ACS Applied Materials & Interfaces*, *4*(5), 2618–2629. doi:10.1021/am300292v

Theron, J., Walker, J. A., & Cloete, T. E. (2008). Nanotechnology and water treatment: Applications and emerging opportunities. *Critical Reviews in Microbiology*, *34*(1), 43–69. doi:10.1080/10408410701710442

Vargas-Bernal, R., Rodrguez-Miranda, E., & Herrera-Prez, G. (2012). Evolution and expectations of enzymatic biosensors for pesticides. In S. P. Soundararajan (Ed.), *Pesticides: Advances in Chemical and Botanical Pesticides*. InTech. doi:10.5772/46227

Venkatesan, J., & Kim, S.-K. (2010). Chitosan composites for bone tissue engineering—An overview. *Marine Drugs*, **8**(8), 2252–2266. doi:10.3390/md8082252

Wang, L., Lu, C., Zhang, B., Zhao, B., Wu, F., & Guan, S. (2014). Fabrication and characterization of flexible silk fibroin films reinforced with graphene oxide for biomedical applications. *RSC Advances*, *4*(76), 40312–40320. doi:10.1039/C4RA04529G

Wang, Y.-F., Chen, Y.-C., Li, D.-K., & Chuang, M.-H. (2011). Technetium-99m-labeled autologous serum albumin: A personal-exclusive source of serum component. *Journal of Biomedicine & Biotechnology, 2011*, 413802. doi:10.1155/2011/413802

Wang, Y., Ma, R., Hu, K., Kim, S., Fang, G., Shao, Z., & Tsukruk, V. V. (2016). Dramatic enhancement of graphene oxide/silk nanocomposite membranes: Increasing toughness, strength, and young's modulus via annealing of interfacial structures. *ACS Applied Materials & Interfaces, 8*(37), 24962–24973. doi:10.1021/acsami.6b08610

Wolinsky, J. B., Colson, Y. L., & Grinstaff, M. W. (2012). Local drug delivery strategies for cancer treatment: Gels, nanoparticles, polymeric films, rods, and wafers. *Journal of Controlled Release, 159*(1), 14–26. doi:10.1016/j.jconrel.2011.11.031

Xie, Q., Xu, Z., Hu, B., He, X., & Zhu, L. (2017). Preparation of a novel silk microfiber covered by AgCl nanoparticles with antimicrobial activity. *Microscopy Research and Technique, 80*(3), 272–279. doi:10.1002/jemt.22683

Xu, Z., Shi, L., Yang, M., & Zhu, L. (2019). Preparation and biomedical applications of silk fibroin-nanoparticles composites with enhanced properties: A review. *Materials Science and Engineering: C, 95*, 302–311. doi:10.1016/J.MSEC.2018.11.010

Ye, P., Yu, B., Deng, J., She, R.-F., & Huang, W.-L. (2017). Application of silk fibroin/chitosan/nano-hydroxyapatite composite scaffold in the repair of rabbit radial bone defect. *Experimental and Therapeutic Medicine, 14*(6), 5547–5553. doi:10.3892/etm.2017.5231

Zhang, H., Ma, X., Cao, C., Wang, M., & Zhu, Y. (2014). Multifunctional iron oxide/silk-fibroin (Fe_3O_4–SF) composite microspheres for the delivery of cancer therapeutics. *RSC Advances, 4*(78), 41572–41577. doi:10.1039/C4RA05919K

Zhao, F., Yao, D., Guo, R., Deng, L., Dong, A., & Zhang, J. (2015). Composites of polymer hydrogels and nanoparticulate systems for biomedical and pharmaceutical applications. *Nanomaterials, 5*(4), 2054–2130. doi:10.3390/nano5042054

Zhao, G., Qing, H., Huang, G., Genin, G. M., Lu, T. J., Luo, Z., … Zhang, X. (2018). Reduced graphene oxide functionalized nanofibrous silk fibroin matrices for engineering excitable tissues. *NPG Asia Materials, 10*(10), 982–994. doi:10.1038/s41427-018-0092-8

Zhao, R., Jia, T., Shi, H., & Huang, C. (2019). A versatile probe for serum albumin and its application for monitoring wounds in live zebrafish. *Journal of Materials Chemistry B, 7*(17), 2782–2789. doi:10.1039/C9TB00219G

Zhao, Z., Li, Y., & Xie, M.-B. (2015). Silk fibroin-based nanoparticles for drug delivery. *International Journal of Molecular Sciences, 16*(3), 4880. doi:10.3390/IJMS16034880

Zhao, Z., Li, Y., Xie, M.-B., & Bing, Y. (2015). Silk fibroin-based nanoparticles for drug delivery. *International Journal of Molecular Sciences, 16*, 4880–4903. doi:10.3390/ijms16034880

Zheng, Z. Z., Liu, M., Guo, S. Z., Wu, J. B., Lu, D. S., Li, G., … Kaplan, D. L. (2015). Incorporation of quantum dots in silk biomaterials for fluorescence imaging. *Journal of Materials Chemistry B, 3*(31), 6509–6519. doi:10.1039/C5TB00326A

Zhuang, Y., Hou, H., Zhao, X., Zhang, Z., & Li, B. (2009). Effects of collagen and collagen hydrolysate from jellyfish (*Rhopilema esculentum*) on mice skin photoaging induced by UV irradiation. *Journal of Food Science*, **74**(6), H183–H188. doi:10.1111/j.1750-3841.2009.01236.x

4 Sources of Natural Polymers from Plants with Green Nanoparticles

Satya Eswari Jujjavarapu, K. Chandrasekhar, Sweta Naik, Aditya L Toppo, and Veena Thakur

CONTENTS

4.1 INTRODUCTION

There is presently great interest in the current information regarding polymer related matters. Recently, increased production of petroleum-derived polymers for domestic and industrial purposes has been relentlessly damaging the environment (Kadier et al., 2018). Specialist predictions about the future availability of fossil fuels (such as oil, natural gas and coal) as energy sources that are not renewable, varies from one to three generations. After disposal, these petroleum-derived polymers remain in the environment for many years and subsequently damage the ecological system (Patel, Pandit, & Chandrasekhar, 2017). For example, plastic bags will still be around for several centuries. People use them for about 15 to 30 seconds or a few minutes (Chandrasekhar & Venkata Mohan, 2014a). If we throw it away, the plastic bag can either end up in a landfill, it can be burned with all other trash, or it can be blown away into the water (Mohan & Chandrasekhar, 2011). However, that plastic will return to humans in the fish we eat 25 years from now. It takes at least 450 years for one plastic bottle to decompose. If we accidentally drop our empty plastic bottle in the forest, you will still be able to pick it up in the year 2400. Most petroleum polymers do not biodegrade; they just fragment into tiny microplastics that are not visible to the naked eye (Kakarla, Kuppam, Pandit, Kadier, & Velpuri, 2017).

It is nearly impossible to buy anything that is not wrapped in plastic. Consequently, we throw away a lot of fossil-fuel-based polymer packaging – 140 million tons a year to be precise. Most of it ends up in landfill, a smaller percentage gets burned, and only a tiny part gets recycled (Pandit, Sarode, Sargunaraj, & Chandrasekhar, 2018). We might see videos or pictures on the news of birds with plastic in their stomachs. We have become accustomed to the idea that even on the remotest of all islands, where nobody lives, the beaches are full of petroleum-derived polymer waste. Approximately every minute, we throw a garbage-truck-full of petroleum-derived polymer material into the ocean (Chandrasekhar & Venkata Mohan, 2014b). Alternatively, if we recycled all our plastic, you would imagine we would end up with a cleaner planet, but according to expert reports, our use of petroleum-derived polymers will be four times higher by 2050, so even if we recycle everything we use, it won't be enough. We will still have to produce a lot more plastic. More importantly, new plastic is cheaper than recycled plastic. That means that, even if you are a model citizen who recycles their plastic, most companies won't even use it. Producing this much plastic is plain dumb. We are human beings. Surely we can think of packaging that is reusable, reliable or just not plastic? Surely we can invent our way out of this plastic waste issue (Johnston et al., 2019)? When you want your takeaway coffee, you can bring your thermos. You don't need a bottle for your shampoo, and you can bring your bags to the supermarket. But we won't get there easily. Not with recycling alone, not with cleaning up alone. It's time to stop using plastic and to invent one hell of a solution (Valentino, Morgan-Sagastume, Campanari, & Werker, 2017). Keeping in mind the worsening ecological circumstances caused by several factors, including progress in science and technology, a growing population, global warming, etc., researchers all over the world have recently focused on biopolymers from renewable resources with much success. Henceforth, biodegradable polymers are considered to be a possible answer to address this issue, and environmentally friendly and easily biodegradable polymers have garnered significant attention in recent years (Laycock et al., 2017).

These biodegradable polymers are broadly classified into two different classes, that is, natural polymers and synthetic polymers, based on their origin. Natural polymers are acquired from natural sources; synthetic polymers are the polymer substances produced through numerous chemical synthesis processes. The best example of a natural polymer is starch; it is produced by green plants as an energy reserve, stored in plant cells for future requirements (van Beilen & Poirier, 2008). It is a highly biodegradable, low-cost and abundantly available natural polysaccharide (plant-based natural polymer), broadly distributed in the form of minute granules as the chief standby polymer/carbohydrate in the trunks, roots, grains and fruits of all types of leafy plants (Mukherjee, Lerma-Reyes, Thompson, & Schrick, 2019). Furthermore, it is produced from atmospheric carbon dioxide and water in the presence of chlorophyll and sunlight by a process called photosynthesis. Due to its readily biodegradable nature, biocompatibility, low cost and renewability, this plant-based natural polymer has gained much attention in recent years.

4.2 NATURAL POLYMERS

A polymer is a large macromolecule formed when monomer units are repeatedly linked together by chemical bonds. These compounds are produced by polymerization and they occur both in natural form as well as in synthetic form. Natural polymers are more in demand because they are biodegradable, with a few exceptions, and also biocompatible, nontoxic and, moreover, readily available. In addition, they are capable of chemical modification (Hon, 2017; Teodorescu, Bercea, & Morariu, 2018). Based on their source of origin, polymers are classified into two types, i.e., either polysaccharide or protein-derived, coming from plant or animal sources, respectively.

Polymers produced by living organisms have been utilized by humans for a very long time, as they are biodegradable in nature and do not cause harm. On the other hand, polymers produced by chemicals have adverse effects on the atmosphere as well as on human health. Biopolymers and their composites occupy a unique position in the new world of biomaterials. Various applications of these composites in the fields of food technology, medicine and the pharmaceuticals are increasing day by day, which is leading to advances in biomaterial technology (Liu, Xie, Li, Zhou, & Chen, 2015; Kunduru, Basu, & Domb, 2016; Ivanova, Bazaka, & Crawford, 2014). Natural polymers and their types are illustrated in Figure 4.1.

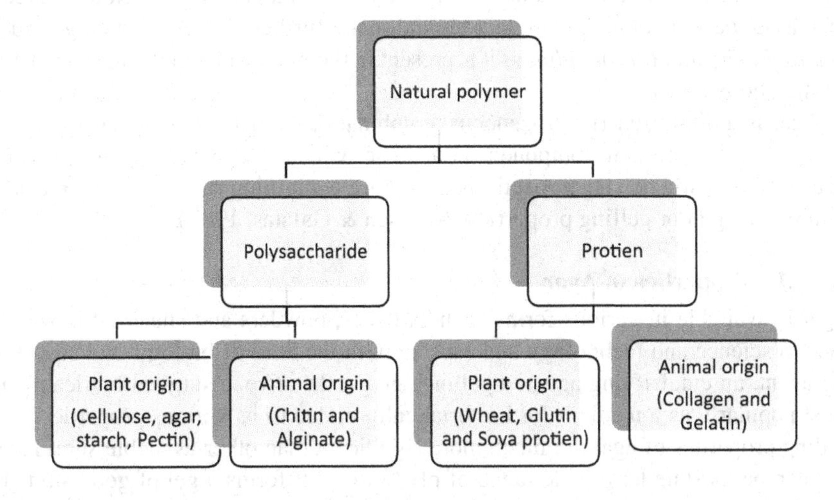

FIGURE 4.1 Natural polymers and their types.

4.2.1 The Need for Plant Polymers

The following properties of plant polymers mean that they are in demand for composite production:

1. Biodegradability, as they are derived from living organisms they are biodegradable in nature and do not cause harm to human beings.
2. Biocompatibility and nontoxicity, as most plant material is made only of carbohydrates and of reiterating monomer (simple sugars) units, which are nontoxic.
3. Cost-effective: Their production cost is cheaper compared to synthetic material.
4. Safe: These polymers are derived from natural sources, hence do not produce any side effects.
5. Easy availability: Due to their applications in various fields, these are produced on a large scale.

However, there is no single polymer that possesses all the desirable properties such as extraordinary mechanical strength, a high oxygen barrier, high thermogravimetric properties, chemical stability and high biodegradability (van Beilen & Poirier, 2007; Snell, Singh, & Brumbley, 2015). Therefore, a biocomposite material can be designed to fulfill all of the above requirements. New applications are emerging in addition to traditional uses, particularly in the automotive and building industries. In the place of synthetic fibers, plant fibers, such as flax and hemp, can be reinforced with PVC, PE or PP-type polymers. Some plant-derived polymers are described in the following sections.

4.2.2 Agar and Its Composites

Agar is a substance that is gelatinous and can be obtained from various sources, such as *Gelidium amansii* (Gelidaceae) and many further classes of red algae such as *Gracilaria*, and *Pterocladia*, as it is present in their cell walls in the form of structural carbohydrates.

Agar is a dissimilar (heterogeneous) combination of agarose and agaropectin in which the predominant component is agarose, which is a straight-chain polymer of agarobiose, whereas agaropectin is a mixture containing molecules of an acidic nature having poor gelling properties (Armisen & Galatas, 1987.).

4.2.2.1 Properties of Agar

Agar is available in various forms such as flakes, powders and sheets. It is widely used in science and technology and especially in the field of biology as a suspending agent, an emulsifying agent, a gelling agent, a laxative, a surgical lubricant and most popularly as a medium for bacterial culture, due to its various properties. The gelling properties of agar are much more significant than other available substances and can be used under a wide range of pH (5 to 8). It forms a gel of good strength which is resistant to temperatures of up to 85°C. It will not lose its properties if used repeatedly, even after having melted a limited number of times (Praiboon, Chirapart, Akakabe, Bhumibhamon, & Kajiwara, 2006; Lee et al., 2017). It is widely accepted as a herbal polymer, which means it is environmentally friendly and has many advantages over non-herbal alternatives, being biocompatible, biodegradable, economical and nontoxic. Different algal sources of agar are shown in Figure 4.2 and the structure of agar is presented in Figure 4.3.

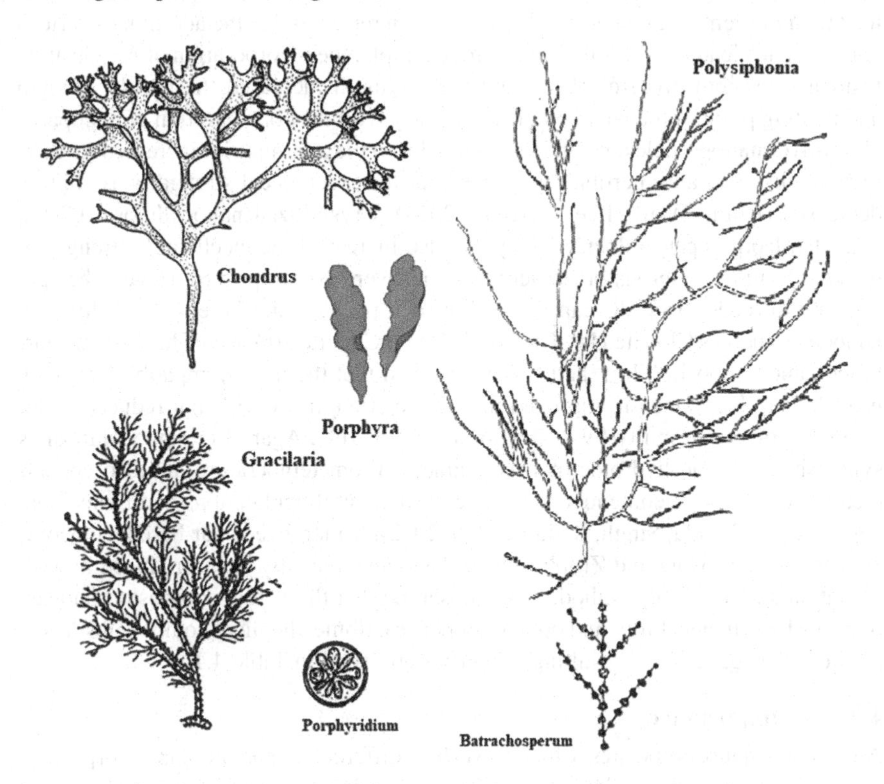

FIGURE 4.2 Algal sources of agar.

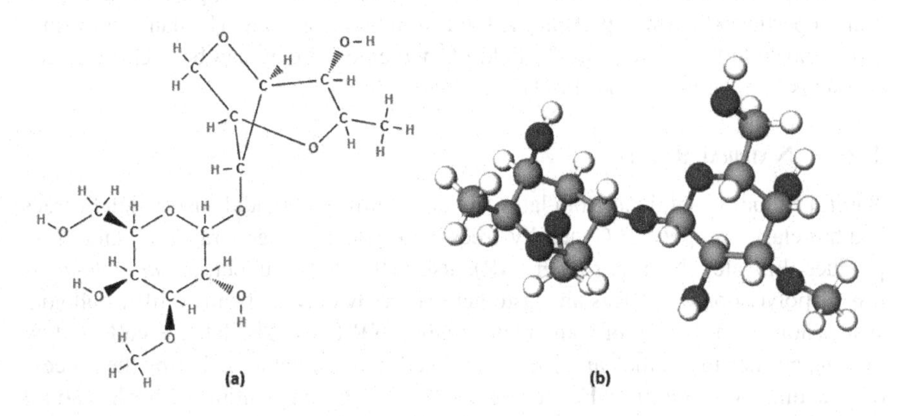

FIGURE 4.3 Structure of agar (a) 2-dimensional, (b) 3-dimensional.

4.2.2.2 Agar-Based Nanocomposites

Nanocomposites have two components: the standard matrix material and additional nanomolecules that contribute to the properties of the nanocomposite; for example, the addition of a nanoparticle can increase mechanical strength, electrical and thermal conductivity, etc. Agar nanocomposites are those in which the main polymer or matrix material is agar, in addition to which there are several modifications that will lead to improvement in its properties. It has been reported that the addition of various nanoparticles leads to several changes, for example, the incorporation of Ag-Cu alloy nanoparticles into glycerol plasticized agar solution increased the tensile strength and melting point of the nanocomposite linearly with the concentration of nanoparticles. Silver nanoparticles, when incorporated into the agar matrix, increased its thermal stability and antimicrobial properties but the mechanical strength was slightly decreased (Rhim, Wang, Lee, & Hong, 2014). Crystallized nanocellulose (CNC), extracted from paper mulberry bast pulp, also increased the mechanical strength of the agar film; its water vapor barrier properties were also improved. Hence, the agar films reinforced with CNC can be used in food packaging. The effects of different nanoclays such as Cloisite Na^+, Cloisite 30B and Cloisite 20A were checked on agar-based nanocomposite films, and it was observed that the most compatible nanoclay was Cloisite Na^+, since it improved the tensile strength by 18% but reduced water (H_2O) vapor penetrability by 24% (Rhim et al., 2014). Agar–SiO_2 nanocomposites synthesized via the hydrothermal technique at room temperature are also considered to be a suitable contender for both environmental applications and biomedical applications (Shukla, Singh, Reddy, & Jha, 2012). Mineralized agar-based nanocomposites (Zn-carbonate and Zn-phosphate) have also recently been produced by precipitation and a casting method. It was observed that these nanocomposites showed improved thermal stability and optical properties. Some chemical components incorporated into agar and the resulting properties are listed in Table 4.1.

4.2.2.3 Importance

Agar-based nanocomposites which have had different nanomaterials incorporated into them are used for various applications. They are used widely for packaging of food and other material since they have increased mechanical strength and antimicrobial properties which will prevent the food from being spoiled. The nanocomposites also contribute to advances in the field of medicine, as they can be used in wound dressings (Cooper & Linton, 1952).

4.2.3 Natural Rubber

Natural rubber is obtained from latex sap or the milky colloidal suspension of trees and has elastic properties. Generally, trees belonging to the genera *Hevea* and *Ficus* produce this latex. Natural rubber (NR), also called India rubber, or *caoutchouc*, is *cis*-1,4-poly(isoprene). NR is an elastomer originally derived from a milky colloidal suspension from the sap of some plants, called NR latex. The latex is collected by making an incision in the bark, and it is refined into a suitable rubber by the process of vulcanization (Surya & Hayeemasae, 2019). The composition of fresh, natural rubber latex in the form of an emulsion and dry natural rubber is listed in Table 4.2.

TABLE 4.1
Chemical Components Incorporated into Agar and Its Properties

S. No.	Component or Plasticizer Added	Film Characteristics and Improvements
1	Arabinoxylan/ glycerol	The addition of arabinoxylan improves moisture barrier efficiency but decreases mechanical properties of the film.
2	Starch/glycerol	The addition of starch degrades surface resistance to water and mechanical properties of the film.
3	Silver (Ag) nanoparticles	The film exhibits good mechanical stability, H_2O vapor and gas barrier in addition to strong antimicrobial activity.
4	Nanoclay/glycerine	Incorporation of clay (up to 10%) raises ductility and decreases H_2O vapor penetrability.
5	Grapefruit seed extract (GSE)	The addition of GSE increases the color, UV fence, dampness, H_2O solubility & H_2O vapor penetrability, but decreases the surface hydrophobicity, ductility and elastic modulus of the film. The film exhibits distinctive antimicrobial activity.
6	Banana powder and Silver (Ag) nanoparticles/ glycerol	The addition of banana powder increases UV light absorption, H_2O vapor fencing and antioxidant activity, but decreases the mechanical properties of bilayer film. The composite film exhibits distinctive antimicrobial activity and mechanical properties.
7	Fish gelatin and TiO_2 nanoparticles	The addition of TiO_2 decreases water vapor permeability and increases tensile strength, UV light barrier property, swelling ratio and H_2O content of the film.

Source: Cooper & Linton (1952).

The independent movement of polymers is prevented by this process. The chemical structure of natural rubber is shown in Figure 4.4.

4.2.3.1 Composites of Natural Rubber and Its Importance

1. Natural rubber/cellulose nanocrystals (NR/CNCs) form true bio-composites from renewable resources. Cellulose nanocrystals form biocomposites with natural rubber, which shows improved thermomechanical properties. They have greater tensile strength, and CNCs act as strengthening agent and crosslinking agents in the natural rubber medium (Manaila, Stelescu, & Craciun, 2018).
2. NR and plasticized starch (PS, extracted from potato)–based eco-mixtures have been prepared by peroxide cross-linking. Some environmentally friendly mixtures based on NR and PS were procured and characterized for use as filler. All these mixtures were attained by peroxide cross-linking in the company of a polyfunctional monomer with the help of cross-linking agents such as trimethylolpropane and trimethacrylate (Masłowski, Miedzianowska, & Strzelec, 2017).
3. Recently, the development of high-performance materials made from available and cheap natural resources is increasing worldwide. A better

TABLE 4.2
Composition of Fresh, Natural Rubber Latex in the Form of an Emulsion and Dry Natural Rubber

	NR Latex	Dry NR	Deproteinized Dry NR
Component	Total solid content	Rubber content	Rubber content
	Dry rubber content	Resinous substances	Proteinaceous substances
	Resinous substances	Proetinaceous substances	Ash
	Inorganic salts	Inorganic salts	Volataile matter content
	Ash, sugars, water	Sugars	Dirt content
		Cu and Mg	
		Water	
Composition (percentage)	36	93–95	96
	33	2	0.12
	1–2.5	2–3	0.15
	1–1.5	<0.2	0.3
	0.5	<0.2	0.01
	<1	2–3 ppm	
	60	0.5	

Source: Dafe et al. (2017).

solution for this issue is mixtures of NR. Complexes of NR, comprising barley, corn and wheatgrass as bio-filters, were investigated. Elastomers that are packed with cereal grass signify a systematic and technical novelty. The utilization of this straw as a plaster intended for elastomer mixtures extends the choice of useful properties and decreases production costs. These bio-composites are eco-friendly and increase the opportunity for the use of straw, which is generally considered as one of the most challenging agronomic leftovers. The rubber blends comprise lignocellulose resources which establish typically favorable kinetics of cross-linking. By

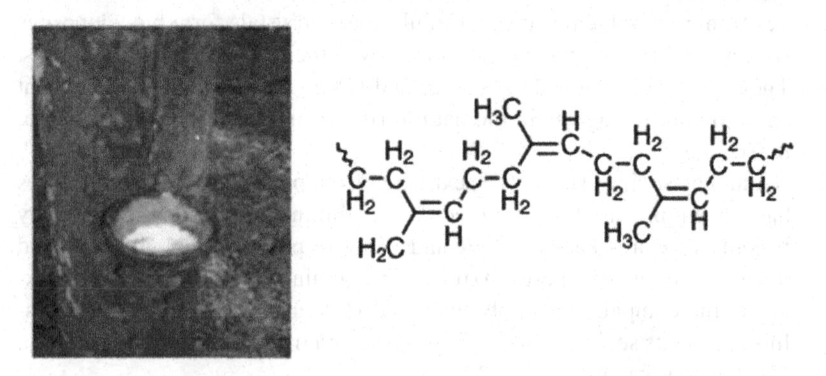

FIGURE 4.4 (a) Chemical structure of natural rubber [*cis*-1,4-poly(isoprene)] (b) Natural rubber latex from a *Hevea* tree.

adding an appropriate amount of plaster, altered NR vulcanizes, by which this composite obtains great mechanical properties, barrier properties, and the capability to wet under the impact of compression stress (Du et al., 2019).

4. Porous starch is used as a plaster for the rubber industry in the place of carbon black (CB). Starch is an economical, plentiful, renewable and eco-friendly plaster for rubber strengthening through appropriate alteration. As soon as the fraction of dodecenyl succinic anhydride porous starch (DDSA-PS)/CB was augmented, the Payne effect of rubber mixtures was lessened to a huge degree in addition to a drop in the mechanical properties and wear resistance of vulcanized rubbers. In particular, the progressing confrontation and hysteretic assets were enriched, replicated as the decrease of heat developed, by a combination of altered permeable starch into the final NR mixtures. As a result, DDSA-PS may be used as a capable plaster for the rubber industry, partially in place of CB (Xu, Cui, Fu, & Lin, 2018).
5. Cross-linked rubber mixtures present recyclable, healable and adaptable capabilities. Conservative cross-linking provides rubbers with exceptional mechanical properties while simultaneously cracking them for thermosetting with the hope of recycling and self-curing. From the viewpoint of maintainable improvement of material, it is critical, and significant, to incorporate all these attractive assets into cross-linked marketable rubbers. An eco-friendly and healable epoxidized natural rubber (ENR)/citric acid–modified bentonite (CABt) complex was prepared. CABt with several carboxyls on its surface serves as cross-linking material in the direction of covalently cross-link epoxidized natural rubber with interchangeable β-hydroxyl ester linkages, along with well-organizing and strengthening epoxidized natural rubber. ENR/CABt mixtures might change the linkage topology at a higher temperature because of the transesterification reactions of β-hydroxyl ester linkages among CABt and epoxidized natural rubber. In addition, the small cross-linking degree of the network and the inherent stickiness of the epoxidized natural rubber medium enable the chain dispersal and transesterification reactions of β-hydroxyl ester-links, which recycle and heal ENR/CABt complexes. Cross-linked rubber complexes are biodegradable and self-healing in nature (Kulkarni, Butte, & Rathod, 2012).

Applications include

1. Footwear, battery boxes, foam mattresses, balloons, toys
2. Customer products such as footballs or golf balls and other sports or entertainment items, erasers
3. Catheters and surgical gloves
4. Non-tyre rubber substances such as industrial goods transmission belts, elevator belts, pipes and tubes, industrial coatings and bridge bearings
5. Heavy industry equipment like shock mounts, vibration isolators, gaskets, seals, rolls, hoses and tubing

4.2.4 Pectin and Its Composites

Pectins are natural heteropolysaccharides of plant cell walls, mainly consisting of a lined polysaccharide of α-1,4-linked D-galacturonic acid residues and 1,2-linked L-rhamnose residues. Pectin exists in the chief cell wall and central lamella of numerous plants. The composition of pectin varies according to the plant species, the part of the plant and also, sometimes, and with time, within a plant (Lara-Espinoza, Carvajal-Millán, Balandrán-Quintana, López-Franco, & Rascón-Chu, 2018). Pectin generally contributes to the structure of plant tissue, its quantity and nature being important for consistency in fruit and vegetables throughout their development and maturation and, similarly, in their storage and processing. It controls the flow of liquid in the plant. Pectin also offers a significant source of nutritional fiber. In addition, it is used as a gelling agent, a solidifying agent, an additive and an emulsifier. It has great importance in the medical field, where it is used as a transporter for controlled drugs or bioactive release, and also for various drug delivery methods.

4.2.4.1 Composites of Pectin

Pectin can interact with many beneficial compounds to form composites. The functional group present in pectin allows interaction with a variety of molecules. The structure of pectin is illustrated in Figure 4.5.

The interactions between polymers are widely studied because of their importance and applications. Nowadays, there is more demand for products composed of bio-based materials, which are environmentally friendly, and are cheaper and safer compared to existing products. To develop such a material, the interactions of different polysaccharides must be studied (Bierhalz, da Silva, & Kieckbusch, 2012). Composites of pectin and their applications are listed in Table 4.3.

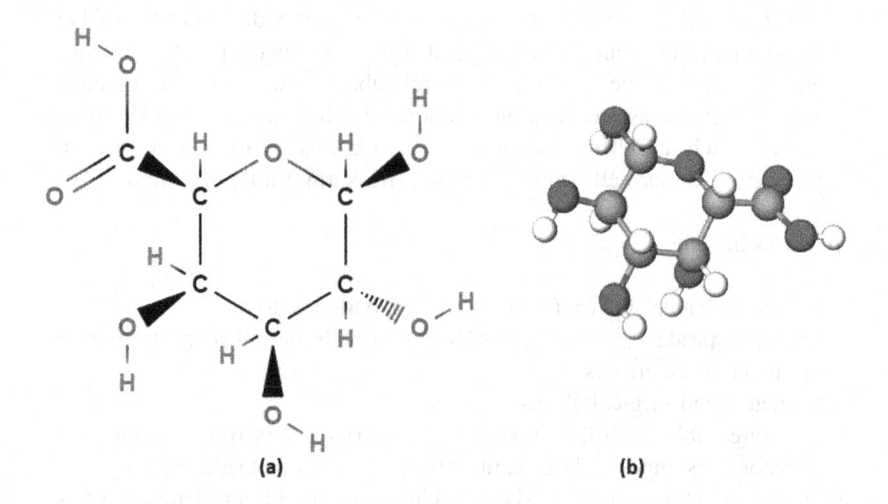

FIGURE 4.5 Structure of pectin (a) 2-dimensional, (b) 3-dimensional.

TABLE 4.3
Composites of Pectin and Their Applications

Composite	Application
Pectin/ Alginate	Full applications in the food industry, as well as manufacturing of cold-setting fruit gels, maintenance of acidic suspensions such as salad cream or mayonnaise. Pharmaceutical industry applications: encapsulation of active ingredients, drug delivery systems
Pectin/ Chitosan	Food industry Biomedicine Drug delivery systems
Pectin/ Gelatin	Food industry Bone regeneration medicine Composites for wound
Pectin/ Protein	Acetaminophen release Esterified pectin as acidified milk stabilizer

Source: Farris et al. (2011).

Different composites of pectin include the following:

1. *Pectin/Alginate*: By mixing these two gelling agents, a composite of pectin can be prepared, but its structural properties mainly depend upon the mannuronic and guluronic acid proportion of the alginate and the degree of esterification of the pectin. When the ratio of mannuronic acid and guluronic acid is small and there is a high degree of pectin esterificaton, the strongest gels are prepared, but in the case of an equal ratio, the interaction is found to be optimum (Bierhalz et al., 2012).
2. *Pectin/Gelatin*: Gelatin is the most-studied food polymer, and interaction between pectin and gelatin has also been reported by several researchers. The interactions between these two are segregative as well as associative. Gelation occurs when the viscosity is increased, followed by the formation of the network and the cessation of phase separation. The interaction of polymers in the complex depends upon various factors such as rigidity, solubility and density (Farris et al., 2011).
3. *Pectin/Chitosan*: Composites formed by the pectin and chitosan polysaccharides are environmentally friendly, as they are nontoxic, biocompatible and biodegradable. These polymers are also known as polyelectrolyte complexes. Various interactions, including Van der Waals, electrostatic, hydrogen, hydrophobic and coordinate bonding, occur between the pectin and the chitosan, resulting in very strong intermolecular bonds. Many factors, such as temperature, pH, concentration, ionic condition of the medium, and charge density, play a vital role in the constancy of these gels (Sriamornsak & Puttipipatkhachorn, 2004).
4. *Pectin/Protein complex*: The mechanism involved in the composite formation of pectin and protein is associative or segregative, depending upon the ionic and structural characters of the two components. Phase separation is a

result of the associative process that occurs between two oppositely charged polymers. High steric exclusion and strong electrostatic repulsion result in the segregative phase. Physicochemical properties, such as ionic strength, pH, the ratio of pectin to protein and charge density, are responsible for the interactions between these two biopolymers.

4.2.5 Starch

Starch is a naturally occurring polymer that is obtained from plants and is composed of amylase and amylopectin carbohydrates. These assemblies are formed from glucopyranose units that are coupled to each other with α-1,4 glycosidic links in a twisting line or branched chains. The fact that it is obtained from natural sources means that it is a biodegradable, renewable, cost-effective and sustainable material, which makes it suitable for multiple applications (Alvarez-Lorenzo, Blanco-Fernandez, Puga, & Concheiro, 2013). The structural components of starch are given in Figure 4.6.

4.2.5.1 Composites of Starch

4.2.5.1.1 Sugar-Palm-Derived Cellulose Composites with Sugar Palm Starch-Based Biofilms

Biofilms can be prepared using sugar palm as a plant source. Packaging materials prepared from sugar palm–based films have been reported to show low resistance to water and poor mechanical strength, which restricts its usage in food packaging (Das & Pal, 2015). The formation of a composite using sugar palm–derived cellulose shows increased tensile strength, keeping its biodegradable nature (Birch & Schiffman, 2014).

4.2.5.1.2 Composites of Thermoplastic Corn Starch and Cellulose-Based Fillers

Composites of this type use corn as a plant source. The addition of cellulose fillers increases the surface water resistance and tensile strength of the prepared composite but decreases elongation values when compared with composites that do not have cellulose fillers (Barakat, 2011; Baracat et al., 2012).

4.2.5.1.3 Thermoplastic Starch-Based Composites Reinforced with Rapeseed Fibers

Composites of this type use rapeseed as a plant source. When composites are prepared using rapeseed fibers, the elasticity of the material diminishes and shows no considerable effect on the water uptake ability of the material (Jensen, Rolin, & Ipsen, 2010).

4.2.5.1.4 Thermoplastic Sago Starch/Kenaf Core Fiber Biocomposites

Composites of this type use sago palm as a plant source. The increased fiber content in this composite results in increased tensile strength. In addition, the thermal stability of the biocomposites increases with the addition of kenaf core fibers to the sago starch matrices. The ability of such biocomposites to absorb water also decreases with the addition of fibers (Wu, Degner, & McClements, 2014).

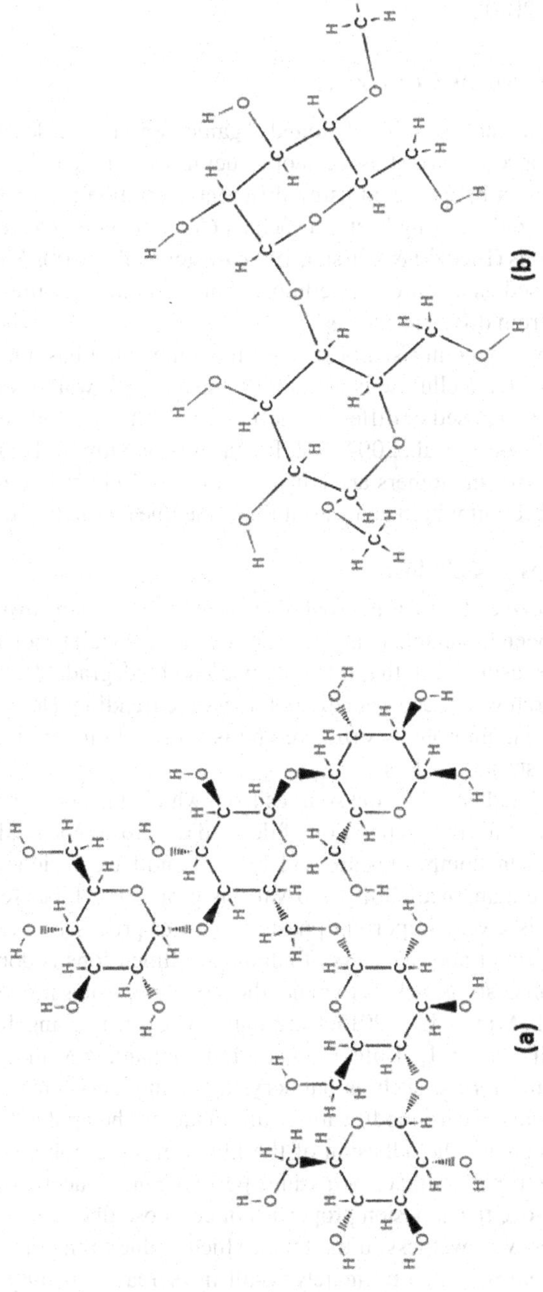

FIGURE 4.6 Structural components of starch (a) amylopectin, (b) amylose.

4.2.5.1.5 Composite of Starch Reinforced with Date Palm and Flax Fibers

The addition of fibers increased the tensile strength, biodegradation, thermal stability and water uptaking ability, all properties improving with increasing fiber content (Saravanan & Rao, 2010).

4.2.6 Cellulose and Its Composites

Cellulose is one of the most abundantly found organic polysaccharides present in the environment. Cellulose is present as a major structural component in the plant cell wall from where it can be extracted using different extraction protocols. Cellulose can also be synthesized by using bacteria such as *Gluconacetobacter xylinus* under specialized conditions (Bledzki & Gassan, 1999; Foster et al., 2018). Algae and tunicates can also be used as a source of cellulose, but there are structural differences found in cellulose from different species.

Cellulose is normally comprised of glucose monomers and has a molecular formula of $(C_6H_{10}O_5)_n$. Hemicellulose is also found in some cell walls associated with cellulose. They are composed of different short chains of hexose and pentose sugars which are branched (Taira et al., 2004; Rol, Belgacem, Gandini, & Bras, 2019).

In cellulose, the two monomers are joined by the β1–4 glucosidic bond and the hydrogen bonding present within chains stabilizes the linear structure of cellulose.

4.2.6.1 Properties of Cellulose

The chemical nature of cellulose is hydrophobic because of the many hydroxyl groups present in it; it has been found that many organic solvents also fail to solubilize it. The major advantages of using it are that it is nontoxic and biodegradable, which makes it suitable for research where eco-friendly polymers are trending. Hemicellulose, on the other hand, has an amorphous structure and is weaker in nature then cellulose, that is, their tensile strength is less.

Cellulose is classified based on the sources from which it is derived. Many of the forms are further used for the preparation of derived cellulose (Nie et al., 2018).

Cellulose-based nanocomposites are widely used and have various applications because of their abundant availability and various properties. Fiber reinforcement in nanocomposites is a very important phase. In earlier processes, artificial fibers were used, but the use of natural fibers of nanoscale dimensions is done nowadays to reduce production costs as well as to make the nanocomposites nontoxic and biodegradable (Nguyen, Min, & Lee, 2015; Carbinatto, de Castro, Evangelista, & Cury, 2014; Tummalapalli et al., 2016; Dafe, Etemadi, Dilmaghani, & Mahdavinia, 2017). Cellulose in its many forms, such as nanocrystals, nanofibers, etc., is reinforced into polymer materials, where modifications are made by the application of different treatments to increase the adhesion of the fibers into the polymer. Chemicals such as sodium hydroxide, silane, potassium permanganate, acetic acid, etc. are well-known to improve the adhesion properties of cellulose fibers. Silane treatment increases the tendency of wetness in the fiber, which is due to the presence of several hydrocarbon chains, which ultimately result in increased affinity towards the polymer matrix. Acetylation is another promising process which results in surface modification that further results in hydrophobic nanocellulose fibers. The structures of cellulose and hemicellulose are given in Figure 4.7. The surface can also be modified so that antimicrobial activity is increased, for which graphene oxide is used along with bacterial cellulose (Rascón-Chu et al., 2018; A. Korma et al., 2016; Lomelí-Ramírez, Barrios-Guzmán, García-Enriquez, Rivera-Prado, & Manríquez-González, 2014; Bilbao-Sainz, Bras, Williams, Sénechal, & Orts, 2011; Lomelí-Ramírez et al., 2014).

4.2.6.2 Application of Cellulose and Its Composites

The field with the most potential for the application of cellulose-based nanocomposites is the paper and packaging industry because of the good strength and hydrophobic nature of the nanocomposites. Nanocellulose composite films are also known for their conductive properties, which makes them useful in the electronics industry. They also have immense applications in the biomedical field due to their high-surface-area-to-volume ratio; one example is in dialysis membranes. The classifications of cellulose and its applications are given in Figure 4.8. Some novel inventions related to cellulose-based nanocomposites are self-healing materials, drug-delivering nanopills, improved protective textiles, etc. (Espigulé et al., 2013; Bodîrlău, Teacă, & Spiridon, 2013; Sarifuddin, Ismail, & Ahmad, 2012).

4.2.7 Carbohydrates

Carbohydrates (sugars) are a very important class of naturally formed organic compounds that contain hydrates of carbon, as the name suggests. They are made up of simple sugars and are the body's primary energy source. The general formula is $C_n(H_2O)_n$. Based upon the value of 'n', they are divided into monosaccharides, disaccharides and polysaccharides. The most common monosaccharides are glucose, galactose and fructose. Among the disaccharides are sucrose, lactose and maltose, and among the polysaccharides are storage polysaccharides such as starch and glycogen and structural polysaccharides such as cellulose and chitin (Li, Tabil, & Panigrahi, 2007; Klemm, Heublein, Fink, & Bohn, 2005; Kumar Trivedi, Patil, Mohan Tallapragada, Mohan Tallapragada Effect of Biofield, & Mohan Tallapragada, 2015; Klemm et al., 2005; Dahman, 2017).

Among the familiar carbohydrates are numerous sugars, starches and cellulose, all of which are significant for the preservation of life in plants as well as in animals. They provide energy through oxidation and serve as a form of stored chemical energy (Li et al., 2007).

It is well known that starch is a supremely plentiful form of storage polysaccharide in plants. As a result, it is a low-cost source for syrups containing glucose, fructose or maltose, which are generally employed in food and in food-related industries. Moreover, the sugars produced can be fermented to yield valuable chemicals such as bioethanol (Yang et al., 2016).

4.2.7.1 Composites of Carbohydrates and Their Importance

1. Keratin has been extracted from human hair and blended with chitosan to fabricate keratin–chitosan-based porous scaffolds for biomedical and tissue engineering applications (Das, 2017.).

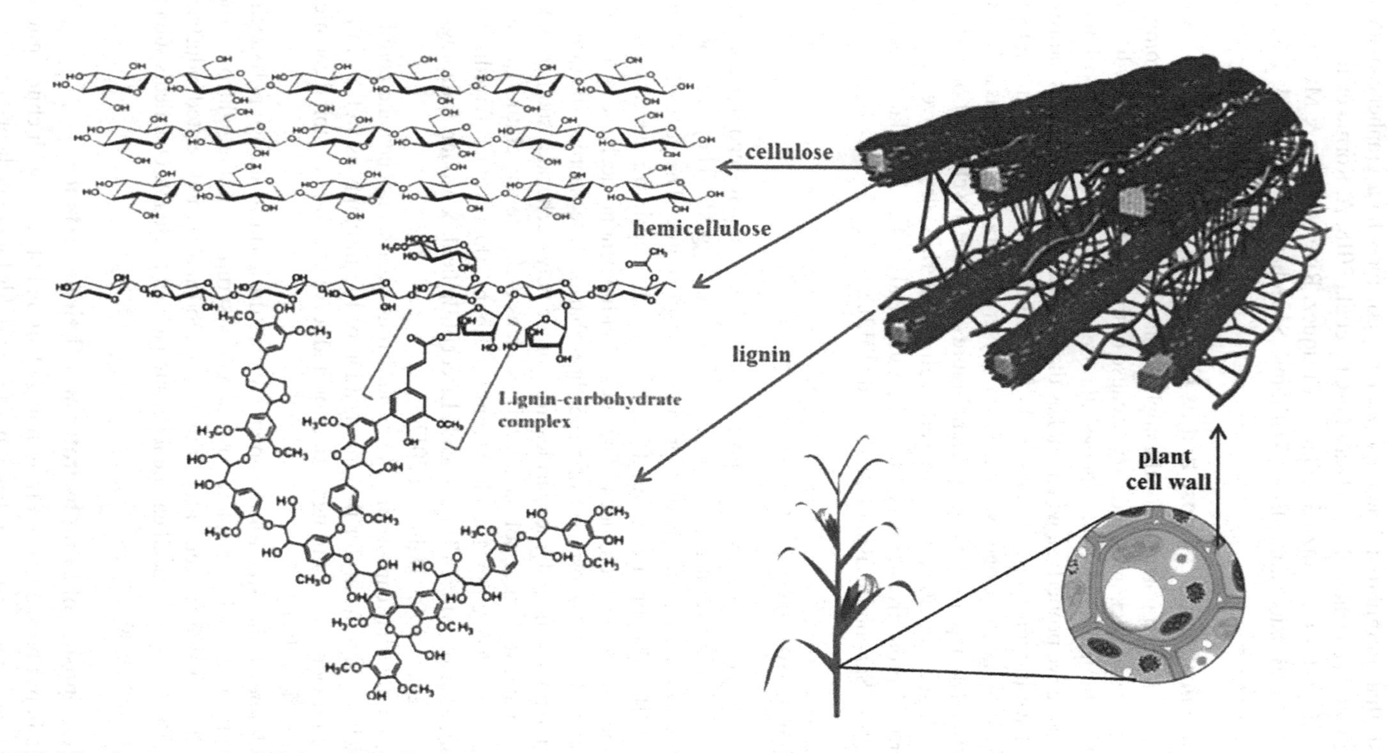

FIGURE 4.7 Structure of cellulose and hemicelluloses.

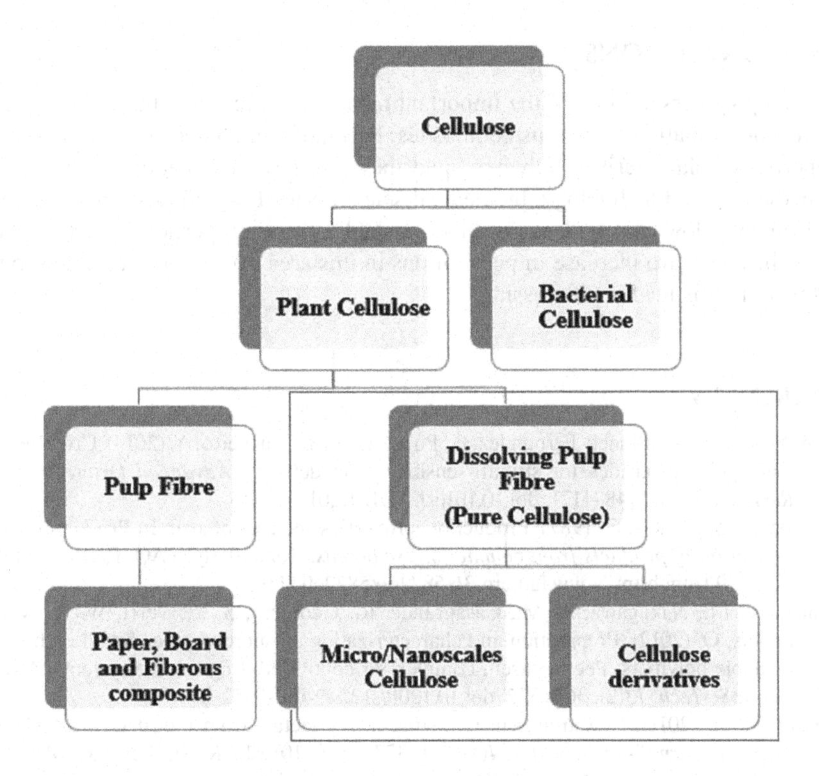

FIGURE 4.8 Classification of cellulose and its applications.

2. Compound hydrogels have been prepared from tea cellulose. Here, chitosan and guar gum might increase the thermal stability and mechanical property of the compound hydrogels, whereas soluble starch may well increase the equilibrium growth ratio. The chitosan and guar gum might enhance pervasion resistance and be useful for release control of the hydrogels. Not only chitosan but also κ-carrageenan, guar gum and soluble starch possess great biocompatibility and non-cytotoxic properties (Liu & Huang, 2016).
3. Alginate/polycaprolactone (PCL) complex nanofibers can selectively improve cells with high drug resistance. Alginate fibers were combined with PCL fibers using the co-electro-spinning method. Unlike alginate, PCL is a regularly employed bio-substantial in framework construction. The microenvironment of frameworks can be influenced with regulating alginate/PCL fiber ratios, by which diverse cell populations can be selected because of their adhesion ability. Liver CSCs (cancer stem cells) were supplemented from Hep G2 cells, a liver-cell line in this particular study (A. Korma et al., 2016; Hu, Lin, & Hong, 2019).
4. Starch films can be incorporated with tea polyphenols and used for active food packaging (Feng et al., 2018).
5. Carbohydrate composites are used for bio-based food packaging and other materials that can be used as a substitute for plastic.

4.3 CONCLUSIONS

Plant biopolymers are one of the important natural resources and have high potential for the formation of various composites. Formation of composites enhances the properties of plant-derived polymers, and they also have applications in the fields of medicine, food technology, biomedical science, etc. They are eco-friendly, easily available, low-cost and easily biodegradable; these properties attract further research. The sharp increase in publications in this area over the last decade shows that work in this field is progressing.

REFERENCES

Alvarez-Lorenzo, C., Blanco-Fernandez, B., Puga, A. M., & Concheiro, A. (2013). Crosslinked ionic polysaccharides for stimuli-sensitive drug delivery. *Advanced Drug Delivery Reviews, 65*(9), 1148–1171. doi:10.1016/J.ADDR.2013.04.016

Armisen, R., & Galatas, F. (1987). Production, properties and uses of agar. In *Production and utilization of products from commercial seaweeds*. Rome, Italy: FAO. Retrieved July 22, 2019 from http://www.fao.org/3/x5822e/x5822e03.htm

Baracat, M. M., Nakagawa, A. M., Casagrande, R., Georgetti, S. R., Verri, W. A., & de Freitas, O. (2012). Preparation and characterization of microcapsules based on biodegradable polymers: Pectin/casein complex for controlled drug release systems. *AAPS PharmSciTech, 13*(2), 364–372. doi:10.1208/s12249-012-9752-0

Barakat, M. A. (2011). New trends in removing heavy metals from industrial wastewater. *Arabian Journal of Chemistry, 4*(4), 361–377. doi:10.1016/J.ARABJC.2010.07.019

Bierhalz, A. C. K., da Silva, M. A., & Kieckbusch, T. G. (2012). Natamycin release from alginate/pectin films for food packaging applications. *Journal of Food Engineering, 110*(1), 18–25. doi:10.1016/J.JFOODENG.2011.12.016

Bilbao-Sainz, C., Bras, J., Williams, T., Sénechal, T., & Orts, W. (2011). HPMC reinforced with different cellulose nano-particles. *Carbohydrate polymers, 86*(4), 1549–1557. doi:10.1016/j.carbpol.2011.06.060

Birch, N. P., & Schiffman, J. D. (2014). Characterization of self-assembled polyelectrolyte complex nanoparticles formed from chitosan and pectin. *Langmuir, 30*(12), 3441–3447. doi:10.1021/la500491c

Bledzki, A., & Gassan, J. (1999). Composites reinforced with cellulose based fibres. *Progress in Polymer Science, 24*(2), 221–274. doi:10.1016/S0079-6700(98)00018-5

Bodîrlău, R., Teacă, C. A., & Spiridon, I. (2013). Green composites comprising thermoplastic corn starch and various cellulose-based fillers. *BioResources, 9*(1), 39–53. Retrieved July 23, 2019 from https://bioresources.cnr.ncsu.edu/resources/green-composites-comprising-thermoplastic-corn-starch-and-various-cellulose-based-fillers/

Carbinatto, F. M., de Castro, A. D., Evangelista, R. C., & Cury, B. S. F. (2014). Insights into the swelling process and drug release mechanisms from cross-linked pectin/high amylose starch matrices. *Asian Journal of Pharmaceutical Sciences, 9*(1), 27–34. doi:10.1016/J.AJPS.2013.12.002

Chandrasekhar, K., & Venkata Mohan, S. (2014a). Bio-electrohydrolysis as a pretreatment strategy to catabolize complex food waste in closed circuitry: Function of electron flux to enhance acidogenic biohydrogen production. *International Journal of Hydrogen Energy, 39*(22), 11411–11422. doi:10.1016/J.IJHYDENE.2014.05.035

Chandrasekhar, K., & Venkata Mohan, S. (2014b). Induced catabolic bio-electrohydrolysis of complex food waste by regulating external resistance for enhancing acidogenic biohydrogen production. *Bioresource Technology, 165*, 372–382. doi:10.1016/J.BIORTECH.2014.02.073

Cooper, K. E., & Linton, A. H. (1952). The importance of the temperature during the early hours of incubation of agar plates in assays. *Journal of General Microbiology, 7*(1–2), 8–17. doi:10.1099/00221287-7-1-2-8

Dafe, A., Etemadi, H., Dilmaghani, A., & Mahdavinia, G. R. (2017). Investigation of pectin/starch hydrogel as a carrier for oral delivery of probiotic bacteria. *International Journal of Biological Macromolecules, 97*, 536–543. doi:10.1016/j.ijbiomac.2017.01.060

Dahman, Y. (2017). *Nanotechnology and functional materials for engineers*. Amsterdam: Elsevier.

Das, D., & Pal, S. (2015). Modified biopolymer-dextrin based crosslinked hydrogels: Application in controlled drug delivery. *RSC Advances, 5*(32), 25014–25050. doi:10.1039/C4RA16103C

Das, P. (2017). *Fabrication and characterization of hydroxyapatite based biocompatible composite scaffold for bone tissue engineering*. Master's thesis. BRAC University, Dhaka Bangladesh. Retrieved from http://dspace.bracu.ac.bd/xmlui/bitstream/handle/10361/9089/14276002_MNS.pdf?sequence=1&isAllowed=y

Du, X., Zhang, Y., Pan, X., Meng, F., You, J., & Wang, Z. (2019). Preparation and properties of modified porous starch/carbon black/natural rubber composites. *Composites Part B: Engineering, 156*, 1–7. doi:10.1016/J.COMPOSITESB.2018.08.033

Espigulé, E., Puigvert, X., Vilaseca, F., Méndez, J. A., Mutjé, P., & Girones, J. (2013). Thermoplastic starch-based composites reinforced with rape fibers: Water uptake and thermomechanical properties. *BioResources, 8*(2), 2620–2630. Retrieved from https://ojs.cnr.ncsu.edu/index.php/BioRes/article/view/BioRes_08_2_2620_Espigule_Thermoplastic_Starch_Composites

Farris, S., Schaich, K. M., Liu, L., Cooke, P. H., Piergiovanni, L., & Yam, K. L. (2011). Gelatin–pectin composite films from polyion-complex hydrogels. *Food Hydrocolloids, 25*(1), 61–70. doi:10.1016/J.FOODHYD.2010.05.006

Feng, M., Yu, L., Zhu, P., Zhou, X., Liu, H., Yang, Y., … Chen, P. (2018). Development and preparation of active starch films carrying tea polyphenol. *Carbohydrate Polymers, 196*, 162–167. doi:10.1016/j.carbpol.2018.05.043

Foster, E. J., Moon, R. J., Agarwal, U. P., Bortner, M. J., Bras, J., Camarero-Espinosa, S., … Youngblood, J. (2018). Current characterization methods for cellulose nanomaterials. *Chemical Society Reviews, 47*(8), 2609–2679. doi:10.1039/C6CS00895J

Hon, D. N.-S. (2017). Functional natural polymers: A new dimensional creativity in lignocellulosic chemistry. In *Chemical modification of lignocellulosic materials* (pp. 1–10). Routledge. doi:10.1201/9781315139142-1

Hu, W.-W., Lin, C.-H., & Hong, Z.-J. (2019). The enrichment of cancer stem cells using composite alginate/polycaprolactone nanofibers. *Carbohydrate polymers, 206*, 70–79. doi:10.1016/j.carbpol.2018.10.087

Ivanova, E. P., Bazaka, K., & Crawford, R. J. (2014). Natural polymer biomaterials: Advanced applications. In *New functional biomaterials for medicine and healthcare* (pp. 32–70). Elsevier. doi:10.1533/9781782422662.32

Jensen, S., Rolin, C., & Ipsen, R. (2010). Stabilisation of acidified skimmed milk with HM pectin. *Food Hydrocolloids, 24*(4), 291–299. doi:10.1016/J.FOODHYD.2009.10.004

Johnston, B., Kowalczuk, M., Hill, D., Tchuenbou-Magaia, F., Jonah, I., & Radecka, I. (2019). From trash to treasure – turning plastic waste into biodegradable polymers using bacteria. *Access Microbiology, 1*(1A). doi:10.1099/acmi.ac2019.po0462

Kadier, A., Jiang, Y., Lai, B., Rai, P. K., Chandrasekhar, K., Mohamed, A., & Kalil, M. S. (2018). Biohydrogen production in microbial electrolysis cells from renewable resources. In *Bioenergy and Biofuels* (pp. 331–356). CRC Press. doi:10.1201/9781351228138-12

Kakarla, R., Kuppam, C., Pandit, S., Kadier, A., & Velpuri, J. (2017). Algae—the potential future fuel: Challenges and prospects. In *Microbial applications Vol.1* (pp. 239–251). Springer International Publishing. doi:10.1007/978-3-319-52666-9_11

Kirby, A. R., MacDougall, A. J., & Morris, V. J. (2006). Sugar beet pectin–protein complexes. *Food Biophysics, 1*(1), 51–56. doi:10.1007/s11483-006-9005-4

Klemm, D., Heublein, B., Fink, H.-P., & Bohn, A. (2005). Cellulose: Fascinating biopolymer and sustainable raw material. *Angewandte Chemie International Edition, 44*(22), 3358–3393. doi:10.1002/anie.200460587

Korma, S. A., Kamal-Alahmad, Niazi, S., Ammar, A.-F., Zaaboul, F., & Zhang, T. (2016). Chemically modified starch and utilization in food stuffs. *International Journal of Nutrition and Food Sciences, 5*(4), 264. doi:10.11648/j.ijnfs.20160504.15

Kulkarni, V. S., Butte, K. D., & Rathod, S. S. (2012). Natural polymers: A comprehensive review. *International Journal of Research in Pharmaceutical and Biomedical Sciences, 3*(4), 1597–1613. Retrieved July 23, 2019 from https://www.researchgate.net/publication/236217541_Natural_Polymers-_A_comprehensive_Review

Kumar Trivedi, M., Patil, S., Mohan Tallapragada, R., Mohan Tallapragada Effect of Biofield, R., & Mohan Tallapragada, R. R. (2015). Treatment on the physical and thermal characteristics of aluminium powders. *Industrial Engineering Management, 4*(1), 1000151. doi:10.4172/2169-0316.1000151ï

Kunduru, K. R., Basu, A., & Domb, A. J. (2016). Biodegradable polymers: Medical applications. In *Encyclopedia of polymer science and technology* (pp. 1–22). John Wiley & Sons. doi:10.1002/0471440264.pst027.pub2

Lara-Espinoza, C., Carvajal-Millán, E., Balandrán-Quintana, R., López-Franco, Y., & Rascón-Chu, A. (2018). Pectin and pectin-based composite materials: Beyond food texture. *Molecules, 23*(4), 942. doi:10.3390/molecules23040942

Laycock, B., Nikolić, M., Colwell, J. M., Gauthier, E., Halley, P., Bottle, S., & George, G. (2017). Lifetime prediction of biodegradable polymers. *Progress in Polymer Science, 71*, 144–189. doi:10.1016/J.PROGPOLYMSCI.2017.02.004

Lee, W.-K., Lim, Y.-Y., Leow, A. T.-C., Namasivayam, P., Abdullah, J. O., & Ho, C.-L. (2017). Factors affecting yield and gelling properties of agar. *Journal of Applied Phycology, 29*(3), 1527–1540. doi:10.1007/s10811-016-1009-y

Li, X., Tabil, L. G., & Panigrahi, S. (2007). *Journal of Polymers and the Environment*. Kluwer Academic-Plenum-Human Sciences Press. Retrieved from http://agris.fao.org/agris-search/search.do?recordID=US201300747680

Liu, X., Xie, F., Li, X., Zhou, S., & Chen, L. (2015). Food polymers functionality and applications. *International Journal of Polymer Science, 2015*, 813628. doi:10.1155/2015/813628

Liu, Z., & Huang, H. (2016). Preparation and characterization of cellulose composite hydrogels from tea residue and carbohydrate additives. *Carbohydrate polymers, 146*, 226–233. doi:10.1016/j.carbpol.2016.03.100

Lomelí-Ramírez, M. G., Barrios-Guzmán, A. J., García-Enriquez, S., Rivera-Prado, J. de J., & Manríquez-González, R. (2014). Chemical and mechanical evaluation of bio-composites based on thermoplastic starch and wood particles prepared by thermal compression. *BioResources, 9*(2), 2960–2974. Retrieved from https://ojs.cnr.ncsu.edu/index.php/BioRes/article/view/5338

Manaila, E., Stelescu, M., & Craciun, G. (2018). Degradation studies realized on natural rubber and plasticized potato starch based eco-composites obtained by peroxide cross-linking. *International Journal of Molecular Sciences, 19*(10), 2862. doi:10.3390/ijms19102862

Masłowski, M., Miedzianowska, J., & Strzelec, K. (2017). Natural rubber biocomposites containing corn, barley and wheat straw. *Polymer Testing, 63*, 84–91. doi:10.1016/J.POLYMERTESTING.2017.08.003

Mohan, S. V., & Chandrasekhar, K. (2011). Solid phase microbial fuel cell (SMFC) for harnessing bioelectricity from composite food waste fermentation: Influence of electrode assembly and buffering capacity. *Bioresource Technology, 102*(14), 7077–7085. doi:10.1016/J.BIORTECH.2011.04.039

Mukherjee, T., Lerma-Reyes, R., Thompson, K. A., & Schrick, K. (2019). Making glue from seeds and gums: Working with plant-based polymers to introduce students to plant biochemistry. *Biochemistry and Molecular Biology Education, 47*(4), 468–475. doi:10.1002/bmb.21252

Nguyen, T. B. L., Min, Y. K., & Lee, B.-T. (2015). Nanoparticle biphasic calcium phosphate loading on gelatin-pectin scaffold for improved bone regeneration. *Tissue Engineering. Part A, 21*(7–8), 1376–1387. doi:10.1089/ten.TEA.2014.0313

Nie, S., Zhang, K., Lin, X., Zhang, C., Yan, D., Liang, H., & Wang, S. (2018). Enzymatic pretreatment for the improvement of dispersion and film properties of cellulose nanofibrils. *Carbohydrate Polymers, 181*, 1136–1142. doi:10.1016/J.CARBPOL.2017.11.020

Pandit, S., Sarode, S., Sargunaraj, F., & Chandrasekhar, K. (2018). Bacterial-mediated biofouling: Fundamentals and control techniques. In *Biotechnological applications of quorum sensing inhibitors* (pp. 263–284). Singapore: Springer Singapore. doi:10.1007/978-981-10-9026-4_13

Patel, V., Pandit, S., & Chandrasekhar, K. (2017). Basics of methanogenesis in anaerobic digester. In *Microbial applications Vol.2* (pp. 291–314). Cham: Springer International Publishing. doi:10.1007/978-3-319-52669-0_16

Praiboon, J., Chirapart, A., Akakabe, Y., Bhumibhamon, O., & Kajiwara, T. (2006). Physical and chemical characterization of agar polysaccharides extracted from the Thai and Japanese species of gracilaria. *ScienceAsia, 32*, 11–17. doi:10.2306/scienceasia1513-1874.2006.32(s1).011

Rascón-Chu, A., Díaz-Baca, J. A., Carvajal-Millan, E., Pérez-López, E., Hotchkiss, A. T., González-Ríos, H., … Campa-Mada, A. C. (2018). Electrosprayed core–shell composite microbeads based on pectin-arabinoxylans for insulin carrying: Aggregation and size dispersion control. *Polymers, 10*(2). doi:10.3390/POLYM10020108

Rhim, J.-W., Wang, L.-F., Lee, Y., & Hong, S.-I. (2014). Preparation and characterization of bio-nanocomposite films of agar and silver nanoparticles: Laser ablation method. *Carbohydrate Polymers, 103*, 456–465. doi:10.1016/J.CARBPOL.2013.12.075

Rol, F., Belgacem, M. N., Gandini, A., & Bras, J. (2019). Recent advances in surface-modified cellulose nanofibrils. *Progress in Polymer Science, 88*, 241–264. doi:10.1016/J.PROGPOLYMSCI.2018.09.002

Saravanan, M., & Rao, K. P. (2010). Pectin–gelatin and alginate–gelatin complex coacervation for controlled drug delivery: Influence of anionic polysaccharides and drugs being encapsulated on physicochemical properties of microcapsules. *Carbohydrate Polymers, 80*(3), 808–816. doi:10.1016/J.CARBPOL.2009.12.036

Sarifuddin, N., Ismail, H., & Ahmad, Z. (2012). Effect of fiber loading on properties of thermoplastic sago starch/kenaf core fiber biocomposites. *BioResources, 7*(3), 4294–4306. Retrieved from https://ojs.cnr.ncsu.edu/index.php/BioRes/article/view/BioRes_07_3_4294_Sarifuddin_IA_Fiber_Loading_Thermoplastic_Sago_Kenaf_Biocomposite

Shukla, M. K., Singh, R. P., Reddy, C. R. K., & Jha, B. (2012). Synthesis and characterization of agar-based silver nanoparticles and nanocomposite film with antibacterial applications. *Bioresource Technology, 107*, 295–300. doi:10.1016/J.BIORTECH.2011.11.092

Snell, K. D., Singh, V., & Brumbley, S. M. (2015). Production of novel biopolymers in plants: Recent technological advances and future prospects. *Current Opinion in Biotechnology, 32*, 68–75. doi:10.1016/J.COPBIO.2014.11.005

Sriamornsak, P., & Puttipipatkhachorn, S. (2004). Chitosan-pectin composite gel spheres: Effect of some formulation variables on drug release. *Macromolecular Symposia, 216*(1), 17–22. doi:10.1002/masy.200451203

Surya, I., & Hayeemasae, N. (2019). Reinforcement of natural rubber and epoxidized natural rubbers with fillers. *Simetrikal Journal of Engineering and Technology, 1*(1), 12–21. Retrieved from https://talenta.usu.ac.id/jet/article/view/682

Taira, T., Saito, Y., Niki, T., Iguchi-Ariga, S. M. M., Takahashi, K., & Ariga, H. (2004). DJ-1 has a role in antioxidative stress to prevent cell death. *EMBO Reports, 5*(2), 213–218. doi:10.1038/sj.embor.7400074

Teodorescu, M., Bercea, M., & Morariu, S. (2018). Biomaterials of poly(vinyl alcohol) and natural polymers. *Polymer Reviews, 58*(2), 247–287. doi:10.1080/15583724.2017.1403928

Tummalapalli, M., Berthet, M., Verrier, B., Deopura, B. L., Alam, M. S., & Gupta, B. (2016). Drug loaded composite oxidized pectin and gelatin networks for accelerated wound healing. *International Journal of Pharmaceutics, 505*(1–2), 234–245. doi:10.1016/j.ijpharm.2016.04.007

Valentino, F., Morgan-Sagastume, F., Campanari, S., & Werker, A. (2017). Carbon recovery from wastewater through bioconversion into biodegradable polymers. *New Biotechnology, 37*, 9–23. doi:10.1016/J.NBT.2016.05.007

van Beilen, J. B., & Poirier, Y. (2007). Prospects for biopolymer production in plants. In A. Fiechter & C. Sautter (Eds.), *Green Gene Technology* (pp. 133–151). Berlin-Heidelberg: Springer. doi:10.1007/10_2007_056

van Beilen, J. B., & Poirier, Y. (2008). Production of renewable polymers from crop plants. *The Plant Journal, 54*(4), 684–701. doi:10.1111/j.1365-313X.2008.03431.x

Wu, B., Degner, B., & McClements, D. J. (2014). Soft matter strategies for controlling food texture: Formation of hydrogel particles by biopolymer complex coacervation. *Journal of Physics: Condensed Matter, 26*(46), 464104. doi:10.1088/0953-8984/26/46/464104

Xu, C., Cui, R., Fu, L., & Lin, B. (2018). Recyclable and heat-healable epoxidized natural rubber/bentonite composites. *Composites Science and Technology, 167*, 421–430. doi:10.1016/J.COMPSCITECH.2018.08.027

Yang, X.-N., Xue, D.-D., Li, J.-Y., Liu, M., Jia, S.-R., Chu, L.-Q., … Zhong, C. (2016). Improvement of antimicrobial activity of graphene oxide/bacterial cellulose nanocomposites through the electrostatic modification. *Carbohydrate Polymers, 136*, 1152–1160. doi:10.1016/J.CARBPOL.2015.10.020

5 Sources of Natural Polymers from Microorganisms with Green Nanoparticles

K. Chandrasekhar, Satya Eswari Jujjavarapu, Prasun Kumar, Gopalakrishnan Kumar, Potla Durthi Chandrasai, Enamala Manoj Kumar, and Murthy Chavali

CONTENTS

5.1 SOURCE OF NATURAL POLYMERS FROM MICROBES WITH GREEN NANOPARTICLES

Over a period of several decades, increased production of petroleum-derived plastic as an alternative source for clay, metal, and wood materials has been severely damaging the environment (Chandrasekhar et al., 2018; Deval et al., 2017; Ram Kumar Pandian et al., 2009; Venkateswar Reddy et al., 2017). These petroleum-derived polymers take decades to degrade or decompose in the environment; moreover, they produce toxic components in the environment at the time of decomposition (Chandrasekhar & Venkata Mohan, 2012; Srikanth et al., 2012; Venkateswar Reddy et al., 2012; Venkateswar Reddy & Venkata Mohan, 2012a; Venkateswar Reddy & Venkata Mohan, 2012b). The greenhouse effect in the environment is increasing day to day due to the rapid consumption of fossil fuels which release carbon dioxide, the chief greenhouse gas that harms the nature (Chandrasekhar et al., 2015a; Chandrasekhar et al., 2018; Chandrasekhar et al., 2015b; Chandrasekhar et al., 2017; Deval et al., 2017; Enamala et al., 2018; Kadier et al., 2018b; Kadier et al., 2017; Kadier et al., 2016a; Kadier et al., 2016b; Kadier et al., 2015; Kakarla et al., 2017; Kumar et al., 2017c; Kumar et al., 2018b; Reddy et al., 2011a; Reddy et al., 2011b; Saratale et al., 2017; Sivagurunathan et al., 2017; Venkata Mohan et al., 2019; Venkata Mohan et al., 2013). Due to limited resources, year by year, these petroleum reserves are rapidly being drained. Furthermore, improved production of petroleum-derived plastics has been recognized as a severe risk to nature due to the non-ecofriendly nature of plastic. Being non-biodegradable in the environment, disposal of these petroleum-derived plastic materials has become a significant concern (Amulya et al., 2014; Pandit et al., 2018; Periyasamy Sivagurunathan et al., 2018; Venkateswar Reddy et al., 2014; Venkateswar Reddy et al., 2015), which has led to the exploration of new polymers that can be removed from the environment in an eco-friendly manner. Biodegradable (i.e. can be microbiologically breakdown to the final products) polymers are considered to be a possible answer (Amulya et al., 2014; Kumar et al., 2018c; Kumar & Kim, 2018; Naseem et al., 2016; Ram Kumar Pandian et al., 2009; Venkateswar Reddy et al., 2017). Consequently, it is necessary to find an alternative to traditional petroleum-based polymers, and eco-friendly and readily biodegradable polymers have gained considerable attention during the past two decades (Venkata Mohan et al., 2010; Venkateswar Reddy et al., 2016).

Recently, nanocomposites composed of biopolymer matrices have been considered as suitable and attractive substitutes for traditional resources selected as scaffolding material for bone replacement. An impeccable scaffolding material is defined as having manageable biodegradability, outstanding biocompatibility, cytocompatibility, suitable microstructure, and mechanical properties (El-Meliegy et al., 2018; Yunus Basha et al., 2015). These biopolymers are massive macromolecules comprising only monomer units. These macromolecules are high molecular-weight biopolymers with variable material characteristics depending on the nature of their monomer composition (Rovera et al., 2018). Moreover, apart from being biodegradable polymers, some of these macromolecules are also biocompatible in nature and without much difficulty can be used for numerous applications (Ram Kumar Pandian et al., 2009; Venkata Mohan & Venkateswar Reddy, 2013; Venkateswar Reddy et al., 2016; Venkateswar Reddy & Venkata Mohan, 2012a).

Therefore, biopolymers can be widely used in different industries such as packaging, agriculture, and medicine for a wide variety of applications. It is a fact that the massive production of these perishable polymers and their extensive usage is essential to protect the environment (Khan et al., 2015; Kumar & Kim, 2018; Ul-Islam et al., 2015). These biodegradable polymers can be broadly categorized into two different classes, that is, natural polymers (P_N) and synthetic polymers (P_S) based on their origin. It is well known that P_N are acquired from natural sources. Besides, P_S are produced through various chemical synthesis methods. P_N can be further categorized into four diverse classes based on their source of origin, comprising agricultural, animal (gelatin and collagen), marine (chitin which is converted into chitosan), and bacteriological bases (including polyhydroxyalkanoates [PHA] and polylactic acid [PLA]) (Dmour & Taha, 2018; Gokarneshan, 2019; Khan et al., 2015; Kirschweng et al., 2017; Kumar & Kim, 2018; Ul-Islam et al., 2015).

Natural polymers encompass polysaccharides containing starch, alginate, chitosan, and hyaluronic acid byproducts or proteins containing collagen, soy, fibrin gels, and silk (Halake et al., 2016; Shi et al., 2016). The innate extracellular environment of the body is typically prepared by natural polymers, together with collagen, fibrinogen, and elastin (Armentano et al., 2010). These natural polymers are biocompatible, biodegradable, and also have a cell recognition characteristic that improves the nature of cell adhesion. In spite of numerous benefits, synthetic polymers do not have cell adhesion signaling property (Akilbekova et al., 2018; George et al., 2019; Halake et al., 2014; Hassan et al., 2019). Therefore, the focus of several researchers has shifted to investigating, identifying, and synthesizing natural polymers for several industrial applications, including bone-tissue engineering applications. On the other hand, these natural polymers have some drawbacks, including poor mechanical properties, immunogenicity, and partial availability (Armentano et al., 2010; Yunus Basha et al., 2015).

Natural polymer–ceramic mixtures are also considered ideal materials to use as bone-graft scaffolds. At the moment, commercially available biopolymer composites are mostly constructed by using a collagen biopolymer (Halake et al., 2014; Udawattha et al., 2018; Zou et al., 2019). Researchers have observed and stated that chitosan-based composite polymer materials possess mechanical strength similar to that of cortical bone. The natural polymer composite materials of chitosan and gelatin-based scaffolds possess greater mechanical strength with acceptable porosity (George et al., 2019; Udawattha et al., 2018; Yunus Basha et al., 2015). PHA is formed entirely by microbial fermentation process while PLA is, to some extent, synthesized. In the case of PLA, the lactic acid (LA) monomers are created by a microbial fermentation process and then polymerized using a chemical process (in the presence of a chemical catalyst) (George et al., 2019). P_N can be degraded entirely by microbes—a group of enzymes involved in this breakdown process, where the long-chain polymers are broken down into their monomers (Amulya et al., 2014; Figueiredo et al., 2018; Raschip et al., 2011). The shortened polymer chain is later metabolized as a carbon source to support microbial growth and metabolic activities (Mei et al., 2017; Ning et al., 2018). Alternatively, P_S can be synthesized with bio-derived monomers or synthetic monomers derived from petroleum hydrocarbons.

Recently, nanocomposites composed of biopolymer media and bio-ceramic plasters have been considered as good choices for conventional resources in the medical field (Madhusudana Rao et al., 2018; Rao et al., 2014). Usually, polymer nanocomposite materials are produced from a mixture of polymers and organic/inorganic fillers at the nanometer scale (Kumar et al., 2019; Madhusudana Rao et al., 2018; Papageorgiou et al., 2019; Rao et al., 2016; Rao et al., 2017). The interaction between the polymer matrix and nanostructures is considered to be the basis for improved mechanical, as well as functional, properties of the nanocomposite materials over traditional microcomposites (Dhivya et al., 2018). During the past three decades, there has been constant growth in research to investigate biopolymers and nanocomposite material, and to improve material properties by employing nanometric-engineered structures, due to the inherent properties of nanocomposites such as high surface-to-area-volume ratio (Dhivya et al., 2018; Rao et al., 2016; Saha et al., 2016). The research of the past four decades has constantly investigated and worked on nanocomposites and investigated their potential applications in various fields. Since 1990, researchers have published articles about nanocomposites (regarding nanocomposite material synthesis and applications). According to the ISI web of science reports, from 1990 to 2019 around 129,016 records have been published regarding nanocomposites (see Figure 5.1 and Figure 5.2).

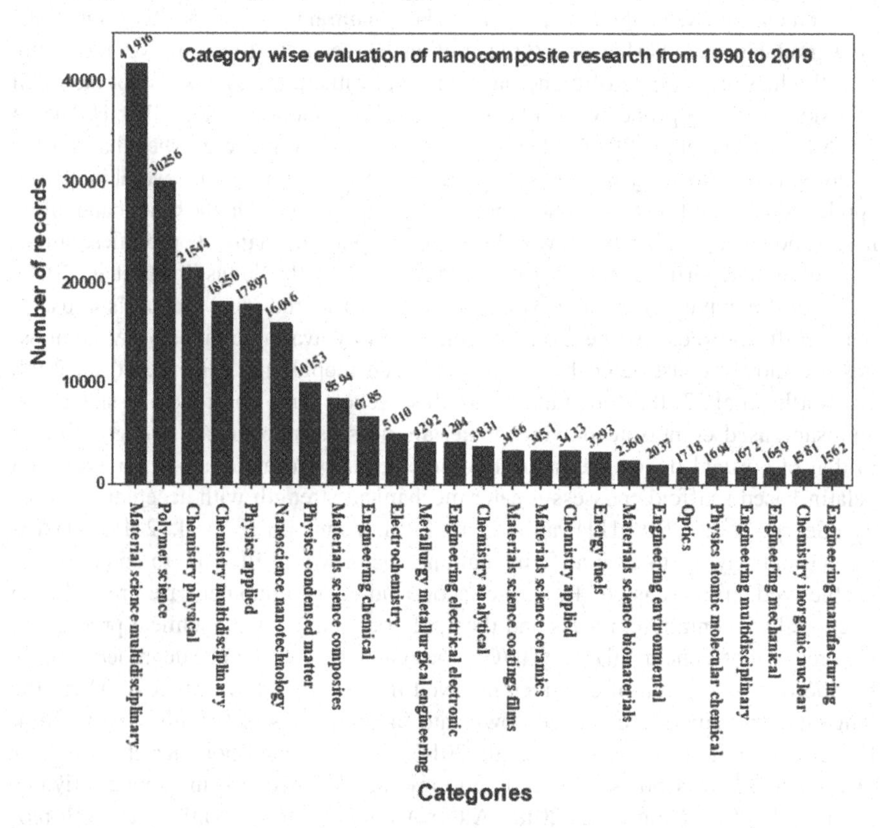

FIGURE 5.1 Bar graph representation regarding category-wise evaluation of nanocomposites research from 1990 to 2019 (adapted from ISI Web of Science Keyword: Nanocomposite on 25 May 2019).

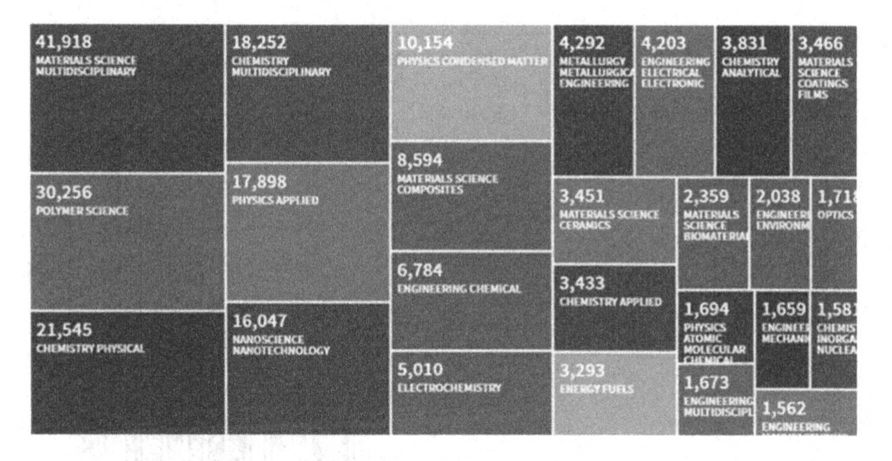

FIGURE 5.2 Tree-map representation regarding category wise evaluation of nanocomposite research from 1990 to 2019 (adapted from ISI Web of Science Keyword: Nanocomposite on 25 May 2019).

Observed records regarding nanocomposites are from different categories, such as materials science multidisciplinary (41,961 records), polymer science (30,256 records), chemistry physical (21,544 records), chemistry multidisciplinary (18250 records), physics applied (17,897 records), nanoscience nanotechnology (16,046 records), physics condensed matter (10,153 records), materials science composites (8594 records), engineering chemical (6785 records), electrochemistry (5010 records), metallurgy metallurgical engineering (4292 records), engineering electrical electronic (4204 records), chemistry analytical (3831 records), materials science coatings films (3466 records), materials science ceramics (3451 records), chemistry applied (3433 records), energy fuels (3293 records), materials science biomaterials (2360 records), engineering environmental (2037 records), optics (1718 records), physics atomic molecular chemical (1694 records), engineering multidisciplinary (1672 records), engineering mechanical (1659 records), chemistry inorganic nuclear (1581 records), engineering manufacturing (1562 records) etc.

According to ISI Web of Science reports, the first record regarding a nanocomposite was in 1990. In the same year, four records were made regarding nanocomposites, and the number has kept on increasing year on year. The highest number of records, 15736, was in 2018 (see Figure 5.3 and Figure 5.4). The detailed information has been clearly illustrated both in bar graphs (see Figure 5.3) and in tree-maps (see Figure 5.4).

Country-wise evaluation of nanocomposite research from 1990 to 2019 has been collected from ISI Web of Science and studied to understand country-wise performance regarding nanocomposite research work. According to ISI Web of Science reports, the highest number of records has been made in China (44,722 records), followed by the Unitecd States (18,235 records). India occupied third place in this regard with 12,056 records, followed by South Korea (7716 records), Iran (6711 records), Germany (4824 records), Japan (4476 records), France (4422 records), Italy (4038 records), Spain (3352 records), occupying fourth, fifth, sixth, seventh, eighth, ninth, and tenth positions respectively (see Figure 5.5 and Figure 5.6). The detailed

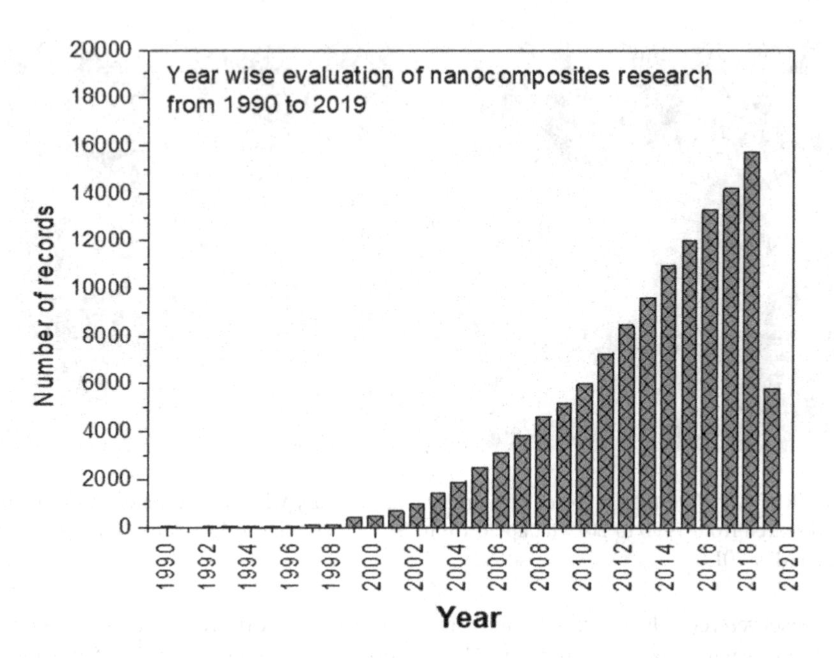

FIGURE 5.3 Bar graph representation regarding year-wise evaluation of nanocomposite research from 1990 to 2019 (adapted from ISI Web of Science Keyword: Nanocomposite on 25 May 2019).

information has been clearly illustrated both in bar graphs (see Figure 5.5) and tree-maps (see Figure 5.6).

The present chapter deals with different types of microbial biopolymers such as dextran, xanthan, alginate, cellulose, hyaluronic acid, polyhydroxyalkanoates, and pullulan (see Table 5.1) by providing a detailed discussion regarding production

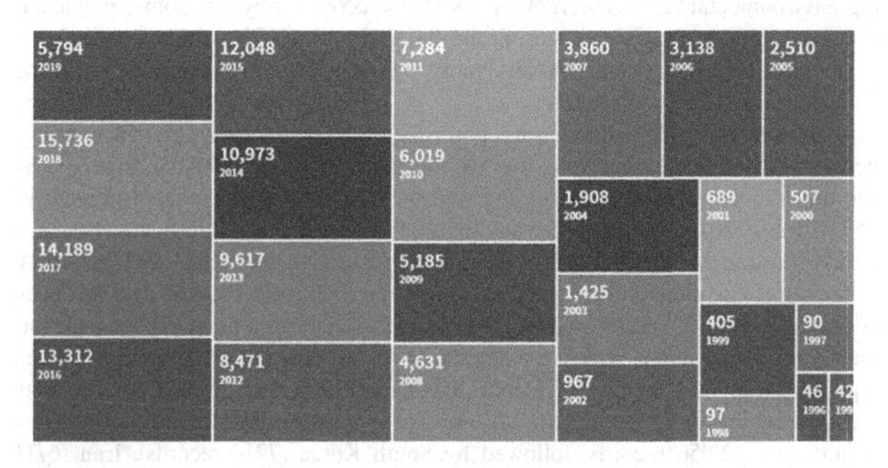

FIGURE 5.4 Tree-map representation regarding year-wise evaluation of nanocomposite research from 1990 to 2019 (adapted from ISI Web of Science Keyword: Nanocomposite on 25 May 2019).

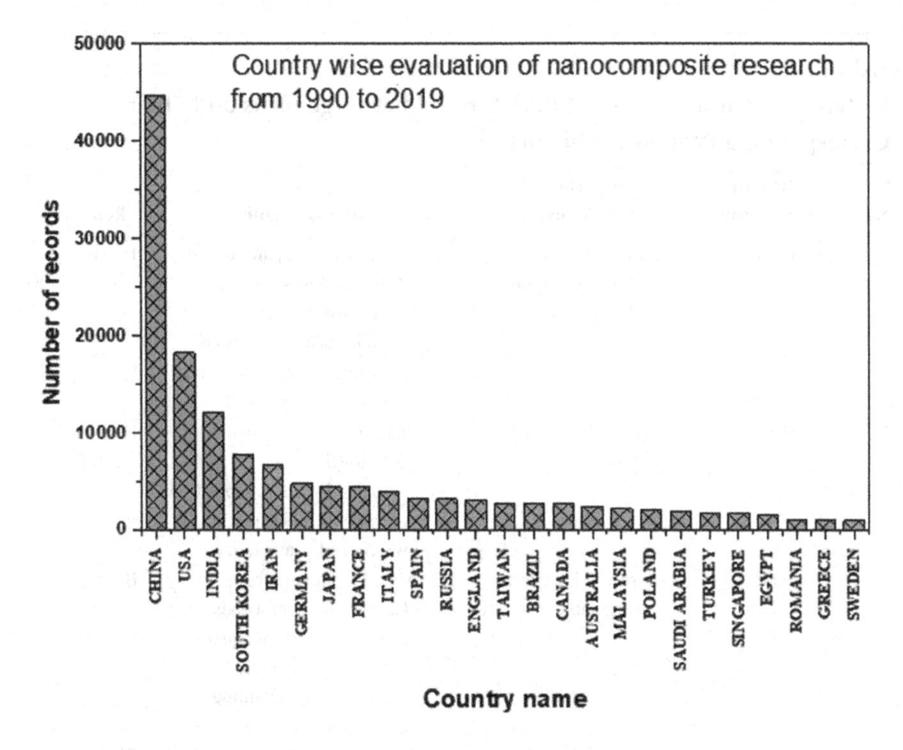

FIGURE 5.5 Bar graph representation regarding the country-wise evaluation of nanocomposite research from 1990 to 2019 (adapted from ISI Web of Science Keyword: Nanocomposite on 25 May 2019).

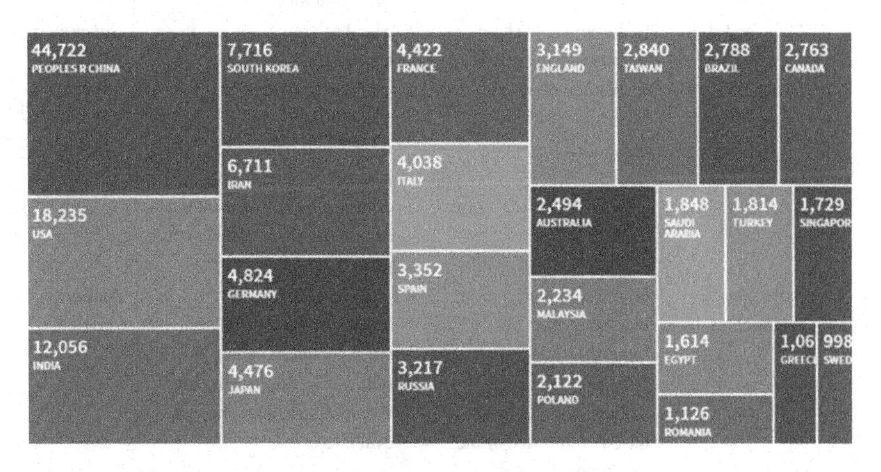

FIGURE 5.6 Tree-map representation regarding the country-wise evaluation of nanocomposite research from 1990 to 2019 (adapted from ISI Web of Science Keyword: Nanocomposite on 25 May 2019).

TABLE 5.1
Different Natural Polymer Material from Microorganisms and Their Corresponding Process Highlights

S. No.	Type of Biopolymer	Properties of the Biopolymer	Process Highlights	Reference
1.	Dextran	Hydrophilic, biocompatible, and biodegradable polysaccharide	Examine the preparation of dextran- chitosan/ nano-hydroxyapatite complex scaffolds with a different weight percentage of nano-hydroxyapatite.	El-Meliegy et al. (2018)
2.	Xanthan	Biodegradable, natural polymer	The surfaces with diverse functionalities were prepared by serial deposition of thin layers on silicon wafers and characterized.	Bueno et al. (2018)
3.	Alginate	Hetero-polysaccharide Biodegradable in nature	Aimed to prepare the Ca-alginate microbeads by electrospraying an aqueous alginate solution into calcium ions containing distilled water.	Bae et al. (2019)
4.	Bacterial cellulose	Bacterial cellulose shows antibacterial, healing, lipophilic, biomimetic properties.	Aimed to introduce a new technique to functionalize bacterial cellulose with antimicrobial properties by *in situ* method.	Chen et al. (2019)
5.	Microbial hyaluronic Acid	Hyaluronic acid shows hydrophilicity, good dispersion ability, biocompatibility, and non-toxicity.	Aimed to prepare biocompatible cryogel composites with great mechanical strength based on HA and halloysite nanotubes of numerous combinations, and their uses as a scaffold for diverse cell growing media.	Suner et al. (2019)
6.	Polyhydroxy-alkanoates	Natural polymer, biodegradable, biocompatible, non-toxic in nature. However, properties will change based on the size of the monomer.	Aimed to evaluate the bioconversion of different carbon source including poly (*cis*-1,4-Isoprene) into Polyhydroxyalkanoates was evaluated.	Andler et al. (2019)
7.	Pullulan	High flexibility and optical properties, which are due to its fully amorphous organization	Nanoindentation experiments on biopolymer nanocomposite coatings on a plastic film.	Rovera et al. (2018)

routes and their potential applications in different fields including the food industry, pharma, and the medical field (e.g. cardiac stent development).

5.1.1 Dextran and Its Composites

Dextran is a hydrophilic polysaccharide belonging to the class of homopolysaccharides. Under the hydrolysis process, dextran provides only a single kind of monomer component, for example, cellulose, alginate, etc. (Hussain et al., 2017). It is a neutral polysaccharide, produced and accumulating on the cell surface of bacteria (LA bacteria*)*. It contains D-anhydroglucosyl units combined by α one to six linkage in the key chains with α one to three linkage at the branch, formed by chemo/biological methods. It is a neutral, water-soluble, hydrophilic, biodegradable, and biocompatible natural polymer material produced through diverse bacterial strains in the presence of sucrose as a carbon source, by the action of a series of enzymes (see Table 5.1). Due to these properties, it is an appropriate material to use as scaffolding material in tissue engineering (El-Meliegy et al., 2018). It has been stated that it is a significant polymer in the medical field for several applications, such as for the progress of drug delivery structures, hydrogel preparation, and wound coverings, among others (Almeida et al., 2013; Hussain et al., 2017). It is greatly biocompatible and biodegradable in the environment. Besides, it can also be used as a molecular mesh for parting and cleansing of large molecules (Hussain et al., 2017; Safinya & Ewert, 2012). Several research reports are available regarding the applications of biopolymer composite scaffold materials (such as dextran, hyaluronic acid, chitosan, etc.) for tissue engineering. These biopolymers are readily biodegradable by the act of several enzymes and won't form any poisonous end-products. Furthermore, some of the natural polymers also having antimicrobial, hemostatic properties which make them more suitable to the field of tissue engineering.

5.1.1.1 Dextran Blend With Sulfate-Coated Chitosan

Chitosan is a natural biopolymer, recently gaining much interest in tissue engineering. It is easily broken down inside the human body by human enzymes and releases non-toxic by-products. Additionally, this biopolymer also has other properties, such as being antimicrobial, hemostatic, and having osteoconductivity. Hence it can be widely used in tissue engineering concepts (El-Meliegy et al., 2018). This polymer, encompassing glucosamine and N-acetylglucosamine brought about from the deacetylation of chitin, is extensively useful in the food and pharmaceutical industries due to its non-toxicity, outstanding biocompatibility, and biodegradable properties.

In recent years, several research articles have been published regarding the usage of combined scaffolds for tissue restoration as a foundation material. Nevertheless, chitosan solubilizes in solutions that contain pH less than pH 6.5. It is not soluble in solutions that contain pH around 7 (Li et al., 2014). The chitosan byproducts of N, N dimethyl hexadecyl carboxymethyl chitosan, were produced by using carboxymethyl chitosan. Innovative amphiphilic chitosan derived liposome comprising definite functional groups have been developed (Li et al., 2014; Zou et al., 2013). Zou et al. (2015) produced dextran sulfate layered amphiphilic chitosan derived nanoliposome by a microfluidization method together with a film vaporization method. A dextran sulfate layered amphiphilic chitosan derivative–based nanoliposome exhibited high zeta potential once mixed with a replicated intestinal liquid (Zou et al., 2015).

5.1.1.2 Dextran Mixture Using Curcumin

Curcumin is a color compound (low molecular weight phenolic pigment) which is extracted from turmeric, which has several possible uses in the medical field, as an anti-inflammatory agent, antidiabetic, and antineoplastic. It can stop cancer in a variety of cell lines (Hussain et al., 2017; Konatham et al., 2011; Sahu et al., 2008; Tummalapalli et al., 2016). Encapsulation of this curcumin in a nanoparticle-supporting drug-transport method through marking ligands can reduce the quantity and improve the efficacy of the drug (Hussain et al., 2017; Yu et al., 2014). Due to improved infusion and adhesion properties, these nanoparticles can be uses for precise and directed drug delivery to a target (Hussain et al., 2017).

5.1.2 Xanthan and Its Composites

Xanthan is a composite exopolysaccharide produced by a microbe in the microenvironment, which is responsible for numerous illnesses in plants (Bhatia et al., 2015; Bueno et al., 2015; Kumar et al., 2018a; Kumar et al., 2017b; Takeuchi et al., 2009; Trombino et al., 2019). These properties, and its biochemical symmetry, make it appropriate material to use as scaffolding material in tissue engineering (Bueno et al., 2015; Hanna & Saad, 2019; Petri, 2015). It has been stated to be a significant polymer in the medical field for several applications such as in the medical field, hydrogel preparation, the food industry, and wound coverings, among others (Kim et al., 2015; Kim et al., 2017; Kumar et al., 2018a). This xanthan is extensively utilized as a solidifying and stabilizing agent in a varied range of food and industrial products (Petri, 2015). Xanthan will demonstrate unique properties when it is blended with other polymers and forms new composites such as a xanthan blend with galactomannan (Vendruscolo et al., 2005), a xanthan blend with konjac glucomannan (Alvarez-Manceñido et al., 2008), a xanthan blend with chitosan (Eftaiha et al., 2010), a xanthan blend with galactomannan (Jian et al., 2012), a xanthan blend with lignin (Raschip et al., 2011), and a xanthan blend with starch (Shalviri et al., 2010) etc.

5.1.2.1 Xanthan Gum Mixture Using Poly(acrylamide)

The solubilization and reaction nature of xanthan is improved once the xanthan is converted into carboxymethyl xanthan, due to the presence of carboxymethyl group (Bhatia et al., 2015; Bizotto & Sabadini, 2008; Kumar et al., 2017a; Kumar et al., 2009; Maia et al., 2012; Mundargi et al., 2007; Pandey & Mishra, 2012). Badwaika et al. (2016) prepared a carboxymethyl xanthan glue embedded with poly(acrylamide) copolymers by the free radical polymerization method. This copolymer can be used as a vehicle for drug delivery in the medical industry (Hussain et al., 2017). Poly (ethylene oxide), polyacrylamide, and polyisobutylene are low physical strength polymers. Hence, these polymers can be mixed with exopolysaccharides like xanthan gum to enhance their physical strength (Tan et al., 2012).

5.1.2.2 Xanthan Mixture Using Polypyrrole

Polypyrrole (PPy) is a polymer with high ion-exchange capacity, hydrophobicity (Molina et al., 2010; Şahin et al., 2008; Slimane et al., 2009; Yuvaraj et al., 2008), and stability (Yuvaraj et al., 2008). However, PPy has limited applications because of partial biocompatibility and solubility issues (Mirhosseini et al., 2008). In the last decade, biocompatible xanthan/PPy supports have been prepared by the electropolymerization method (Bueno et al., 2015; Hussain et al., 2017).

5.1.2.3 Xanthan Gum Mixture Using Polyaniline

Polyaniline is a familiar conducting material (conducting polymer) with decent stability and can be easily produced by following simple methods (Naskar et al., 2010; Ptaszek et al., 2009). However, its solubility issues and non-biodegradable nature are two key facts that limit the use of this material in various fields (Mazumder et al., 2012). A few researchers have considered preparation of polyaniline/xanthan glue nanocomposite liquid through oxidative polymerization of aniline in associated xanthan glue. However, this nanocomposite can disturb the hydrophilic property of polyaniline. It also enhances the conductive property and thermal stability of the nanocomposites by augmenting xanthan into polyaniline (Hussain et al., 2017).

5.1.3 Bacterial Alginate and Its Composites

Alginate fits into a class of straight-chain copolymers of (1,4)-linked β-D-mannuronic and α-l-guluronic acids. The ratio of β-D-mannuronic and α-l-guluronic acids and their structure describes the physical properties, biocompatibility, and biodegradability of alginate under typical physiological circumstances (Li et al., 2017; Salem et al., 2019; Trandafilovic et al., 2012; Vigués et al., 2018; Wichai et al., 2019). Recently, alginate-stabilized iron oxide nanoparticles have been employed to identify liver cancer (Ma et al., 2007). Trandafilovic et al. (2012) utilized the alginate with the intention of attaining measured synthesis of ZnO nanoparticles and examined their structure and antimicrobial behavior. From the antimicrobial tests, authors have stated that the ZnO–alginate nanocomposite consists of robust action in contradiction of the pathogens *Staphylococcus aureus* and *Escherichia coli* (Trandafilovic et al., 2012). In another study, bacterial cellulose (BC) and alginate were used as a potential substrate to prepare a novel blend of membrane material. To fabricate this membrane, the blend solution was coagulated with $CaCl_2$ aqueous solution in a Teflon container. To achieve a nanoporous structure, a supercritical carbon dioxide (CO_2) drying method was employed. Finally, the properties and morphology of the BC and blend membranes were studied and noticed that the blend membrane with 80%/20% of BC and alginate respectively, showed a similar configuration and improved water adsorption capability.

5.1.4 Applications of Bacterial Cellulose and Its Composites

Cellulose is primarily obtained from plants. It is the most abundant and accessible biopolymer on earth (Parkhey et al., 2017; Torres et al., 2019). In spite of its intensive uses in several areas, plant cellulose has several disadvantages, amongst which the most significant is its lack of purity and the need for extreme pretreatments before it can be used in numerous applications (Atila et al., 2015; Konwarh et al., 2013; Torres et al., 2019). Besides, bacterial cellulose (BC) is a purified form of cellulose material consisting of microfibrils. These microfibrils are organized in a three-dimensional web-design, which is formed by numerous kinds of microbial species mostly fitting into the genus *Acetobacter*. This web-shaped style of microfibrils provides BC with a spongy geometry and great mechanical strength (Abral et al., 2019; Khan et al., 2015). BC possesses greater crystallinity, liquid absorption capability, and mechanical strength over plant cellulose, making it the best biomaterial for several applications in various industries including the medical industry (Atila et al., 2015; Parkhey et al., 2017; Torres et al., 2019). These properties and its biochemical symmetry make it a suitable material to use as scaffolding material in tissue engineering. It has been stated to be a significant polysaccharide in the medical field for several applications such as for the progress of drug delivery structures, hydrogel preparation, and wound coverings, among others. Nevertheless, these BC materials also have a few limitations, such as low production rate and the need for purification after production, which confine its applications in numerous areas (Khan et al., 2015; Ul-Islam et al., 2015). BC produced by limited but definite microorganisms is a natural exopolysaccharide and pure in nature. BC also possess several key properties such as adhesiveness, UV protection, longer retention of moisture content, they generate aerobic conditions, etc. Hence, this nanostructured matrix is being widely explored for numerous industrial applications (Abral et al., 2019). The characteristics of BC, for example, biocompatibility, stability, and biodegradability make it suitable for numerous medical applications, together with a suitable biomaterial, for artificial skin, supporting material intended for tissue engineering, and wound covering (Torres et al., 2019). Nevertheless, unpolished BC lacks some properties, such as physical strength and moisture retention capacity, which limits its claims in numerous fields. Consequently, the production of BC mixtures has been steered to resolve these limitations. Different types of artificial BC-complex approaches have been industrialized depending on the type and significant uses of the product (Shah et al., 2013). BC mixtures are chiefly manufactured through *in situ* adding of support ingredients to artificial BC media or the ex situ infiltration of such resources into BC microfibrils. The production of BC composites with several ingredients ranges from organic polymers to inorganic nanoparticles (Chantereau et al., 2019; Ye et al., 2019). These composite materials are widely used for several applications such as tissue redevelopment, curing of serious injuries, enzyme immobilization, drug delivery, and the creation of health devices that might substitute connective tissues. Numerous key devices, counting biosensors, bio-catalysts, E-papers, electrical tools, and optoelectronic expedients, are equipped from BC composites with conductive ingredients (Abral et al., 2019; Chantereau et al., 2019; Shah et al., 2013).

5.1.4.1 Scaffold

The 3-D scaffold plays a crucial role in the accomplishment of tissue redevelopment which offers the structural support for cell adhesion, distribution, propagation, diversity, and shape of the evolving tissue. However, the pure BC material won't be degraded by way of mammalian cells because of the lack of cellulose enzymes. Nevertheless, modified BC material is biodegradable in water, saline, mammalian cell fluids, etc. During the initial stage of research on biodegradable scaffolds, development was hampered by numerous limitations, including the rapid biodegradation of scaffold material associated with cell growth. Development of stable scaffold material using biocompatible and extremely resistant materials will help to overcome this issue. In recent years, the expansion of BC-grounded platforms is being paid a great deal of attention as a result of BC's biocompatible properties and the ability to use it as an embedded material. Besides, it has been noticed that native BC cannot provide cell development, as it is partially biocompatible in nature. Nevertheless, this issue can be addressed by soaking BC scaffold material in serum and electrolytic liquids, for example, sodium hydroxide (NaOH) solution (Watanabe et al., 1998).

5.1.4.2 Cornea

In 2010, Wang et al. (2010) fabricated a biocompatible and elastic cornea artificially in the laboratory. According to their investigation, they attempted to grow a synthetic cornea with BC-grounded complex material by a polyvinyl alcohol combination. In this combination, polyvinyl alcohol is highly stable, whereas BC is biocompatible and holds great mechanical strength. As a result, a complex was prepared from BC and polyvinyl alcohol by means of the freeze–thaw process. The mechanical properties, light transmittance, and stability are some of the key features of the composite used in the preparation of the artificial cornea (Wang et al., 2010).

5.1.4.3 Heart Valves

The unique performance of BC and BC complex materials attracted several researchers to develop and fabricate various biocompatible materials that can be widely used in medical fields, such as in the improvement of synthetic heart valves. At present, two kinds of regulator substitutes are in use. The first is porcine valves, which offer an inadequate lifetime. The second is mechanical heart valves with extended lifetimes, but they necessitate lifetime blood-thinning medicine which may cause some side-effects. These problems might be solved by fabricating BC composite-based heart valves. In recent years, researchers prepared a BC-polyvinyl alcohol composite that maintains mechanical properties similar to innate heart leaflets (Millon et al., 2008). In another study, Fink et al. (2010) assessed the hemocompatibility of BC-grounded vascular graft tubes and matched them with commercial grafts of polyethylene terephthalate (PET).

5.1.4.4 Blood Vessels

Recently researchers have focused on the fabrication of artificial tubes similar to blood vessels by using BC and BC composite materials. These artificial tube-like materials are fabricated by using synthetic materials to bring back blood flow in the case of obstacles in the coronary vessels near the heart. In place of artificial materials, BC could be used as a potential alternative material; it provides less of a threat of the development of blood clots and also has remarkable mechanical strength over other materials such as polypropylene, PET, or cellophane. Moreover, these BC composites are highly biocompatible in nature. Therefore, BC will not show any toxic effect upon interaction with blood and will maintain the evolution and also the proliferation of smooth muscles (Ul-Islam et al., 2015).

5.1.4.5 Bone and Cartilage

The thermal stability and the strength of BC composite material have attracted the interest of researchers to use it in the fabrication of bone and cartilage materials. Bone is made up of a collagen matrix supported by hydroxyapatite. Hence, the biocompatibility and structural properties of BC with cartilage provide novel opportunities for the improvement of BC-based implantations in place of injured bone and cartilage. BC-based composite has provided a convenient platform for the reinforcement of bone and cartilage tissues. Nevertheless, a key restriction in the use of BC is that nanofibrils form a thick lattice that can limit cell penetration.

5.1.5 Microbial Hyaluronic Acid and Its Composites

Hyaluronic acid (HA, also known as hyaluronan; exopolysaccharides) comprises a recapping unit of d-glucuronic acid and N-acetyl-d-glucosamine connected through one to four and one to three linkages. It is an unbranched polysaccharide comprise alternating N-acetyl-D-glucosamine and D-glucuronic acid (Gallo et al., 2019; Haridas & Rosemary, 2019). Usually, HA has been widely produced from rooster combs and bovine vitreous humor. On the other hand, it is highly challenging to separate high molecular weight (around 1000 kDa) HA from these sources since it forms a blend with proteoglycans, which makes the process economically not viable. Furthermore, successive extraction processes and purification processes will lead to a decrease in the molecular weight of the hyaluronic acid (Haridas & Rosemary, 2019). Nonetheless, in recent years, it has been widely produced through the fermentation process, with inferior production costs and a lesser amount of environmental pollution, by using streptococcal species (hyaluronic acid produced and stored in the capsules formed inside the streptococci) as a biocatalyst (Widner et al., 2005). Hence, several industries are showing much interest in bacterial fermentation processes with the hope of attaining commercially feasible biopolymers. During this bacterial fermentation process, the extracellular polysaccharide is freely released into the growth medium, hence control of product yield and biopolymer physical characteristics are possible (Haridas & Rosemary, 2019). Moreover, the quantity of product that can be produced by this method is hypothetically limitless. Furthermore, HA production through recombinant systems has gained much interest in recent years, because this process will not produce unnecessary by-products that are toxic in nature and may diminsh the overall efficiciency of the process (Korurer et al., 2014; Vazquez et al., 2013).

Streptococci are anaerobic microorganisms that utilize glucose as a substrate. They produce lactic acid as an intermediate metabolite. Streptococci also produce hyaluronic acid as a mucoid capsule around the cell. This helps the microorganism to escape from the host immune system. During streptococci growth, HA will be formed as a secondary metabolite. However, HA formation is liable through numerous physio-chemical factors that take account of genetic as well as nutritional factors. Streptococcus has the capability of producing HA as a secondary metabolite in aerobic conditions as well as in anaerobic circumstance. Definite strains of streptococcus yield HA during a specific period in their lifespan. The same microorganism produces an enzyme named hyaluronidase which has the capability of degrading the accumulated or already produced HA inside the cell (Tiwari & Bahadur, 2019; Valverde et al., 2019). The selection of a suitable strain for the production of HA is a crucial step and we should carefully select the appropriate strain to get a higher yield. Different types of enzymes involved in HA production through fermentation (biosynthesis) process are phosphoglucomutase, pyrophosphorylase, UDP, glucose dehydrogenase, hyaluronate synthase, pyrophosphorylase, acetyltransferase, mutase, and amidotransferase (Valverde et al., 2019).

HA is a highly biocompatible material with outstanding viscoelasticity and high moisture-holding capability. In recent years, HA has, due to its highly biocompatible nature, gained much importance due to its use in medicine, cosmetics, and nutraceuticals (Tiwari & Bahadur, 2019). It is a hydrophilic polymer; because of hydrogen bonding amid carboxyl and N-acetyl groups of hyaluronic acid with H_2O it can absorb the huge quantity of H_2O and hold up to 1000-fold more H_2O than its solid volume (Hussain et al., 2017; Morra, 2005; Subramaniam et al., 2016). Hyaluronic acid is a highly viscous solution with efficient moisture-holding properties. It exhibits an effective lubricant property and also possesses a wound-healing character in pathological circumstances by fluid replacement properties. Hyaluronic acid displays a bacteriostatic, non-immunogenic, non-inflammatory (Anisha et al., 2013), biodegradable nature along with biocompatible properties. These properties and its biochemical symmetry make it appropriate material to use as scaffolding material in tissue engineering. It has been stated to be an important biopolymer in the medical field for numerous applications such as for the progress of drug delivery structures, hydrogel preparation, and wound coverings, among others (El-Meliegy et al., 2018). However, it has poor mechanical strength and stability. Due to its physical and chemical properties, it is extensively used as a drug delivery vehicle, as a scaffold for tissue engineering (Anisha et al., 2013), as a component in cosmetics that offer anti-aging and wetness-providing properties to the skin (Morra, 2005; Pitarresi et al., 2010; Yeo et al., 2006). HA is a natural biomolecule consisting of vital applications in animals and humans. Hyaluronic acid comprises carboxyl groups and N-acetyl groups, which form hydrogen bonds with water or aqueous solutions when they come into contact. This will impart conformation strength to the polymer material.

5.1.5.1 Hyaluronic Acid Grafted Using Polyacrylic Acid

Grafting of a diverse polymer is an active technique to functionalize natural polysaccharides. There are two different approaches to implanting: "grafting to" and "grafting from." The sodium (Na) salt of polyacrylic acid is an implanting polymer (Hussain et al., 2017). Nakagawaa et al. (2015) established biocompatible calcium (Ca) salt of hyaluronic acid implanted with polyacrylic (Hyaluronic acid-g-polyacrylic acid) for drug delivery. In relation to unmodified hyaluronic acid, the grafted hyaluronic acid (Hyaluronic acid-g-polyacrylic acid) displayed fewer degradation properties. The acrylate hyaluronic acid hydrogel was utilized as a support material to deliver human stem cells for rat calvarial treatment (Subramaniam et al., 2016).

5.1.5.2 Hyaluronic Blend Acid Using Polypyrrole

In recent years, due to its non-immunogenic nature, hyaluronic acid has gained much importance in the medical and biological fields. It is known that molecular imprinting technology has tremendous sensitivity, ease, quick reaction, is cheap, and has in vivo recognition in addition to electrochemical sensors characteristics. Hence it is widely used in sensor fields, medical fields, biological fields, and environmental fields (Bo et al., 2011; Qin et al., 2010). Graphene is a single-atom-thick, highly electric-conductive material for conducting electrons from the biomolecules. It is a 2-D pane of carbon atoms with great thermal stability as well as good mechanical strength. As a result, for the fabrication of biological sensors as well as electrochemical sensors, this emerges as an ideal material (Hussain et al., 2017; Li et al., 2015). Scientists have used an electropolymerization technique to design electrochemical sensors grounded on molecular imprinting film of polypyrrole-sulfonated graphene/hyaluronic acid-multiwall carbon nanotubes by using tryptamine molecules. The function of polypyrrole-sulfonated graphene complex films and the hyaluronic acid are to enhance the conductivity and act as a diffusing agent to solubilize multiwall carbon nanotubes, respectively (Hussain et al., 2017).

5.1.5.3 Hyaluronic Acid Mixture Using Poly(ethylene Glycol)

HA is a crucial component of the extracellular matrix. Hyaluronic acid biomaterials have numerous benefits due to their biocompatible, non-immunogenic nature. Besides, hyaluronic acid side chains can be freely influenced to incorporate numerous functional groups. Hyaluronic acid–based biomaterials are pooled with another component, for example, collagen, gelatin, or poly(ethylene glycol) (PEG), as cell transporters to the scaffold (Hussain et al., 2017). The advantages of the hyaluronic acid composite materials comprise an ability to adjust physiological characteristics of the hyaluronic acid hydrogel as support material. Recently, Jeong et al. (2014) produced HA-grounded PEG-complex hydrogels by employing artificial neural network analysis (to study the relationship between HA and PEG). In this approach, Michael-type adding of thiolated hyaluronic acid and four-arm PEG-vinylsulfone of numerous molecular weights were employed.

5.1.5.4 Hyaluronic Acid Mixture by Poly(Ne-acryloyl l-lysine)

Cancer is a serious health issue that can cause death. Globally, around one in eight women develop metastatic breast cancer in their lifespan. Several researchers are working to solve this issue by developing new anti-cancer drugs. However, these newly invented anti-cancer drugs are being delayed because of the absence of live tumor replicas that thoroughly impersonate the human illness. Hyaluronic acid has high biodegradability, a non-immunogenic nature, and non-inflammatory properties, and hence can be employed to prepare 3-D tumor models (Toole, 2004). Nonetheless, hyaluronic acid hydrogels have very low mechanical strength. The perfect 3-D model has decent mechanical strength identical with tumor tissue, manageable degradability, and precise release of bioactive substances (Hartman et al., 2009; Qi et al., 2014).

5.1.5.5 Hyaluronic Acid Mixture Using Poly(Lactide-Co-Glycolide)

Nowadays, biodegradable substances are widely used in the medical field to prepare medical devices, as a new material in tissue engineering, and also as carriers, etc. Scientists have been exploring research in a wide range of applications of new materials in tissue engineering. As a result, the development of biodegradable polymers has now gained momentum. In this scenario, several researchers have begun investigations into preparing an HA mix with poly(lactide-co-glycolide) as one of the potential biopolymer materials. Poly (lactide-co-glycolide) is a biomaterial and highly biocompatible polymer, with great demand in medical fields. Nevertheless, it has two disadvantages, namely high price and inadequate cellular linkage ability. Fibrin is a natural biopolymer with great importance in fibrin adhesive preparations for medical applications (Gentile et al., 2014; Rahaman et al., 2011). The obtained poly(lactide-co-glycolide) scaffold has interlocked porous microstructure and great porosity level. The strength of the poly (lactide-co-glycolide) scaffold can be upgraded by raising the amount of coating agent, particularly fibrin (Hussain et al., 2017).

5.1.6 Polyhydroxyalkanoates and its Composites

In recent years, biological production of PHA has gained importance for overcoming the harmful effects in the course of conventional petroleum-based plastics degradation (Amulya et al., 2014). PHA is a type of straight-chain polymer that can be produced by a wide variety of microorganisms with similar properties to synthetic polymers (such as polypropylene and polyethylene). Moreover, these PHA are formed from renewable resources (Venkateswar Reddy & Venkata Mohan, 2012a). However, they are totally biodegradable and biocompatible in nature. After their disposal in the environment, they are metabolized by microbes into water and carbon dioxide under oxygen-present circumstance and methane (CH_4) under oxygen-absent circumstances. A wide variety of single cultures comprising *Alcaligenes latus*, *Azotobacter vinelandii*, *Pseudomonas oleovorans*, and *E. coli* were used in industry to produce PHA on a large scale. Nevertheless, high operational cost (around 11% of total production costs) is considered to be one of the major limiting factors. It has been noticed that the production and commercialization of PHA by employing pure culture increases the price by about four to nine times greater than those of conventional plastics (Venkateswar Reddy & Venkata Mohan, 2012a).

Culture selection is considered a key step in PHA production using the mixed microbial population. Mixed microbial populations grown under feast-and-famine circumstances were exposed to an interior growth restraint arising from the alternative substrate accessibility, which forced the organisms into a physiological adaptation (Venkateswar Reddy et al., 2012). In recent years, bio-electrochemical systems gained much interest due to their efficiency in generating electricity with simultaneous waste remediation and being eco-friendly in nature (Chandrasekhar, 2019; Chandrasekhar & Ahn, 2017; Chandrasekhar & Venkata Mohan, 2014a; Chandrasekhar & Venkata Mohan, 2014b; Kadier et al., 2018a; Kumar et al., 2012; Pandit et al., 2017; Pandit; et al., 2019; Saratale et al., 2017;

Venkata Mohan & Chandrasekhar, 2011a; Venkata Mohan & Chandrasekhar, 2011b). With the increasingly apparent advantages of bio-electrochemical systems, such as the utilization of waste organic matter as substrate, and aeration of the cathode champer, it was realized that bio-cathodes could be used for the production of PHA (Kumar et al., 2018b). Even though PHA are biopolymers with high biodegradable behavior and also can be produced from renewable resources, their possible uses are hindered because of their fragility (Kumar et al., 2016; Kumar et al., 2015a; Kumar et al., 2015b; Kumar et al., 2014; Kumar et al., 2013; Patel et al., 2015; Singh et al., 2015a).

Polyhydroxybutyrate (PHB) and polylactic acid composites are well studied and show enhanced mechanical properties. The PHA/polylactic acid blends were examined essentially to discover properties such as miscibility, crystallization, strength, and biodegradability (see Table 5.1). The mixture of PHA and polylactic acid will enhance the physical strength and compensate for the cost of PHA (Mozumder, 2018). The making of PHA and polylactic acid mixtures is achieved by various techniques such as the melt-compounding method, solvent-based methods, polymerization of lactide on the surface of PHA, co-extrusion, etc. (Mozumder, 2018). PHAs/PLA mixtures are widely used in the fabrication of numerous electrical and electronic devices. The composites of PHAs/polylactic acid are also used in pharmaceutical applications (tissue engineering scaffolds and drug delivery vehicles). The properties of PHAs/polylactic acid composites are promising for various applications such as environmental, food preservation, and household applications, and for partial applications in biomedical fields. Rigorous investigation is essential to increase compatibilizing mechanisms in order to prepare innovative materials appropriate for precise applications mainly in biomedical issues (Kumar & Kim, 2018; Kumar et al., 2018c; Kumar et al., 2016; Kumar et al., 2015a; Mozumder, 2018; Singh et al., 2015a).

5.1.7 Pullulan

Pullulan ($(C_6H_{10}O_5)_n$) is a naturally occurring non-ionic biopolymer that has gained much attention in recent years because of its peculiar features. Under controlled conditions, pullulan can be produced by the yeast *Aureobasidium pullulans.* In this fermentation process, the overall product (pullulan) yield is strongly influenced by media composition and operation conditions (Amrita et al., 2015; An et al., 2017; Atila et al., 2016; Rovera et al., 2018; Singh et al., 2016). This pullulan biopolymer is a non-toxic, non-carcinogenic, non-mutagenic, and edible polysaccharide. This non-ionic biopolymer is soluble in water, and insoluble in organic solvents except in dimethylsulfoxide and formamide. This polysaccharide comprises substantial properties, for example, adhesiveness, film and fiber formability, an anti-fungal growth agent, and degradability including significant mechanical property (An et al., 2017; Aydogdu et al., 2016; Singh et al., 2015b; Tabasum et al., 2018). Hence this biopolymer is widely used in several areas including the food industry, the biomedical field and the pharma field (Rovera et al., 2018; Terán Hilares et al., 2019).

Regarding food industry applications, pullulan is widely used as a thickening agent, a stabilizing agent, and also as a prebiotic agent. Regarding biomedical applications, pullulan is used in gene delivery, tissue engineering, wound healing, and diagnostic applications (An et al., 2017; Cheng et al., 2011; K.R & V, 2017; Singh et al., 2016; Tabasum et al., 2018; Wang et al., 2015). Regarding pharmaceutical applications, pullulan is widely employed in capsule manufacturing and drug delivery processes. To overcome a few limitations, this biopolymer can also be used together with supplementary biopolymers (An et al., 2017; Rovera et al., 2018). A few examples are pullulan–soy protein blend (for higher adhesion to the surface of the food), pullulan–caseinate blend (for high thermal stability), pullulan–whey protein blend (for oxygen permeability, water vapor permeability, and appearance), pullulan–hydroxypropyl methylcellulose blend (for thermos-mechanical properties), pullulan–sodium alginate blend (for thermos-mechanical properties), pullulan–gelatin mixture (for mechanical and oxygen permeability) etc. Even though pullulan has many attractive applications in different fields, it has not been effectively as exploited as it deserves. At the moment very few commercial products are available that relate to pullulan biopolymers. One of the limiting factors for this is its cost of production.

5.1.8 Conclusion

Increased production of petroleum-derived plastic materials has been seriously damaging the environment. Improved production of these petroleum-derived plastics has been recognized as a severe risk to nature due to the non-ecofriendly nature of plastic. Disposal of these petroleum-derived plastic materials has become a major concern. Biodegradable polymers can play an important role in the construction of a clean environment by substituting for the extensively used non-biodegradable synthetic plastics. Recently, researchers have noticed that a few of these biodegradable biopolymers are highly biocompatible in nature. In view of this, industrial-scale production of these biodegradable polymers is essential to ensure alternative sources of plastic. Therefore, inventive industrial applications can be devised if researchers, at both academic and industrial levels, are motivated to innovatively develop new strategies enabling the microbial production of new and effective biopolymers such as PHA, PLA, dextran, xanthan, bacterial alginate, bacterial cellulose, cyanophycin, microbial hyaluronic acid and their composites. In such attempts, it is of the utmost importance to aim for biodegradable, eco-friendly, and cost-efficient solutions.

REFERENCES

Abral, H., Kadriadi, Mahardika, M., Handayani, D., Sugiarti, E. 2019. Characterization of disintegrated bacterial cellulose nanofibers/PVA bionanocomposites prepared via ultrasonication. *International Journal of Biological Macromolecules*, **135**, 591–599.

Akilbekova, D., Shaimerdenova, M., Adilov, S., Berillo, D. 2018. Biocompatible scaffolds based on natural polymers for regenerative medicine. *International Journal of Biological Macromolecules*, **114**, 324–333.

Almeida, J.F., Ferreira, P., Alves, P., Lopes, A., Gil, M.H. 2013. Synthesis of a dextran based thermo-sensitive drug delivery system by gamma irradiation. *International Journal of Biological Macromolecules*, **61**, 150–155.

Alvarez-Manceñido, F., Landin, M., Martínez-Pacheco, R. 2008. Konjac glucomannan/xanthan gum enzyme sensitive binary mixtures for colonic drug delivery. *European Journal of Pharmaceutics and Biopharmaceutics*, **69**(2), 573–581.

Amrita, Arora, A., Sharma, P., Katti, D.S. 2015. Pullulan-based composite scaffolds for bone tissue engineering: Improved osteoconductivity by pore wall mineralization. *Carbohydrate Polymers*, **123**, 180–189.

Amulya, K., Venkateswar Reddy, M., Venkata Mohan, S. 2014. Acidogenic spent wash valorization through polyhydroxyalkanoate (PHA) synthesis coupled with fermentative biohydrogen production. *Bioresource Technology*, **158**, 336–342.

An, C., Ma, S.-j., Chang, F., Xue, W.-j. 2017. Efficient production of pullulan by Aureobasidium pullulans grown on mixtures of potato starch hydrolysate and sucrose. *Brazilian Journal of Microbiology*, **48**(1), 180–185.

Andler, R., Vivod, R., Steinbüchel, A. 2019. Synthesis of polyhydroxyalkanoates through the biodegradation of poly(cis-1,4-isoprene) rubber. *Journal of Bioscience and Bioengineering*, **127**(3), 360–365.

Anisha, B.S., Biswas, R., Chennazhi, K.P., Jayakumar, R. 2013. Chitosan–hyaluronic acid/nano silver composite sponges for drug resistant bacteria infected diabetic wounds. *International Journal of Biological Macromolecules*, **62**, 310–320.

Armentano, I., Dottori, M., Fortunati, E., Mattioli, S., Kenny, J.M. 2010. Biodegradable polymer matrix nanocomposites for tissue engineering: A review. *Polymer Degradation and Stability*, **95**(11), 2126–2146.

Atila, D., Keskin, D., Tezcaner, A. 2015. Cellulose acetate based 3-dimensional electrospun scaffolds for skin tissue engineering applications. *Carbohydrate Polymers*, **133**, 251–261.

Atila, D., Keskin, D., Tezcaner, A. 2016. Crosslinked pullulan/cellulose acetate fibrous scaffolds for bone tissue engineering. *Materials Science and Engineering: C*, **69**, 1103–1115.

Aydogdu, H., Keskin, D., Baran, E.T., Tezcaner, A. 2016. Pullulan microcarriers for bone tissue regeneration. *Materials Science and Engineering: C*, **63**, 439–449.

Badwaik, H.R., Sakure, K., Alexander, A., Ajazuddin, Dhongade, H., Tripathi, D.K. 2016. Synthesis and characterisation of poly(acryalamide) grafted carboxymethyl xanthan gum copolymer. *International Journal of Biological Macromolecules*, **85**, 361–369.

Bae, S.B., Nam, H.C., Park, W.H. 2019. Electrospraying of environmentally sustainable alginate microbeads for cosmetic additives. *International Journal of Biological Macromolecules*, **133**, 278–283.

Bhatia, M., Ahuja, M., Mehta, H. 2015. Thiol derivatization of Xanthan gum and its evaluation as a mucoadhesive polymer. *Carbohydrate Polymers*, **131**, 119–124.

Bizotto, V.C., Sabadini, E. 2008. Poly(ethylene Oxide) × polyacrylamide. Which one is more efficient to promote drag reduction in aqueous solution and less degradable? *Journal of Applied Polymer Science*, **110**(3), 1844–1850.

Bo, Y., Yang, H., Hu, Y., Yao, T., Huang, S. 2011. A novel electrochemical DNA biosensor based on graphene and polyaniline nanowires. *Electrochimica Acta*, **56**(6), 2676–2681.

Bueno, P.V.A., Hilamatu, K.C.P., Carmona-Ribeiro, A.M., Petri, D.F.S. 2018. Magnetically triggered release of amoxicillin from xanthan/Fe3O4/albumin patches. *International Journal of Biological Macromolecules*, **115**, 792–800.

Bueno, V.B., Takahashi, S.H., Catalani, L.H., de Torresi, S.I.C., Petri, D.F.S. 2015. Biocompatible xanthan/polypyrrole scaffolds for tissue engineering. *Materials Science and Engineering: C*, **52**, 121–128.

Chandrasekhar, K. 2019. Chapter 3.5 - Effective and nonprecious cathode catalysts for oxygen reduction reaction in microbial fuel cells. in: *Microbial Electrochemical Technology*, (Eds.) S.V. Mohan, S. Varjani, A. Pandey, Elsevier, pp. 485–501.

Chandrasekhar, K., Ahn, Y.-H. 2017. Effectiveness of piggery waste treatment using microbial fuel cells coupled with elutriated-phased acid fermentation. *Bioresource Technology*, **244**, 650–657.

Chandrasekhar, K., Amulya, K., Mohan, S.V. 2015a. Solid phase bio-electrofermentation of food waste to harvest value-added products associated with waste remediation. *Waste Management*, **45**, 57–65.

Chandrasekhar, K., Kadier, A., Kumar, G., Nastro, R.A., Jeevitha, V. 2018. Challenges in microbial fuel cell and future scope. in: *Microbial Fuel Cell: A Bioelectrochemical System that Converts Waste to Watts*, (Ed.) D. Das, Springer International Publishing. Cham, pp. 483–499.

Chandrasekhar, K., Lee, Y.J., Lee, D.W. 2015b. Biohydrogen production: Strategies to improve process efficiency through microbial routes. *International Journal of Molecular Sciences*, **16**(4), 8266–8293.

Chandrasekhar, K., Pandit, S., Kadier, A., Dasagrandhi, C., Velpuri, J. 2017. Biohydrogen production: Integrated approaches to improve the process efficiency. in: *Microbial Applications Vol.1*: *Bioremediation and Bioenergy*, (Eds.) V.C. Kalia, P. Kumar, Springer International Publishing. Cham, pp. 189–210.

Chandrasekhar, K., Venkata Mohan, S. 2012. Bio-electrochemical remediation of real field petroleum sludge as an electron donor with simultaneous power generation facilitates biotransformation of PAH: Effect of substrate concentration. *Bioresource Technology*, **110**, 517–525.

Chandrasekhar, K., Venkata Mohan, S. 2014a. Bio-electrohydrolysis as a pretreatment strategy to catabolize complex food waste in closed circuitry: Function of electron flux to enhance acidogenic biohydrogen production. *International Journal of Hydrogen Energy*, **39**(22), 11411–11422.

Chandrasekhar, K., Venkata Mohan, S. 2014b. Induced catabolic bio-electrohydrolysis of complex food waste by regulating external resistance for enhancing acidogenic biohydrogen production. *Bioresource Technology*, **165**, 372–382.

Chantereau, G., Brown, N., Dourges, M.-A., Freire, C.S.R., Silvestre, A.J.D., Sebe, G., Coma, V. 2019. Silylation of bacterial cellulose to design membranes with intrinsic anti-bacterial properties. *Carbohydrate Polymers*, **220**, 71–78.

Chen, J., Chen, C., Liang, G., Xu, X., Hao, Q., Sun, D. 2019. In situ preparation of bacterial cellulose with antimicrobial properties from bioconversion of mulberry leaves. *Carbohydrate Polymers*, **220**, 170–175.

Cheng, K.-C., Demirci, A., Catchmark, J.M., Puri, V.M. 2011. Effects of initial ammonium ion concentration on pullulan production by Aureobasidium pullulans and its modeling. *Journal of Food Engineering*, **103**(2), 115–122.

Deval, A.S., Parikh, H.A., Kadier, A., Chandrasekhar, K., Bhagwat, A.M., Dikshit, A.K. 2017. Sequential microbial activities mediated bioelectricity production from distillery wastewater using bio-electrochemical system with simultaneous waste remediation. *International Journal of Hydrogen Energy*, **42**(2), 1130–1141.

Dhivya, S.M., Sathiya, S.M., Mugesh, S., Murugan, M., Rajan, M.A.J. 2018. Evaluation of antibacterial potential of CS/Fe3O4 nanocomposites based plate count method. *Advanced Science Letters*, **24**(8), 5503–5507.

Dmour, I., Taha, M.O. 2018. Chapter 2 - Natural and semisynthetic polymers in pharmaceutical nanotechnology. in: *Organic Materials as Smart Nanocarriers for Drug Delivery*, (Ed.) A.M. Grumezescu, William Andrew Publishing, pp. 35–100.

Eftaiha, A.a.F., Qinna, N., Rashid, I.S., Al Remawi, M.M., Al Shami, M.R., Arafat, T.A., Badwan, A.A. 2010. Bioadhesive controlled metronidazole release matrix based on Chitosan and Xanthan Gum. *Marine Drugs*, **8**(5), 1716–1730.

El-Meliegy, E., Abu-Elsaad, N.I., El-Kady, A.M., Ibrahim, M.A. 2018. Improvement of physico-chemical properties of dextran-chitosan composite scaffolds by addition of nano-hydroxyapatite. *Scientific Reports*, **8**(1), 12180.

Enamala, M.K., Enamala, S., Chavali, M., Donepudi, J., Yadavalli, R., Kolapalli, B., Aradhyula, T.V., Velpuri, J., Kuppam, C. 2018. Production of biofuels from microalgae: A review on cultivation, harvesting, lipid extraction, and numerous applications of microalgae. *Renewable and Sustainable Energy Reviews*, **94**, 49–68.

Figueiredo, P., Lintinen, K., Hirvonen, J.T., Kostiainen, M.A., Santos, H.A. 2018. Properties and chemical modifications of lignin: Towards lignin-based nanomaterials for biomedical applications. *Progress in Materials Science*, **93**, 233–269.

Fink, H., Faxälv, L., Molnár, G.F., Drotz, K., Risberg, B., Lindahl, T.L., Sellborn, A. 2010. Real-time measurements of coagulation on bacterial cellulose and conventional vascular graft materials. *Acta Biomaterialia*, **6**(3), 1125–1130.

Gallo, N., Nasser, H., Salvatore, L., Natali, M.L., Campa, L., Mahmoud, M., Capobianco, L., Sannino, A., Madaghiele, M. 2019. Hyaluronic acid for advanced therapies: Promises and challenges. *European Polymer Journal*, **117**, 134–147.

Gentile, P., Chiono, V., Carmagnola, I., Hatton, P.V. 2014. An overview of poly(lactic-co-glycolic) Acid (PLGA)-based biomaterials for bone tissue engineering. *International Journal of Molecular Sciences*, **15**(3), 3640–3659.

George, A., Shah, P.A., Shrivastav, P.S. 2019. Natural biodegradable polymers based nano-formulations for drug delivery: A review. *International Journal of Pharmaceutics*, **561**, 244–264.

Gokarneshan, N. 2019. 19 - Application of natural polymers and herbal extracts in wound management. in: *Advanced Textiles for Wound Care (Second Edition)*, (Ed.) S. Rajendran, Woodhead Publishing, pp. 541–561.

Halake, K., Birajdar, M., Kim, B.S., Bae, H., Lee, C., Kim, Y.J., Kim, S., Kim, H.J., Ahn, S., An, S.Y., Lee, J. 2014. Recent application developments of water-soluble synthetic polymers. *Journal of Industrial and Engineering Chemistry*, **20**(6), 3913–3918.

Halake, K., Kim, H.J., Birajdar, M., Kim, B.S., Bae, H., Lee, C., Kim, Y.J., Kim, S., Ahn, S., An, S.Y., Jung, S.H., Lee, J. 2016. Recently developed applications for natural hydrophilic polymers. *Journal of Industrial and Engineering Chemistry*, **40**, 16–22.

Hanna, D.H., Saad, G.R. 2019. Encapsulation of ciprofloxacin within modified xanthan gum-chitosan based hydrogel for drug delivery. *Bioorganic Chemistry*, **84**, 115–124.

Haridas, N., Rosemary, M.J. 2019. Effect of steam sterilization and biocompatibility studies of hyaluronic acid hydrogel for viscosupplementation. *Polymer Degradation and Stability*, **163**, 220–227.

Hartman, O., Zhang, C., Adams, E.L., Farach-Carson, M.C., Petrelli, N.J., Chase, B.D., Rabolt, J.F. 2009. Microfabricated electrospun collagen membranes for 3-D cancer models and drug screening applications. *Biomacromolecules*, **10**(8), 2019–2032.

Hassan, A.M., Ayoub, M., Eissa, M., Musa, T., Bruining, H., Farajzadeh, R. 2019. Exergy return on exergy investment analysis of natural-polymer (Guar-Arabic gum) enhanced oil recovery process. *Energy*, **181**, 162–172.

Hussain, A., Zia, K.M., Tabasum, S., Noreen, A., Ali, M., Iqbal, R., Zuber, M. 2017. Blends and composites of exopolysaccharides; properties and applications: A review. *International Journal of Biological Macromolecules*, **94**, 10–27.

Jeong, C.G., Francisco, A.T., Niu, Z., Mancino, R.L., Craig, S.L., Setton, L.A. 2014. Screening of hyaluronic acid–poly(ethylene glycol) composite hydrogels to support intervertebral disc cell biosynthesis using artificial neural network analysis. *Acta Biomaterialia*, **10**(8), 3421–3430.

Jian, H., Zhu, L., Zhang, W., Sun, D., Jiang, J. 2012. Galactomannan (from Gleditsia sinensis Lam.) and xanthan gum matrix tablets for controlled delivery of theophylline: In vitro drug release and swelling behavior. *Carbohydrate Polymers*, **87**(3), 2176–2182.

Kadier, A., Chandrasekhar, K., Kalil, M.S. 2017. Selection of the best barrier solutions for liquid displacement gas collecting metre to prevent gas solubility in microbial electrolysis cells. *International Journal of Renewable Energy Technology*, **8**(2), 93–103.

Kadier, A., Jiang, Y., Lai, B., Rai, P.K., Chandrasekhar, K., Mohamed, A., Kalil, M.S. 2018a. Biohydrogen production in microbial electrolysis cells from renewable resources. in: *Bioenergy and Biofuels*, (Ed.) O. Konur, CRC Press, pp. 331–356.

Kadier, A., Kalil, M.S., Abdeshahian, P., Chandrasekhar, K., Mohamed, A., Azman, N.F., Logroño, W., Simayi, Y., Hamid, A.A. 2016a. Recent advances and emerging challenges in microbial electrolysis cells (MECs) for microbial production of hydrogen and value-added chemicals. *Renewable and Sustainable Energy Reviews*, **61**, 501–525.

Kadier, A., Kalil, M.S., Chandrasekhar, K., Mohanakrishna, G., Saratale, G.D., Saratale, R.G., Kumar, G., Pugazhendhi, A., Sivagurunathan, P. 2018b. Surpassing the current limitations of high purity H2 production in microbial electrolysis cell (MECs): Strategies for inhibiting growth of methanogens. *Bioelectrochemistry*, **119**(Supplement C), 211–219.

Kadier, A., Simayi, Y., Abdeshahian, P., Azman, N.F., Chandrasekhar, K., Kalil, M.S. 2016b. A comprehensive review of microbial electrolysis cells (MEC) reactor designs and configurations for sustainable hydrogen gas production. *Alexandria Engineering Journal*, **55**(1), 427–443.

Kadier, A., Simayi, Y., Chandrasekhar, K., Ismail, M., Kalil, M.S. 2015. Hydrogen gas production with an electroformed Ni mesh cathode catalysts in a single-chamber microbial electrolysis cell (MEC). *International Journal of Hydrogen Energy*, **40**(41), 14095–14103.

Kakarla, R., Kuppam, C., Pandit, S., Kadier, A., Velpuri, J. 2017. Algae—the potential future fuel: Challenges and prospects. in: *Microbial Applications Vol.1: Bioremediation and Bioenergy*, (Eds.) V.C. Kalia, P. Kumar, Springer International Publishing. Cham, pp. 239–251.

Khan, S., Ul-Islam, M., Khattak, W.A., Ullah, M.W., Park, J.K. 2015. Bacterial cellulose-poly(3,4-ethylenedioxythiophene)-poly(styrenesulfonate) composites for optoelectronic applications. *Carbohydrate Polymers*, **127**, 86–93.

Kim, J., Hwang, J., Kang, H., Choi, J. 2015. Chlorhexidine-loaded xanthan gum-based biopolymers for targeted, sustained release of antiseptic agent. *Journal of Industrial and Engineering Chemistry*, **32**, 44–48.

Kim, J., Hwang, J., Seo, Y., Jo, Y., Son, J., Choi, J. 2017. Engineered chitosan–xanthan gum biopolymers effectively adhere to cells and readily release incorporated antiseptic molecules in a sustained manner. *Journal of Industrial and Engineering Chemistry*, **46**, 68–79.

Kirschweng, B., Tátraaljai, D., Földes, E., Pukánszky, B. 2017. Natural antioxidants as stabilizers for polymers. *Polymer Degradation and Stability*, **145**, 25–40.

Konatham, S., Reddy, B., Aukunuru, J. 2011. Enhanced liver delivery and sustained release of curcumin with drug loaded nanoparticles after intravenous administration in rats. *Asian Journal of Pharmaceutical Research and Health Care*, **3**(4), 99–108.

Konwarh, R., Karak, N., Misra, M. 2013. Electrospun cellulose acetate nanofibers: The present status and gamut of biotechnological applications. *Biotechnology Advances*, **31**(4), 421–437.

Korurer, E., Kenar, H., Doger, E., Karaoz, E. 2014. Production of a composite hyaluronic acid/gelatin blood plasma gel for hydrogel-based adipose tissue engineering applications. *Journal of Biomedical Materials Research A*, **102**(7), 2220–9.

Kumar, A., Deepak, Sharma, S., Srivastava, A., Kumar, R. 2017a. Synthesis of xanthan gum graft copolymer and its application for controlled release of highly water soluble Levofloxacin drug in aqueous medium. *Carbohydrate Polymers*, **171**, 211–219.

Kumar, A., Rao, K.M., Han, S.S. 2018a. Application of xanthan gum as polysaccharide in tissue engineering: A review. *Carbohydrate Polymers*, **180**, 128–144.

Kumar, A., Rao, K.M., Kwon, S.E., Lee, Y.N., Han, S.S. 2017b. Xanthan gum/bioactive silica glass hybrid scaffolds reinforced with cellulose nanocrystals: Morphological, mechanical and in vitro cytocompatibility study. *Materials Letters*, **193**, 274–278.

Kumar, A., Singh, K., Ahuja, M. 2009. Xanthan-g-poly(acrylamide): Microwave-assisted synthesis, characterization and in vitro release behavior. *Carbohydrate Polymers*, **76**(2), 261–267.

Kumar, A., Zo, S.M., Kim, J.H., Kim, S.-C., Han, S.S. 2019. Enhanced physical, mechanical, and cytocompatibility behavior of polyelectrolyte complex hydrogels by reinforcing halloysite nanotubes and graphene oxide. *Composites Science and Technology*, **175**, 35–45.

Kumar, A.K., Reddy, M.V., Chandrasekhar, K., Srikanth, S., Mohan, S.V. 2012. Endocrine disruptive estrogens role in electron transfer: bio-electrochemical remediation with microbial mediated electrogenesis. *Bioresource Technology*, **104**, 547–556.

Kumar, G., Sivagurunathan, P., Pugazhendhi, A., Thi, N.B.D., Zhen, G., Chandrasekhar, K., Kadier, A. 2017c. A comprehensive overview on light independent fermentative hydrogen production from wastewater feedstock and possible integrative options. *Energy Conversion and Management*, **141**(Supplement C), 390–402.

Kumar, P., Chandrasekhar, K., Kumari, A., Sathiyamoorthi, E., Kim, B. 2018b. Electro-fermentation in aid of bioenergy and biopolymers. *Energies*, **11**(2), 343.

Kumar, P., Jun, H.-B., Kim, B.S. 2018c. Co-production of polyhydroxyalkanoates and carotenoids through bioconversion of glycerol by Paracoccus sp. strain LL1. *International Journal of Biological Macromolecules*, **107**, 2552–2558.

Kumar, P., Kim, B.S. 2018. Valorization of polyhydroxyalkanoates production process by co-synthesis of value-added products. *Bioresource Technology*, **269**, 544–556.

Kumar, P., Mehariya, S., Ray, S., Mishra, A., Kalia, V.C. 2015a. Biodiesel industry waste: A potential source of bioenergy and biopolymers. *Indian Journal of Microbiology*, **55**(1), 1–7.

Kumar, P., Patel, S.K.S., Lee, J.-K., Kalia, V.C. 2013. Extending the limits of Bacillus for novel biotechnological applications. *Biotechnology Advances*, **31**(8), 1543–1561.

Kumar, P., Ray, S., Kalia, V.C. 2016. Production of co-polymers of polyhydroxyalkanoates by regulating the hydrolysis of biowastes. *Bioresource Technology*, **200**, 413–419.

Kumar, P., Ray, S., Patel, S.K.S., Lee, J.-K., Kalia, V.C. 2015b. Bioconversion of crude glycerol to polyhydroxyalkanoate by Bacillus thuringiensis under non-limiting nitrogen conditions. *International Journal of Biological Macromolecules*, **78**, 9–16.

Kumar, P., Singh, M., Mehariya, S., Patel, S.K.S., Lee, J.-K., Kalia, V.C. 2014. Ecobiotechnological approach for exploiting the abilities of bacillus to produce co-polymer of polyhydroxyalkanoate. *Indian Journal of Microbiology*, **54**(2), 151–157.

Li, J., He, J., Huang, Y. 2017. Role of alginate in antibacterial finishing of textiles. *International Journal of Biological Macromolecules*, **94**, 466–473.

Li, W., Peng, H., Ning, F., Yao, L., Luo, M., Zhao, Q., Zhu, X., Xiong, H. 2014. Amphiphilic chitosan derivative-based core–shell micelles: Synthesis, characterisation and properties for sustained release of Vitamin D3. *Food Chemistry*, **152**, 307–315.

Li, Z., Chen, L., Meng, S., Guo, L., Huang, J., Liu, Y., Wang, W., Chen, X. 2015. Field and temperature dependence of intrinsic diamagnetism in graphene: Theory and experiment. *Physical Review B*, **91**(9), 094429.

Ma, H.-l., Qi, X.-r., Maitani, Y., Nagai, T. 2007. Preparation and characterization of superparamagnetic iron oxide nanoparticles stabilized by alginate. *International Journal of Pharmaceutics*, **333**(1), 177–186.

Madhusudana Rao, K., Kumar, A., Han, S.S. 2018. Polysaccharide based hydrogels reinforced with halloysite nanotubes via polyelectrolyte complexation. *Materials Letters*, **213**, 231–235.

Maia, A.M.S., Silva, H.V.M., Curti, P.S., Balaban, R.C. 2012. Study of the reaction of grafting acrylamide onto xanthan gum. *Carbohydrate Polymers*, **90**(2), 778–783.

Mazumder, M.A.J., Fitzpatrick, S.D., Muirhead, B., Sheardown, H. 2012. Cell-adhesive thermogelling PNIPAAm/hyaluronic acid cell delivery hydrogels for potential application as minimally invasive retinal therapeutics. *Journal of Biomedical Materials Research - Part A*, **100 A**(7), 1877–1887.

Mei, L., Shen, B., Xue, J., Liu, S., Ma, A., Liu, F., Shao, H., Chen, J., Chen, Q., Liu, F., Ying, Y., Ling, P. 2017. Adipose tissue–derived stem cells in combination with xanthan gum attenuate osteoarthritis progression in an experimental rat model. *Biochemical and Biophysical Research Communications*, **494**(1), 285–291.

Millon, L.E., Guhados, G., Wan, W. 2008. Anisotropic polyvinyl alcohol—Bacterial cellulose nanocomposite for biomedical applications. *Journal of Biomedical Materials Research Part B: Applied Biomaterials*, **86B**(2), 444–452.

Mirhosseini, H., Tan, C.P., Hamid, N.S.A., Yusof, S. 2008. Effect of Arabic gum, xanthan gum and orange oil contents on ζ-potential, conductivity, stability, size index and pH of orange beverage emulsion. *Colloids and Surfaces A: Physicochemical and Engineering Aspects*, **315**(1), 47–56.

Mohan, S.V., Chandrasekhar, K., Chiranjeevi, P., Babu, P.S. 2013. Biohydrogen Production from Wastewater. in: Biohydrogen, (Eds.) A. Pandey, J.-S. Chang, P.C. Hallenbeck, C. Larroche, Elsevier. Amsterdam, pp. 223–257.

Molina, J., Fernández, J., del Río, A.I., Lapuente, R., Bonastre, J., Cases, F. 2010. Stability of conducting polyester/polypyrrole fabrics in different pH solutions. Chemical and electrochemical characterization. *Polymer Degradation and Stability*, **95**(12), 2574–2583.

Morra, M. 2005. Engineering of biomaterials surfaces by hyaluronan. *Biomacromolecules*, **6**(3), 1205–1223.

Mozumder, M.S.I. 2018. Polyhydroxyalkanoate and polylactic acid composite. in: *Reference Module in Materials Science and Materials Engineering*, (Ed.) S. Hashmi, Elsevier.

Mundargi, R.C., Patil, S.A., Aminabhavi, T.M. 2007. Evaluation of acrylamide-grafted-xanthan gum copolymer matrix tablets for oral controlled delivery of antihypertensive drugs. *Carbohydrate Polymers*, **69**(1), 130–141.

Nakagawa, Y., Nakasako, S., Ohta, S., Ito, T. 2015. A biocompatible calcium salt of hyaluronic acid grafted with polyacrylic acid. *Carbohydrate Polymers*, **117**, 43–53.

Naseem, A., Tabasum, S., Zia, K.M., Zuber, M., Ali, M., Noreen, A. 2016. Lignin-derivatives based polymers, blends and composites: A review. *International Journal of Biological Macromolecules*, **93**, 296–313.

Naskar, B., Dan, A., Ghosh, S., Moulik, S.P. 2010. Characteristic physicochemical features of the biopolymer inulin in solvent added and depleted states. *Carbohydrate Polymers*, **81**(3), 700–706.

Ning, C., Zhou, Z., Tan, G., Zhu, Y., Mao, C. 2018. Electroactive polymers for tissue regeneration: Developments and perspectives. *Progress in Polymer Science*, **81**, 144–162.

Pandey, S., Mishra, S.B. 2012. Microwave synthesized xanthan gum-g-poly(ethylacrylate): An efficient Pb2+ ion binder. *Carbohydrate Polymers*, **90**(1), 370–379.

Pandit, S., Chandrasekhar, K., Jadha, D.A., Ghangrekar, M.M., Das, D. 2019. Contaminant removal and energy recovery in microbial fuel cells. in: *Microbial Biodegradation of Xenobiotic Compounds*, (Ed.) Y.-C. Chang, Vol. 1, CRC Press. Published January 30, 2019, pp. 76.

Pandit, S., Chandrasekhar, K., Kakarla, R., Kadier, A., Jeevitha, V. 2017. Basic principles of microbial fuel cell: Technical challenges and economic feasibility. in: *Microbial Applications Vol.1*: *Bioremediation and Bioenergy*, (Eds.) V.C. Kalia, P. Kumar, Springer International Publishing. Cham, pp. 165–188.

Pandit, S., Sarode, S., Chandrasekhar, K. 2018. Fundamentals of bacterial biofilm: Present state of art. in: *Quorum Sensing and its Biotechnological Applications*, (Ed.) V.C. Kalia, Springer Singapore. Singapore, pp. 43–60.

Papageorgiou, D.G., Liu, M., Li, Z., Vallés, C., Young, R.J., Kinloch, I.A. 2019. Hybrid poly(ether ether ketone) composites reinforced with a combination of carbon fibres and graphene nanoplatelets. *Composites Science and Technology*, **175**, 60–68.

Parkhey, P., Gupta, P., Eswari, J.S. 2017. Optimization of cellulase production from isolated cellulolytic bacterium: Comparison between genetic algorithms, simulated annealing, and response surface methodology. *Chemical Engineering Communications*, **204**(1), 28–38.

Patel, S.K.S., Kumar, P., Singh, M., Lee, J.-K., Kalia, V.C. 2015. Integrative approach to produce hydrogen and polyhydroxybutyrate from biowaste using defined bacterial cultures. *Bioresource Technology*, **176**, 136–141.

Periyasamy Sivagurunathan, Abudukeremu Kadier, Mudhoo., A., Gopalakrishnan Kumar, Kuppam Chandrasekhar, Takuro Kobayashi, Xu, K. 2018. Nanomaterials for biohydrogen production. in: *Nanomaterials: Biomedical, Environmental, and Engineering Applications*, (Eds.) S. Kanchi, S. Ahmed, M.I. Sabela, C. M. Hussain, Wiley-Scrivener, pp. 217–237.

Petri, D.F.S. 2015. Xanthan gum: A versatile biopolymer for biomedical and technological applications. *Journal of Applied Polymer Science*, **132**(23), 42035.

Pitarresi, G., Palumbo, F.S., Albanese, A., Fiorica, C., Picone, P., Giammona, G. 2010. Self-assembled amphiphilic hyaluronic acid graft copolymers for targeted release of antitumoral drug. *Journal of Drug Targeting*, **18**(4), 264–276.

Ptaszek, A., Berski, W., Ptaszek, P., Witczak, T., Repelewicz, U., Grzesik, M. 2009. Viscoelastic properties of waxy maize starch and selected non-starch hydrocolloids gels. *Carbohydrate Polymers*, **76**(4), 567–577.

Qi, W., Yuan, W., Yan, J., Wang, H. 2014. Growth and accelerated differentiation of mesenchymal stem cells on graphene oxide/poly-l-lysine composite films. *Journal of Materials Chemistry B*, **2**(33), 5461–5467.

Qin, X., Wang, H., Wang, X., Miao, Z., Chen, L., Zhao, W., Shan, M., Chen, Q. 2010. Amperometric biosensors based on gold nanoparticles-decorated multiwalled carbon nanotubes-poly(diallyldimethylammonium chloride) biocomposite for the determination of choline. *Sensors and Actuators B: Chemical*, **147**(2), 593–598.

Rahaman, M.N., Day, D.E., Sonny Bal, B., Fu, Q., Jung, S.B., Bonewald, L.F., Tomsia, A.P. 2011. Bioactive glass in tissue engineering. *Acta Biomaterialia*, **7**(6), 2355–2373.

Ram Kumar Pandian, S., Deepak, V., Kalishwaralal, K., Muniyandi, J., Rameshkumar, N., Gurunathan, S. 2009. Synthesis of PHB nanoparticles from optimized medium utilizing dairy industrial waste using Brevibacterium casei SRKP2: A green chemistry approach. *Colloids Surf B Biointerfaces*, **74**(1), 266–273.

Rao, K.M., Kumar, A., Haider, A., Han, S.S. 2016. Polysaccharides based antibacterial polyelectrolyte hydrogels with silver nanoparticles. *Materials Letters*, **184**, 189–192.

Rao, K.M., Kumar, A., Han, S.S. 2017. Poly(acrylamidoglycolic acid) nanocomposite hydrogels reinforced with cellulose nanocrystals for pH-sensitive controlled release of diclofenac sodium. *Polymer Testing*, **64**, 175–182.

Rao, K.M., Nagappan, S., Seo, D.J., Ha, C.-S. 2014. pH sensitive halloysite-sodium hyaluronate/poly(hydroxyethyl methacrylate) nanocomposites for colon cancer drug delivery. *Applied Clay Science*, **97–98**, 33–42.

Raschip, I.E., Hitruc, E.G., Oprea, A.M., Popescu, M.-C., Vasile, C. 2011. In vitro evaluation of the mixed xanthan/lignin hydrogels as vanillin carriers. *Journal of Molecular Structure*, **1003**(1), 67–74.

Reddy, M.V., Chandrasekhar, K., Mohan, S.V. 2011a. Influence of carbohydrates and proteins concentration on fermentative hydrogen production using canteen based waste under acidophilic microenvironment. *Journal of Biotechnology*, **155**(4), 387–395.

Reddy, M.V., Devi, M.P., Chandrasekhar, K., Goud, R.K., Mohan, S.V. 2011b. Aerobic remediation of petroleum sludge through soil supplementation: Microbial community analysis. *Journal of Hazardous Materials*, **197**, 80–87.

Rovera, C., Cozzolino, C.A., Ghaani, M., Morrone, D., Olsson, R.T., Farris, S. 2018. Mechanical behavior of biopolymer composite coatings on plastic films by depth-sensing indentation - A nanoscale study. *Journal of Colloid and Interface Science*, **512**, 638–646.

Safinya, C.R., Ewert, K.K. 2012. Materials chemistry: Liposomes derived from molecular vases. *Nature*, **489**(7416), 372–374.

Saha, N.R., Sarkar, G., Roy, I., Bhattacharyya, A., Rana, D., Dhanarajan, G., Banerjee, R., Sen, R., Mishra, R., Chattopadhyay, D. 2016. Nanocomposite films based on cellulose acetate/polyethylene glycol/modified montmorillonite as nontoxic active packaging material. *RSC Advances*, **6**(95), 92569–92578.

Şahin, M., Şahin, Y., Özcan, A. 2008. Ion chromatography-potentiometric detection of inorganic anions and cations using polypyrrole and overoxidized polypyrrole electrode. *Sensors and Actuators B: Chemical*, **133**(1), 5–14.

Sahu, A., Kasoju, N., Bora, U. 2008. Fluorescence study of the curcumin-casein micelle complexation and its application as a drug nanocarrier to cancer cells. *Biomacromolecules*, **9**(10), 2905–2912.

Salem, D.M.S.A., Sallam, M.A.E., Youssef, T.N.M.A. 2019. Synthesis of compounds having antimicrobial activity from alginate. *Bioorganic Chemistry*, **87**, 103–111.

Saratale, R.G., Kuppam, C., Mudhoo, A., Saratale, G.D., Periyasamy, S., Zhen, G., Koók, L., Bakonyi, P., Nemestóthy, N., Kumar, G. 2017. Bioelectrochemical systems using microalgae – A concise research update. *Chemosphere*, **177**, 35–43.

Shah, N., Ul-Islam, M., Khattak, W.A., Park, J.K. 2013. Overview of bacterial cellulose composites: A multipurpose advanced material. *Carbohydrate Polymers*, **98**(2), 1585–1598.

Shalviri, A., Liu, Q., Abdekhodaie, M.J., Wu, X.Y. 2010. Novel modified starch–xanthan gum hydrogels for controlled drug delivery: Synthesis and characterization. *Carbohydrate Polymers*, **79**(4), 898–907.

Shi, Z., Gao, X., Ullah, M.W., Li, S., Wang, Q., Yang, G. 2016. Electroconductive natural polymer-based hydrogels. *Biomaterials*, **111**, 40–54.

Singh, M., Kumar, P., Ray, S., Kalia, V.C. 2015a. Challenges and opportunities for customizing polyhydroxyalkanoates. *Indian Journal of Microbiology*, **55**(3), 235–249.

Singh, R.S., Kaur, N., Kennedy, J.F. 2015b. Pullulan and pullulan derivatives as promising biomolecules for drug and gene targeting. *Carbohydrate Polymers*, **123**, 190–207.

Singh, R.S., Kaur, N., Rana, V., Kennedy, J.F. 2016. Recent insights on applications of pullulan in tissue engineering. *Carbohydrate Polymers*, **153**, 455–462.

Sivagurunathan, P., Kuppam, C., Mudhoo, A., Saratale, G.D., Kadier, A., Zhen, G., Chatellard, L., Trably, E., Kumar, G. 2017. A comprehensive review on two-stage integrative schemes for the valorization of dark fermentative effluents. *Critical Reviews in Biotechnology*, **38**(6), 868–882.

Slimane, A.B., Connan, C., Vaulay, M.-J., Chehimi, M.M. 2009. Preparation and surface analysis of pigment@polypyrrole composites. *Colloids and Surfaces A: Physicochemical and Engineering Aspects*, **332**(2), 157–163.

Srikanth, S., Venkateswar Reddy, M., Venkata Mohan, S. 2012. Microaerophilic microenvironment at biocathode enhances electrogenesis with simultaneous synthesis of polyhydroxyalkanoates (PHA) in bioelectrochemical system (BES). *Bioresource Technology*, **125**, 291–299.

Subramaniam, S., Fang, Y.-H., Sivasubramanian, S., Lin, F.-H., Lin, C.-p. 2016. Hydroxyapatite-calcium sulfate-hyaluronic acid composite encapsulated with collagenase as bone substitute for alveolar bone regeneration. *Biomaterials*, **74**, 99–108.

Sugumaran K.R., Ponnusami V. 2017. Review on production, downstream processing and characterization of microbial pullulan. *Carbohydrate Polymers*, **173**, 573–591.

Suner, S.S., Demirci, S., Yetiskin, B., Fakhrullin, R., Naumenko, E., Okay, O., Ayyala, R.S., Sahiner, N. 2019. Cryogel composites based on hyaluronic acid and halloysite nanotubes as scaffold for tissue engineering. *International Journal of Biological Macromolecules*, **130**, 627–635.

Tabasum, S., Noreen, A., Maqsood, M.F., Umar, H., Akram, N., Nazli, Z.-i.-H., Chatha, S.A.S., Zia, K.M. 2018. A review on versatile applications of blends and composites of pullulan with natural and synthetic polymers. *International Journal of Biological Macromolecules*, **120**, 603–632.

Takeuchi, A., Kamiryou, Y., Yamada, H., Eto, M., Shibata, K., Haruna, K., Naito, S., Yoshikai, Y. 2009. Oral administration of xanthan gum enhances antitumor activity through Toll-like receptor 4. *International Immunopharmacology*, **9**(13), 1562–1567.

Tan, H., Peng, Z., Li, Q., Xu, X., Guo, S., Tang, T. 2012. The use of quaternised chitosan-loaded PMMA to inhibit biofilm formation and downregulate the virulence-associated gene expression of antibiotic-resistant staphylococcus. *Biomaterials*, **33**(2), 365–377.

Terán Hilares, R., Resende, J., Orsi, C.A., Ahmed, M.A., Lacerda, T.M., da Silva, S.S., Santos, J.C. 2019. Exopolysaccharide (pullulan) production from sugarcane bagasse hydrolysate aiming to favor the development of biorefineries. *International Journal of Biological Macromolecules*, **127**, 169–177.

Tiwari, S., Bahadur, P. 2019. Modified hyaluronic acid based materials for biomedical applications. *International Journal of Biological Macromolecules*, **121**, 556–571.

Toole, B.P. 2004. Hyaluronan: From extracellular glue to pericellular cue. *Nature Reviews Cancer*, **4**(7), 528–539.

Torres, F.G., Arroyo, J.J., Troncoso, O.P. 2019. Bacterial cellulose nanocomposites: An all-nano type of material. *Materials Science and Engineering: C*, **98**, 1277–1293.

Trandafilovic, L.V., Bozanic, D.K., Dimitrijevic-Brankovic, S., Luyt, A.S., Djokovic, V. 2012. Fabrication and antibacterial properties of ZnO-alginate nanocomposites. *Carbohydrate Polymers*, **88**(1), 263–269.

Trombino, S., Serini, S., Cassano, R., Calviello, G. 2019. Xanthan gum-based materials for omega-3 PUFA delivery: Preparation, characterization and antineoplastic activity evaluation. *Carbohydrate Polymers*, **208**, 431–440.

Tummalapalli, M., Berthet, M., Verrier, B., Deopura, B.L., Alam, M.S., Gupta, B. 2016. Composite wound dressings of pectin and gelatin with aloe vera and curcumin as bioactive agents. *International Journal of Biological Macromolecules*, **82**, 104–113.

Udawattha, C., De Silva, D.E., Galkanda, H., Halwatura, R. 2018. Performance of natural polymers for stabilizing earth blocks. *Materialia*, **2**, 23–32.

Ul-Islam, M., Khan, S., Ullah, M.W., Park, J.K. 2015. Bacterial cellulose composites: Synthetic strategies and multiple applications in bio-medical and electro-conductive fields. *Biotechnol Journal*, **10**(12), 1847–1861.

Valverde, A., Pérez-Álvarez, L., Ruiz-Rubio, L., Pacha Olivenza, M.A., García Blanco, M.B., Díaz-Fuentes, M., Vilas-Vilela, J.L. 2019. Antibacterial hyaluronic acid/chitosan multilayers onto smooth and micropatterned titanium surfaces. *Carbohydrate Polymers*, **207**, 824–833.

Vazquez, J.A., Rodriguez-Amado, I., Montemayor, M.I., Fraguas, J., Gonzalez Mdel, P., Murado, M.A. 2013. Chondroitin sulfate, hyaluronic acid and chitin/chitosan production using marine waste sources: Characteristics, applications and eco-friendly processes: A review. *Marine Drugs*, **11**(3), 747–774.

Vendruscolo, C.W., Andreazza, I.F., Ganter, J.L.M.S., Ferrero, C., Bresolin, T.M.B. 2005. Xanthan and galactomannan (from M. scabrella) matrix tablets for oral controlled delivery of theophylline. *International Journal of Pharmaceutics*, **296**(1), 1–11.

Venkata Mohan, S., Chandrasekhar, K. 2011a. Self-induced bio-potential and graphite electron accepting conditions enhances petroleum sludge degradation in bio-electrochemical system with simultaneous power generation. *Bioresource Technology*, **102**(20), 9532–9541.

Venkata Mohan, S., Chandrasekhar, K. 2011b. Solid phase microbial fuel cell (SMFC) for harnessing bioelectricity from composite food waste fermentation: Influence of electrode assembly and buffering capacity. *Bioresource Technology*, **102**(14), 7077–7085.

Venkata Mohan, S., Chiranjeevi, P., Chandrasekhar, K., Babu, P.S., Sarkar, O. 2019. Chapter 11 - Acidogenic Biohydrogen Production From Wastewater. in: *Biohydrogen (Second Edition)*, (Eds.) A. Pandey, S.V. Mohan, J.-S. Chang, P.C. Hallenbeck, C. Larroche, Elsevier, pp. 279–320.

Venkata Mohan, S., Venkateswar Reddy, M. 2013. Optimization of critical factors to enhance polyhydroxyalkanoates (PHA) synthesis by mixed culture using Taguchi design of experimental methodology. *Bioresource Technology*, **128**, 409–416.

Venkata Mohan, S., Venkateswar Reddy, M., Venkata Subhash, G., Sarma, P.N. 2010. Fermentative effluents from hydrogen producing bioreactor as substrate for poly(β-OH) butyrate production with simultaneous treatment: An integrated approach. *Bioresource Technology*, **101**(23), 9382–9386.

Venkata Mohan, S.V., Chandrasekhar, K., Chiranjeevi, P., Babu, P.S. 2013. Chapter 10 - Biohydrogen Production from Wastewater A2 - Pandey, Ashok. in: Biohydrogen, (Eds.) J.-S. Chang, P.C. Hallenbecka, C. Larroche, Elsevier. Amsterdam, pp. 223–257.

Venkateswar Reddy, M., Amulya, K., Rohit, M.V., Sarma, P.N., Venkata Mohan, S. 2014. Valorization of fatty acid waste for bioplastics production using Bacillus tequilensis: Integration with dark-fermentative hydrogen production process. *International Journal of Hydrogen Energy*, **39**(14), 7616–7626.

Venkateswar Reddy, M., Mawatari, Y., Onodera, R., Nakamura, Y., Yajima, Y., Chang, Y.-C. 2017. Polyhydroxyalkanoates (PHA) production from synthetic waste using Pseudomonas pseudoflava: PHA synthase enzyme activity analysis from P. pseudoflava and P. palleronii. *Bioresource Technology*, **234**, 99–105.

Venkateswar Reddy, M., Mawatari, Y., Yajima, Y., Satoh, K., Venkata Mohan, S., Chang, Y.-C. 2016. Production of poly-3-hydroxybutyrate (P3HB) and poly(3-hydroxyb utyrate-co-3-hydroxyvalerate) P(3HB–co–3HV) from synthetic wastewater using Hydrogenophaga palleronii. *Bioresource Technology*, **215**, 155–162.

Venkateswar Reddy, M., Mawatari, Y., Yajima, Y., Seki, C., Hoshino, T., Chang, Y.-C. 2015. Poly-3-hydroxybutyrate (PHB) production from alkylphenols, mono and poly-aromatic hydrocarbons using Bacillus sp. CYR1: A new strategy for wealth from waste. *Bioresource Technology*, **192**, 711–717.

Venkateswar Reddy, M., Nikhil, G.N., Venkata Mohan, S., Swamy, Y.V., Sarma, P.N. 2012. Pseudomonas otitidis as a potential biocatalyst for polyhydroxyalkanoates (PHA) synthesis using synthetic wastewater and acidogenic effluents. *Bioresource Technology*, **123**, 471–479.

Venkateswar Reddy, M., Venkata Mohan, S. 2012a. Effect of substrate load and nutrients concentration on the polyhydroxyalkanoates (PHA) production using mixed consortia through wastewater treatment. *Bioresource Technology*, **114**, 573–582.

Venkateswar Reddy, M., Venkata Mohan, S. 2012b. Influence of aerobic and anoxic microenvironments on polyhydroxyalkanoates (PHA) production from food waste and acidogenic effluents using aerobic consortia. *Bioresource Technology*, **103**(1), 313–321.

Vigués, N., Pujol-Vila, F., Marquez-Maqueda, A., Muñoz-Berbel, X., Mas, J. 2018. Electroaddressable conductive alginate hydrogel for bacterial trapping and general toxicity determination. *Analytica Chimica Acta*, **1036**, 115–120.

Wang, D., Chen, F., Wei, G., Jiang, M., Dong, M. 2015. The mechanism of improved pullulan production by nitrogen limitation in batch culture of Aureobasidium pullulans. *Carbohydrate Polymers*, **127**, 325–331.

Wang, W., Li, H.-Y., Zhang, D.-W., Jiang, J., Cui, Y.-R., Qiu, S., Zhou, Y.-L., Zhang, X.-X. 2010. Fabrication of Bienzymatic Glucose Biosensor Based on Novel Gold Nanoparticles-Bacteria Cellulose Nanofibers Nanocomposite. *Electroanalysis*, **22**(21), 2543–2550.

Watanabe, K., Tabuchi, M., Morinaga, Y., Yoshinaga, F. 1998. Structural features and properties of bacterial cellulose produced in agitated culture. *Cellulose*, **5**(3), 187–200.

Wichai, S., Chuysinuan, P., Chaiarwut, S., Ekabutr, P., Supaphol, P. 2019. Development of bacterial cellulose/alginate/chitosan composites incorporating copper (II) sulfate as an antibacterial wound dressing. *Journal of Drug Delivery Science and Technology*, **51**, 662–671.

Widner, B., Behr, R., Von Dollen, S., Tang, M., Heu, T., Sloma, A., Sternberg, D., DeAngelis, P.L., Weigel, P.H., Brown, S. 2005. Hyaluronic Acid Production in <em>Bacillus subtilis</em>. *Applied and Environmental Microbiology*, **71**(7), 3747–3752.

Ye, S., Jiang, L., Su, C., Zhu, Z., Wen, Y., Shao, W. 2019. Development of gelatin/bacterial cellulose composite sponges as potential natural wound dressings. *International Journal of Biological Macromolecules*, **133**, 148–155.

Yeo, Y., Highley, C.B., Bellas, E., Ito, T., Marini, R., Langer, R., Kohane, D.S. 2006. In situ cross-linkable hyaluronic acid hydrogels prevent post-operative abdominal adhesions in a rabbit model. *Biomaterials*, **27**(27), 4698–4705.

Yu, C.Y., Wang, Y.M., Li, N.M., Liu, G.S., Yang, S., Tang, G.T., He, D.X., Tan, X.W., Wei, H. 2014. In vitro and in vivo evaluation of pectin-based nanoparticles for hepatocellular carcinoma drug chemotherapy. *Molecular Pharmaceutics*, **11**(2), 638–644.

Yunus Basha, R., T.S, S.K., Doble, M. 2015. Design of biocomposite materials for bone tissue regeneration. *Materials Science and Engineering: C*, **57**, 452–463.

Yuvaraj, H., Woo, M.H., Park, E.J., Jeong, Y.T., Lim, K.T. 2008. Polypyrrole/γ-Fe2O3 magnetic nanocomposites synthesized in supercritical fluid. *European Polymer Journal*, **44**(3), 637–644.

Zou, L., Peng, S., Liu, W., Chen, X., Liu, C. 2015. A novel delivery system dextran sulfate coated amphiphilic chitosan derivatives-based nanoliposome: Capacity to improve in vitro digestion stability of (–)-epigallocatechin gallate. *Food Research International*, **69**, 114–120.

Zou, L., Zhang, Y., Liu, X., Chen, J., Zhang, Q. 2019. Biomimetic mineralization on natural and synthetic polymers to prepare hybrid scaffolds for bone tissue engineering. *Colloids and Surfaces B: Biointerfaces*, **178**, 222–229.

Zou, P., Helson, L., Maitra, A., Stern, S.T., McNeil, S.E. 2013. Polymeric curcumin nanoparticle pharmacokinetics and metabolism in bile duct cannulated rats. *Molecular Pharmaceutics*, **10**(5), 1977–1987.

6 Enhancement of Polymeric Material Surface Properties Using Various Surface Modification Techniques

M.C. Ramkumar, K. Navaneetha Pandiyaraj, P.V.A. Padmanabhan, P. Gopinath, and R.R. Deshmukh

CONTENTS

6.1 INTRODUCTION

Over the past few decades, we have perceived an extraordinary growth in the use of polymers in technology, in both advanced and consumer-product applications and they are extensively replacing traditional engineering materials such as stainless steel, titanium, etc. Polymeric materials (polypropylene, polyethylene, poly terephthalate) have gained attention among researchers for their anticipated material properties (resistance to corrosion, flexibility, low weight and low cost) [1–3] and are widely employed in numerous industrial applications, for instance, biomedical, food packing, automotive etc. [4, 5]. In particular, polymers play a significant role in the biomedical field, as they are widely used in the manufacture of blood-contacting devices such as stents, blood bags, bone cement, artificial heart valves, orthopaedic bearings, etc. [6–8]. Still, necessary variances among polymers and other materials result in numerous vital challenges. Regardless of their unique properties, the key problem in employing polymeric materials in biomedical applications is surface-induced thrombus formation (i.e. when the polymer interacts with blood components) and also their poor cell adhesion and viability. This is mainly due to their inadequate surface properties, for instance, low surface energy, poor adhesion, hydrophobicity, and absence of functional groups [9, 10]. It is well known that the surface properties (surface chemistry, topography, and wettability) are important factors that determine the biocompatibility (cell and blood compatibility) of polymeric material [11, 12]. Hence, it is mandatory to tailor the surface of polymers as the interaction of the biological environment occurs on the material surfaces rather than on bulk and it is significant in controlling their performance *in vivo* [13, 14]. Furthermore, polymers should comprise of extensive bulk properties, as they have to operate suitably in the biological environment. So, an appropriate surface tailoring method is needed for enhancing the hydrophilicity and surface energy and introducing new functional groups without disturbing their bulk properties [15–17]. However, the technique that modifies the surface properties of polymers should be reliable, reproducible, and should give good product yield.

Various surface modification methods have been used to improve the surface properties, thereby improving the biocompatibility of polymers such as ozone treatment, wet chemical, UV-induced polymerization, gamma irradiation, laser treatment, and plasma-based technologies [18–20]. A detailed description of these techniques used for tailoring the surface properties of polymers will be discussed in the following paragraphs.

6.2 SURFACE MODIFICATION TECHNIQUES

6.2.1 Ozone Treatment

Ozone, generally known as O_3, can be easily generated and does not require any sophisticated instruments. Ozone is comparatively stable up to 70°C and when its temperature is increased, it can decompose into ionic species or other excited species. It is a dipolar molecule commonly employed for its oxidative properties. Ozone is a convenient species and possesses the ability to tailor the surface of polymers. It has numerous advantages such as the fact that this system offers species that are environmentally safe, since O_3 decomposes into O_2 + O [21, 22]. Additionally, the surface modification of polymers using O_3 occurs at atmospheric pressure. E.T. Kang et al. modified the surface of polypyrrole (PPY), poly(3-alkylthiophene), and polyaniline (PAN) films using O_3 treatment. Their results showed that the oxidation occurs instantly and mostly at the carbon atoms, leading to the generation of C=O, C–O, COOH, and O–COOH groups on polymeric film surfaces [23]. Keiji Fujimot et al. incorporated peroxides on the surface of polyurethane (PU) films by exposing them to ozone. They reported that the peroxide formation on the surface of polymers mainly depends on ozone concentration and exposure time. In addition, the ozone oxidation incorporated peroxides a layer deeper than the outer surface. Furthermore, they modified the surface by graft polymerization and finally concluded that the modified films exhibited reduction in platelet adhesion [24].

6.2.2 Wet Chemical Treatment

In wet chemical treatment, the materials are modified with liquid reagents in order to produce reactive functionalities on their surfaces. This particular surface modification method does not require any specific instrument and can be performed in laboratories. In this process polymeric materials (polypropylene, polyethylene, polyester, etc.) are exposed to oxidizing wet chemicals such as nitric acid, chromic acid, or potassium permanganate, resulting in the generation of hydroxyl, carboxylic, and carbonyl groups on the surface of polymers [25–28]. Thus, incorporation of these oxygen functionalities improves polarity, which in turn enhances the wettability and adhesion properties. Three-dimensional substrates can be penetrated easily using wet chemical treatment, compared to other surface modification methods. However, longer exposure leads to damage on the surface of material by creating microscopic pits [29]. M. Herrero et al. used the wet chemical method to incorporate various functionalities on PVC films. They reported that the quality of solvent, reaction time, and temperature have a major role to play in incorporating functionalities. The bacterial adhesion of *E. Coli* subjected to surface-modified PVC films was found to be reduced by 50% [30]. Gabriel et al. aimed to enhance the cell-compatible properties of poly-tetrafluoroethylene (PTFE) films by the wet chemical process. Sodium naphthalene was used to introduce hydroxyl (–OH) functional groups for immobilization of arginine-glycine-aspartate (RGD) peptide. Their cell-compatible results showed that the surface-tailored films exhibited outstanding growth of endothelial cells [31].

6.2.3 UV-Induced Polymerization

The UV-assisted surface modification method appears to be a simple and economical process, compared to other modification techniques [32]. It has been used widely for various biological and industrial applications, that is, sterilization of packing materials and activation of surfaces. In this process, UV light is extensively employed to perform surface graft polymerization, frequently in the presence of a photoinitiator or photosensitizer. Generally, benzophenone and its derivatives have been extensively used as photoinitiators [33–36], as they resulted in higher grafting efficiency and also efficiently initiates various radical-induced graft polymerizations.

Hwang et al. tailored PP membranes via UV-induced graft polymerization of glycidyl methacrylate (GMA) using benzophenone as a photoinitiator. They reported that the degree of grafting was enhanced by increase in the intensity of UV irradiation, reaction time, and concentration of photoinitiator. SEM and AFM analysis clearly showed that the modification technique amended the surface morphology of PP membranes [37]. Salmi-Mani et al. enhanced the surface properties of PET films via a two-step process for improving the antibacterial properties. Initially, the films were grafted by photoinitiator by aminolysis, followed by photopolymerization of acrylamide monomer. They reported that the modified PET films exhibited excellent antimicrobial properties, that is, the bacterial adhesion of *Rhodococcus wratislaviensis* and *Staphylococcus aureus* was reduced by 85% and 97% respectively [38]. Kim et al. prepared copolymers of monoacrylate-poly(ethylene glycol) (PEGMA) and poly(3-hydroxyoctanoate) (PHO) in the presence of chloroform and irradiated with UV light, and examined their blood compatibility using platelet adhesion and protein adsorption analysis. Their result showed that enhancement in blood compatibility was subsequent to grafting of PEGMA. The inhibition of platelet and protein was increased with an increase in the concentration of PEGMA and PHO. Finally they concluded that the prepared copolymers can be employed in various biomedical applications [39]. Song et al. aimed to improve the adhesion property of PET film by UV grafting of acrylic acid on its surfaces. The surface-modified films exhibited excellent hydrophilic properties due to the presence of acrylic acid functionalities. In addition, the adhesive strength of the PET films was increased subsequent to UV grafting of acrylic acid [40]. Feng et al. modified the surface of polycarbonateurethane (PCU) by UV-induced photopolymerization of poly(ethylene glycol) monoacrylates (PEGMAs) at different molecular weights, in order to enhance its hemocompatibility. The authors reported that after grafting of PEGMAs, hydrophilicity was increased substantially and the modified films inhibited adhesion of platelets. They also found that the molecular weight of PEGMAs played a significant role in determining the hemocompatibility [41]. Ramanathan et al. modified the surface properties of various polymers such as polyurethane, (PU), polypropylene (PP), polystyrene (PS), and polysulfone (PSU) in the presence of acrylic acid vapor via UV treatment. The surface-modified films showed excellent hydrophilic property even after 65 days of storage time [42].

6.2.4 Gamma Irradiation

In general, gamma radiation is used for basic studies and for small doses of irradiation with intense penetration. The main sources of gamma radiation are cobalt-60 and cesium-137 radioactive isotopes. It is extensively employed for sterilization of medical appliances, to solve the issues related to the attachment of pathogens. Gamma radiation treatment of polymers has great significance as it supports attaining certain anticipated amendments in the surface properties of polymers. It offers a distinctive way to alter the structural, mechanical, chemical, electrical, and optical properties of polymers resulting in permanent alterations in their macromolecular structure [43]. Gorna et al. modified the surface of polyurethane film with different hydrophilic properties, by exposing it to gamma radiation at a typical dose of 25 kGy, employed for sterilization. The hydrophilic films irradiated exhibited reduction in mechanical strength. They also found that a significant amendment in thermal properties was observed. Furthermore, the roughness of surface-modified films increased to 36–76% and the contact angle decreased to 20–45% [44]. Siddhartha et al. studied the influence of 1.25 MeV gamma radiation on optical and surface properties of polyethylene terephthalate (PET) films. They reported that crystalline size and crystallinity of the polymers were increased and the bandgap of the polymers decreased with the increase in the dose of the gamma radiation. This may be due to the formation of clusters and defects on the surface of materials, whereas the cluster size was found to be dependent on the adsorbed dose [45]. Sinha et al. modified the surface of polycarbonate at a different dose (10^1–10^6 Gy) of gamma radiation from Co 60 source. They found the formation of phenolic groups on the film surfaces at a higher dose of 10^6 Gy. The crystallinity was increased with increasing the dose to higher values. Due to the chain scissions, an increase in mobility of chains was observed, resulting in a decrease in glass transition temperature, and the same was influenced by gamma dose [46].

6.2.5 Laser Treatment

Laser radiation is extensively employed for numerous medical applications and it plays a key role in surface modification of polymers for improvement of its biocompatibility. Laser treatment provides various advantages, compared to other physical and chemical surface modification techniques. Laser treatment also offers specific surface tailoring properties, which is very challenging to accomplish with other traditional techniques. The treated surfaces are free from impurities; in some cases, laser treatment does not alter the bulk properties. It provides the user an exceptionally high degree of control and flexibility. Jelvani et al. altered polyethersulfonate (PES) film surfaces by CO_2 laser treatment to enhance biocompatibility. The wettability was increased for the PES films that were modified by laser radiation below the ablation threshold. The PES films modified at optimum conditions showed enhancement in biocompatibility (i.e. reduced platelet adhesion) [47]. Dadsetan et al. aimed to enhance the blood compatibility of PET by treating them via CO_2-pulsed and KrF excimer lasers. Their results exhibited that the PET films crystallinity was decreased after surface treatment. Platelets were highly adhered to the surface of pristine films, whereas a reduction in the number of adhered and spreading platelets was witnessed on laser-modified PET film surfaces. The authors concluded that the decrease in platelet adhesion was mainly ascribed to amendment in chemical structure, crystallinity, and morphology of PET films induced by laser treatment [48]. Tiaw et al. tailored the surfaces of ultra-thin polycaprolactone (PCL) films by femtosecond laser and excimer laser for tissue engineering applications. The surface modification was performed at different parameters, for example, pulse repetition rate and pulse energy, and was characterized by various techniques. Results revealed that laser treatment amended the surface morphology of the films and the hydrophilicity was enhanced on the surface of films modified at different parameters [49]. Khorasani et al. tailored polydimethylsiloxane (PDMS) film surfaces via CO_2-pulsed laser and their *in vitro* analysis outcome revealed that the surface-tailored films significantly suppressed the adhesion of platelets [50]. Wang et al. investigated the effect of femtosecond laser on PMMA film surfaces at different focus distances and laser fluences. It was found that hydrophilicity can be enhanced by suitably governing the treatment conditions and the amendments were mainly influenced by surface chemistry rather than roughness [51].

Various surface modification techniques employed for surface modification of polymers have been discussed. In spite of their advantages, they possess some limitations while being used in surface modification, such as emission of volatile organic compounds (VOC), intake of solvents, longer processing time, and stability problems [52–54]. In order to overcome these limitations, it is necessary to find a suitable surface modification technique. One such technology is the non-thermal plasma surface modification technique, as it is a promising technique that has gained attention among researchers in recent years.

Plasma-based surface modification is a solvent-free single-step process and it is proficient in incorporating a wide range of functionalities in less time with a uniform surface treatment. Additionally, it possesses different advantages relating to biomedical industries, as listed:

1. It is an eco-friendly process.
2. Plasma modification techniques are reliable, moderately economical, and appropriate for different substrates (ceramics, polymers, and composites).
3. It can easily alter an extensive range of surface properties.
4. Plasma modification technique can be easily scaled up to industry level compared to other conventional techniques.

In this chapter, we focus on the modification of surface properties of polymers using the non-thermal plasma technique. The plasma modification technique is an appropriate process for incorporation of various functional groups [55] and it can also modify samples having complex structures such as scaffolds [56]. In addition, it can also be employed for immobilization of biomolecules or proteins [57–60].

6.3 CLASSIFICATION OF PLASMA

Plasma is a quasi-neutral fluid consisting of ions, electrons, and neutrals, which are in a ground and an excited state. It is also denoted as the fourth state of matter [61]. Moreover, plasma is electrically neutral from a macroscopic point of view. In general, plasma is classified into two types: thermal plasmas and non-thermal or cold plasmas. Thermal plasma or hot plasma is characterized by local thermodynamic equilibrium (LTE), that is, the temperature of an electron (T_e) is almost equal to the temperature of the heavier particle (T_h) ($T_e \approx T_h$) (see Figure 6.1a). They have extensive industrial applications, such as treatment of waste materials, plasma spraying, advanced material processing etc. [62]. Non-thermal plasma or cold plasma has a lower degree of ionization and it is distinguished by lower energy densities and large variations of the temperature of heavier particles and electrons, that is, $T_e >>> T_h$ (Figure 6.1b). These discharges can be created at room temperature and can be used for various industrial applications such as surface modification of textiles, food packaging, and agricultural and biomedical applications [63, 64].

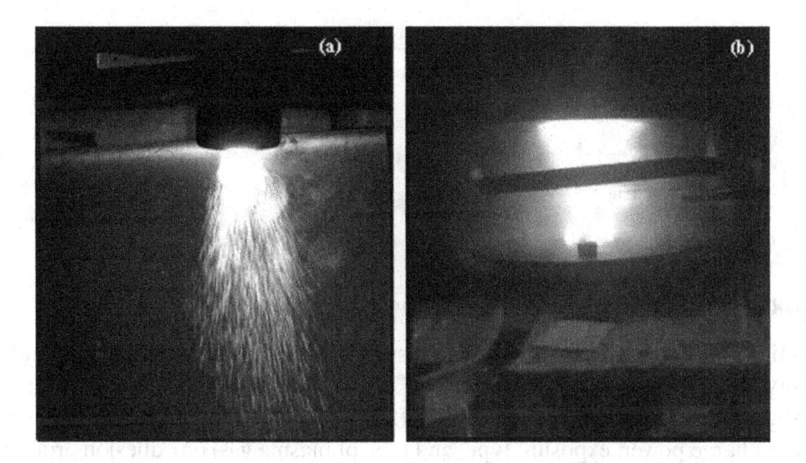

FIGURE 6.1 Photograph of (a) thermal and (b) non-thermal plasma.

6.3.1 Non-Thermal Plasma

Non-thermal plasma has gained attention owing to its various advantages, for instance, low operating temperature, low cost, and lack of solvents. Non-thermal plasmas are employed widely for surface modification as the molecules, atoms, and ions are relatively cold and do not damage the surface of heat-sensitive materials such as biological tissues and polymers [65]. This offers the ability to effectively alter the surface properties of the polymeric materials without disturbing their material bulk [66]. Subsequently, different mechanisms, such as surface activation or functionalization, etching, and crosslinking take place on the material surfaces when they interact with plasma. Furthermore, deposition of polymer-like films on various substrates can also be performed using non-thermal plasma, known as plasma polymerization. Polymer films developed by plasma polymerization are extremely homogeneous and exhibit excellent adhesion to numerous substrates, providing the anticipated physical, chemical, and morphological properties to the polymers [67]. Plasma polymer interaction can be classified into various distinct mechanisms, such as etching, deposition, activation, and functionalization. In the following sections, we will mainly focus on some of the significant work regarding the plasma surface modification technique used to tailor the surface of polymers.

6.3.2 Plasma Treatment

When polymers are made to interact with non-thermal plasma, produced using air, O_2, N_2, or NH_3, oxygen or nitrogen functional groups are incorporated on polymer surfaces [68, 69]. The interaction between the active species produced by plasma and various groups of polymer molecules leads to the generation of hydrophilic functionalities. Furthermore, argon and helium plasmas are used to create free radicals and are subsequently employed for cross-linking or grafting of oxygen functionalities by exposing them to the environment [70, 71]. The schematic representation of plasma treatment is shown in Figure 6.2. Recently, there has been a growing interest in

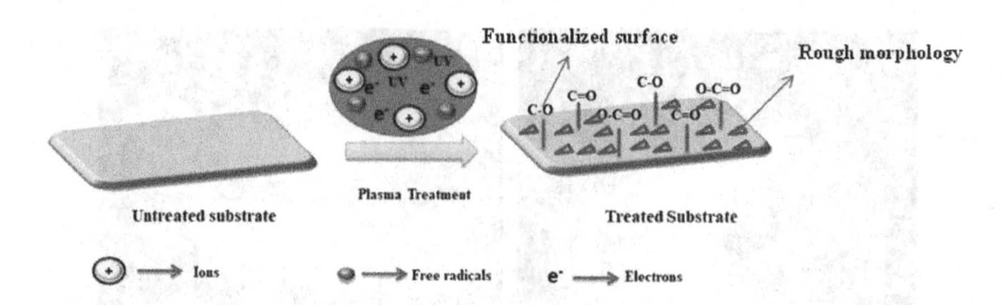

FIGURE 6.2 Schematic representation of plasma treatment.

investigating the relationship between plasma operating parameters and the surface density of functional groups formed [72, 73].

Ataeefard et al. (2009) examined the effect of various plasma operating parameters (discharge power, exposure type, and type of plasma gas) on adhesion properties of LDPE films by employing low-pressure RF plasma. Their results showed incorporation of higher concentration of polar functionalities such as hydroxyl, carboxyl, etc. on plasma-treated LDPE films as confirmed by attenuated total reflectance–Fourier transform infrared (ATR–FTIR) analysis. AFM and SEM analyses clearly showed the drastic change in surface roughness and topography subsequent to plasma treatment. X-ray diffraction (XRD) outcome showed reduction in crystallinity of LDPE films after plasma treatment even for short treatment time. They concluded that while Ar plasma and O_2 plasma alike create functional groups, Ar plasma is more efficient than O_2 plasma; in addition, it increased the roughness of the films, which has a strong impact on adhesive properties [74].

Amanatides et al. (2006) tailored the PET film surfaces using He/O_2 RF discharge plasma. They studied the influence of gas pressure and bias potential on the adhesion of bacteria (*Staphylococcus epidermidis*) in detail. The authors found that the PET films showed an increase in surface roughness subsequent to plasma treatment. However, all the plasma-modified PET films exhibited a substantial reduction in adhesion of bacteria. In addition, they reported the amendment in chemical structure, and the increase in surface free energy was found to be the major reason for reduction in bacterial adhesion on PET surfaces [75]. Aziz et al. tailored the surface of ultra-high molecular weight polyethylene (UHMWPE) in various discharge atmospheres (argon, helium, nitrogen, and air). XPS results undoubtedly revealed the introduction of O_2 functionalities on UHMWPE film surfaces. Human foreskin fibroblast interactions with unmodified and modified samples were studied *in vitro*. The authors found that the plasma-modified samples exhibited more cell attachment and viability than unmodified samples [76]. Esena et al. (2005) used atmospheric pressure air DBD plasma to tailor PET film surfaces. Results showed that the surface roughness enhanced with increase in plasma exposure time. In addition, the surface modification led to an amendment in optical properties and an increase in cell adhesion [77]. Gonzalez et al. (2008) activated the surface of PET and PEN films via atmospheric pressure helium and O_2 plasma. Subsequent to plasma treatment the polarity and adhesion strength were enhanced significantly. XPS results revealed a decrease in concentration of carbon atoms and an increase in concentration of oxygen atoms. Further, it was also found that the variation in surface properties was mainly due to the breakage of methylene and aromatic carbon atoms by subsequent incorporation of oxygen species present in the plasma region, resulting in the creation of ester-like C=O and C–O bonds [78]. Ramires et al. altered polyethylene terephthalate (PET) surfaces using oxygen and ammonia plasma to enhance the biocompatible properties. Furthermore, to investigate the biocompatibility of surface-treated PET films, cytocompatibility and cytotoxicity tests were performed using endothelial cells. The authors reported that the surface-modified PET films exhibited nontoxicity towards the cells. The outcome of cytocompatibility analysis unveiled an enhancement in cell growth and the existence of well-spread cells on plasma-treated PET films. The improvement in cell compatibility was essentially attributed to the incorporation of oxygen functionalities on PET film surfaces [79].

In another study, Vishnuvarthanan et al. examined the influence of various treatment times and plasma power on variations in surface properties of PP films via O_2 plasma. FTIR and AFM outcome clearly confirmed the incorporation of oxygen functionalities as well as marked roughness on PP films surfaces. XRD outcome revealed that the plasma-modified PP films exhibited crystalline nature compared with untreated ones [80]. Various polymers such as LDPE, HDPE, and UHMWPE were modified by Reznickova et al. who used Ar plasma discharge for enhancement of adhesion and spreading properties of various cell lines (vascular smooth muscle cell [VSMC] and L292). Their results portrayed that the plasma modification initiated the ablation on the polymer surface, leading to a drastic change in their surface roughness and morphology, which greatly influence the adhesion and spreading of L929 and VSMC cells [81]. Nakagawa et al. aimed to study the interaction between poly(L-lactide) (PLLA) films and MC3T3-E1cells and the treatment was performed in various atmospheres of CO_2, air, and perfluoropropane (C_3F_8) gas. The PLLA films treated using CO_2 and air plasma showed a decrease in contact angle values; whereas the contact angle was increased for the films treated using C_3F_8 gas. Finally, cell culture studies unveiled that the cell response was substantially higher for PLLA samples treated using air or CO_2, compared to C_3F_8 gas [82]. Khorasani et al. treated poly(D,L-lactic acid-coglycolic aid) (PLGA) and poly(L-lactic acid) films using O_2 plasma. Their results showed that the enhancement in cell adhesion was observed, this way mainly due to the amendments in surface chemistry and hydrophilicity after plasma treatment. Cell culture studies revealed that the B65 nervous cell was highly adhered and proliferated on plasma-modified films compared to untreated films [83]. Junkar et al. employed nitrogen and oxygen plasma to tailor the surface properties of PET films. Results revealed that amendments in surface chemistry, topography, and hydrophilicity were observed. In addition, plasma treatment enhanced the growth of endothelial cells and fibroblast and varied the adhesion properties of platelets. Remarkably, reduction in platelet adhesion was prominent on O_2 plasma-treated film surfaces; whereas nitrogen plasma-treated films exhibited similar adhesion to that of pristine PET films [84]. Slepicka et al. used Ar plasma to alter the surfaces of various polymers such as PET, PLLA, HDPE, and PTFE. Subsequent to plasma treatment, a substantial decrease in contact angle and variation in surface morphology and roughness was observed. Conclusively, plasma treatment has exceptionally enhanced the biocompatibility of PTFE films [85]. Khorasani et al. examined the interaction between the CO_2 plasma-treated PLLA films and B65 nervous tissue cells and witnessed an enhancement in adhesion and proliferation of B65 cells. The authors reported that the O_2 containing plasma treatment greatly increased the hydrophilicity of PLLA samples, leading to an enhancement in attachment and growth of nervous tissue cells [86]. Kuzminova et al. studied the influence of atmospheric pressure DBD plasma in air on PET foils. HUVEC (human umbilical vein endothelial) and Saos-2 human osteoblast-like cells were used to study the cytocompatibility of the plasma-treated films. The authors witnessed that plasma treatment has a positive impact on cell growth on PET foils. This outcome was related to the predominant increase in surface energy and the existence of O_2 functionalities on PET foils [87].

Jaganjac et al. reported the influence of O_2 and fluorine plasma treatment of PET films on the adhesion and proliferation of human microvascular endothelial cells. The presence of oxygen functionalities on PET film surfaces promoted binding of proteins and endothelialization [88]. De Geyter et al. treated the surface of polylactic acid (PLA) samples by DBD at various atmospheres such as air, helium, nitrogen, and argon. The authors reported that argon and air plasma-incorporated oxygen functionalities resulted in enhanced wettability; however, nitrogen plasma treatment resulted in a marked increase in wettability, compared to argon and air plasma [89]. Junkar et al. modified PET films by O_2 and nitrogen plasmas for different exposure times to enhance the adhesion of fucoidan for improvement of antithrombogenic properties. The authors reported that plasma treatment introduced O_2 functionalities on PET films and altered the surface topography even for short exposure times. Finally, they concluded that the fucoidan attachment was enhanced by O_2 plasma treatment, essentially due to the enhancement in surface roughness [90].

6.3.3 Plasma Polymerization

Plasma polymerization is a particular type of strategy that can be used to fabricate thin films (known as plasma polymers) with exceptional physicochemical properties [91]. In cold plasma, gaseous or liquid precursors are transformed into reactive fragments that sequentially recombine to form polymers [92]. This process also incorporates extremely cross-linked functional layers on the surface of the material. The formed polymer in plasma does not contain the chemical composition and structure similar to polymers produced by other conventional methods [93]. The plasma polymer properties are influenced by the flow rate of the monomer, applied power, and working pressure. The polymers obtained from this technique possess outstanding properties, for instance, pinhole-free, greatly cross-linked, insoluble in most organic solvent, etc. [94, 95]. In addition, the films are coherent and adherent to various material surfaces, for example, conventional polymer, glass, and metal surfaces. Figure 6.3 shows the schematic representation of plasma polymerization.

In this section, various functional coatings obtained using plasma polymerization will be discussed in detail. Since there is a substantial amount of literature available on plasma polymerization on polymers, the outcome will be shown for each type of coatings exclusively.

6.3.3.1 Acrylic Acid Coatings

Plasma polymerization of acrylic acid has gained attention among researchers owing to its widespread applications in the biomedical field. As these coatings comprise a

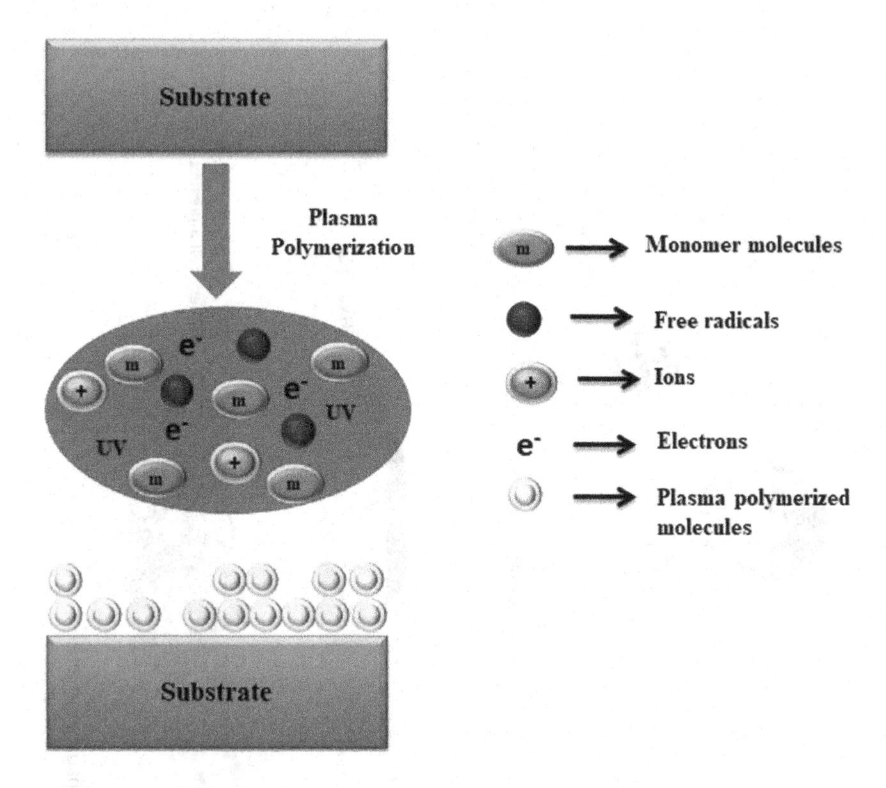

FIGURE 6.3 Schematic representation of plasma polymerization.

higher concentration of carboxylic groups, it can increase the hydrophilic property of a material. It is well known that the hydrophilic property of the material plays a key role in reducing the adsorption of proteins, which results in preventing the thrombus formation [96, 97], and can promote cell attachment and proliferation on the surface of materials. Carboxyl rich coatings have widespread applications in the biomedicine field; for instance, it can act as an active site for covalent immobilization of biomolecules and cell culture substrates [98, 99]. Figure 6.4 shows the schematic of acrylic acid coatings for enhanced cell-compatible properties.

Basarir et al. aimed to improve the electrochemical properties of PP films via plasma polymerization of acrylic acid and the same was performed at different operating conditions such as time, power, and pressure. The authors reported that the plasma polymerization was successfully performed on PP films, confirmed by the existence of carboxylic functionalities, and the modified samples exhibited enhancement in electrochemical properties [100]. Jafari et al. altered PE film surfaces by acrylic acid polymerization at different plasma powers. Results showed that the reduction in contact angle value was mainly due to the existence of COOH/COOR functionalities on PE film surfaces. In addition, deposited coatings were found to be uniform and highly stable [101]. Morent et al. deposited acrylic acid coatings on PP films at different plasma conditions (discharge power and substrate location), in order to obtain a stable coating. The authors witnessed that the films deposited at

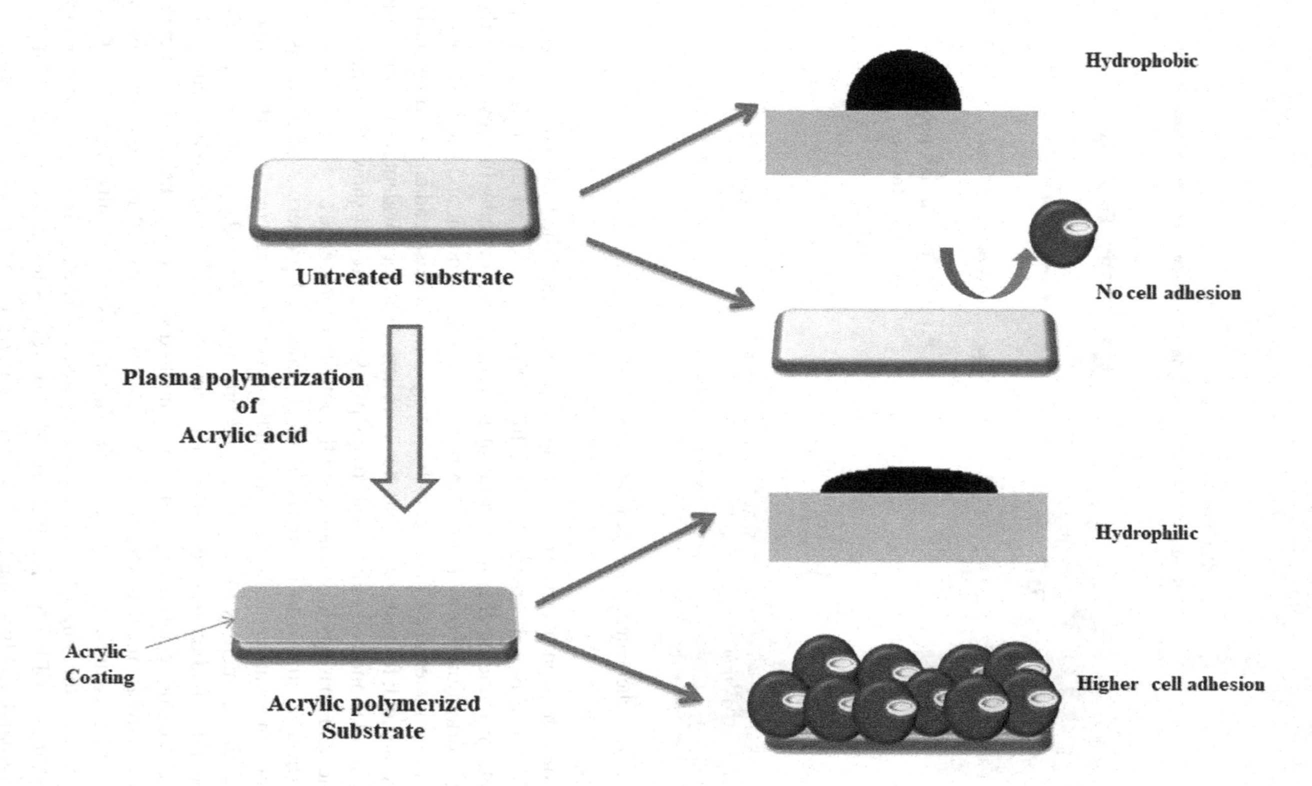

FIGURE 6.4 Schematic of acrylic acid coatings for enhanced cell compatible properties.

lower power possessed a higher concentration of COOH functional groups, whereas the concentration of the same was decreased when the power was increased to higher values. Furthermore, XPS and FTIR results confirmed that the deposited coatings are stable in water [102]. In a similar study, Cools et al. aimed to prepare stable coatings on various substrates (PS, PET, and UHMWPE) via plasma polymerization of acrylic acid using parallel-plate DBD setup at medium pressure. The polymerization was performed at different discharge power and monomer flow rate. The authors found that the coatings prepared were mostly stable. Furthermore, *in vitro* analysis results revealed that the deposited coatings promoted extensive cell-material interactions [103]. Kang et al. polymerized allylamine and acrylic acid on microporous PP membranes. Increasing plasma power resulted in damaged membranes, whereas extending the treatment time led to an increase in plasma coating and micropore blocking. In addition, the hydrophilicity was increased with increase in treatment time. They also reported that the modified films decreased the fouling with bovine serum albumin to 50%. Finally, they concluded that the reduction in BSA was mainly due to enhancement in hydrophilicity on the surface of PP membranes [104].

Topola et al. developed acrylic coatings on PET substrates using a DBD reactor. They found that subsequent to acrylic acid polymerization, hydrophilic property and surface energy of PET films was enhanced. The presence of OH functionalities on modified PET surfaces was confirmed by ATR–FTIR analysis. Furthermore, the authors unveiled that the coatings were found to be uniform and exhibited a smooth surface [105]. Ramkumar et al. inspected the influence of various gaseous plasma (Ar, air, O_2, and Ar + O_2) on plasma polymerization of acrylic acid on LDPE films for improvement of its cell-compatible properties. The incorporation of various functionalities such as C–COOH, C–O, and COOH/COOR was unveiled by XPS analysis and the same was found to be higher on Ar + O_2 plasma-treated acrylic deposited LDPE film surfaces. The modified films exhibited different surface morphology and enhanced hydrophilicity. Additionally, the coatings were stable even after storing for long duration in water. *In vitro* cytocompatibility analysis confirmed that acrylic-deposited LDPE films promoted cell adhesion and proliferation compared to unmodified LDPE films. Also, the cytotoxicity study specifies that the surface-modified films are non-toxic and also a substantial increase in cell viability was observed. The results of RIN-5F cell adhesion, proliferation, and viability on the surface of acrylic-deposited LDPE films are shown in Figure 6.5 [106].

6.3.3.2 Poly(ethylene) Glycol Coatings

Poly(ethylene) glycol (PEG) is a renowned polymer employed for fabrication of anti-fouling coatings on the surface of various substrates and used for numerous biomedical applications. Owing to their anti-fouling character, these coatings can inhibit nonspecific protein and bacterial binding on polymer surfaces. It can also incorporate OH- functionalities on polymeric surfaces, as this group promotes hydrophilic properties on the surface of materials and is found to be responsible for the reduced adsorption of proteins and thus superior blood compatibility [107]. Zou et al. performed plasma polymerization of poly(ethylene glycol) methyl ether methacrylate on the surface of poly(tetrafluoroethylene) films. The authors reported that the surface-tailored films exhibited reduction in contact angle values; this was essentially due to the incorporation of PEGMA functionalities. Protein adsorption

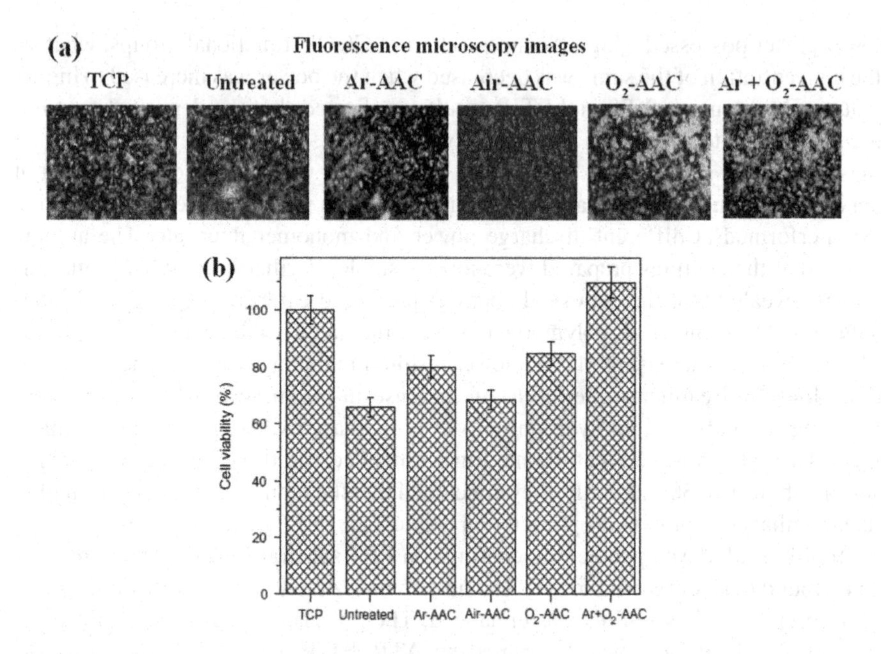

FIGURE 6.5 (a) *In vitro* RIN-5F cell adhesion and (b) cell viability on surface modified LDPE films.

analysis outcome disclosed that the modified films exhibited significant anti-fouling properties [108]. Choi et al. modified various materials, such as polyethylene, glass, and PET by plasma polymerization of PEG. Their results showed that the PEG films exhibited outstanding biocompatibility, unveiled by whole blood flow test. The outcome unveiled that no thrombi and insignificant serum protein adsorption and cellular adhesion was obtained on PEG films surfaces compared to pristine substrates. Furthermore, they investigated the histocompatibility of PEG films, which revealed substantial reduction of tissue adhesion and fibrous proliferation [109]. In another study, Nisol et al. used RF plasma to deposit poly (ethylene) glycol films on the surface of gold and polyvinylfluouride substrates using a tetraglyme monomer. The PEG character was decreased when increasing the discharge power; this was mainly due to the fragmentation of monomer molecule in the plasma region. BSA adsorption outcome revealed that a substantial amount of proteins was adsorbed on the films deposited at higher discharge power, whereas the films deposited at lower discharge power exhibited excellent protein repelling properties [110]. D'Sa et al. aimed to tailor the surface of polystyrene and poly(methyl methacrylate) films by plasma-induced grafting of PEGMA. The authors reported that the plasma power, exposure time, and molecular weight of PEGMA influenced the surface properties of polymers. Protein adsorption analysis showed that the films grafted with low molecular weight revealed some protein adsorption, however no protein was adsorbed on film surfaces grafted using higher molecular weight PEGMA [111].

Sakthi Kumar et al. deposited poly (ethylene glycol) on poly (ethylene terephthalate) film surfaces. The surface-modified films showed a substantial increase in hydrophilic property and smooth topography. Platelet adhesion tests revealed a substantial decrease in adhesion of platelets on PET film surfaces due to the presence of PEG characteristic groups [112]. Jhong et al. aimed to modify the surface of PTFE membranes by plasma copolymerization of PSBMA and PEGMA. The modified membranes showed a reduction in protein adsorption; this can prevent bacterial adhesion to the membranes, resulting in reduced risk of wound infection by microorganisms. Furthermore, the modified membranes possessed excellent blood compatible properties and prevented cell adhesion, Finally, they concluded that the surface-modified membranes can be used for wound healing [113]. Bremmell et al. developed PEG-like coatings on PET film surfaces using diethylene glycol vinyl ether. The authors reported that the surface possesses high PEG-like characteristics, exhibited excellent protein repellent properties, and the adhesion was mainly influenced by applied load and surface chemistry [114]. Pandiyaraj et al. deposited PEO-like coatings on PP film surfaces by employing ethylene glycol dimethyl ether as monomer via cold atmospheric pressure plasma jet-assisted polymerization using a plasma reactor developed inhouse (see Figure 6.6). The PEO coatings were prepared at various discharge potentials and monomer flow rate. The authors found the incorporation of monomeric functionalities on PP film surfaces and the same was dependent on operating conditions. Substantial amendments in surface topography and wettability were witnessed on the surface of modified films. *In vitro* studies evidently revealed that all the PEO-like films possess excellent antifouling properties [115].

Ramkumar et al. tailored LDPE film surfaces by plasma polymerization of PEGMA to enhance their biocompatible properties. The effects of deposition time and applied potential on alteration in surface properties were examined in detail. The authors reported that, subsequent to plasma polymerization of PEGMA, enhancement in hydrophilicity and surface energy was observed. The existence of various functional groups, for instance, C=O, C–O, and O–C=O on the surface of PEGMA films, was confirmed by XPS. In addition, the higher concentration of ester (C–O) functionalities was found on PEGMA films polymerized at higher deposition time and discharge potential. In vitro studies revealed that the coatings displayed both fouling (cell adhesion) and antifouling (decrease in protein adsorption) properties. The authors reported that the dual property was mainly due to the presence of lower concentration of C–O functionalities and incorporation of other functional groups. Figure 6.7 shows the morphology of platelets adhered on the surface of modified LDPE films at different operating conditions [116].

6.3.3.3 Allylamine Coatings

Allylamine is an unsaturated molecule and can be polymerized at low energies. Plasma polymerization of allylamine can be performed to incorporate a higher extent of amine functionalities on the material surfaces for enhancing hydrophilic properties. In addition, it can also be used for immobilization of biomolecules further to improve the biocompatibility of polymers. Aziz et al. aimed to enhance the cell-compatible (cell attachment and proliferation) properties of UHMWPE films by plasma polymerization of allylamine. The deposition was carried out at different operating parameters. The authors reported that plasma polymerization of allylamine changed the surface chemistry of UHMWPE by incorporating various oxygen and nitrogen functionalities, which further led to the enhancement in wettability. The coatings were found to be stable, confirmed by aging tests. Cell compatibility analysis clearly

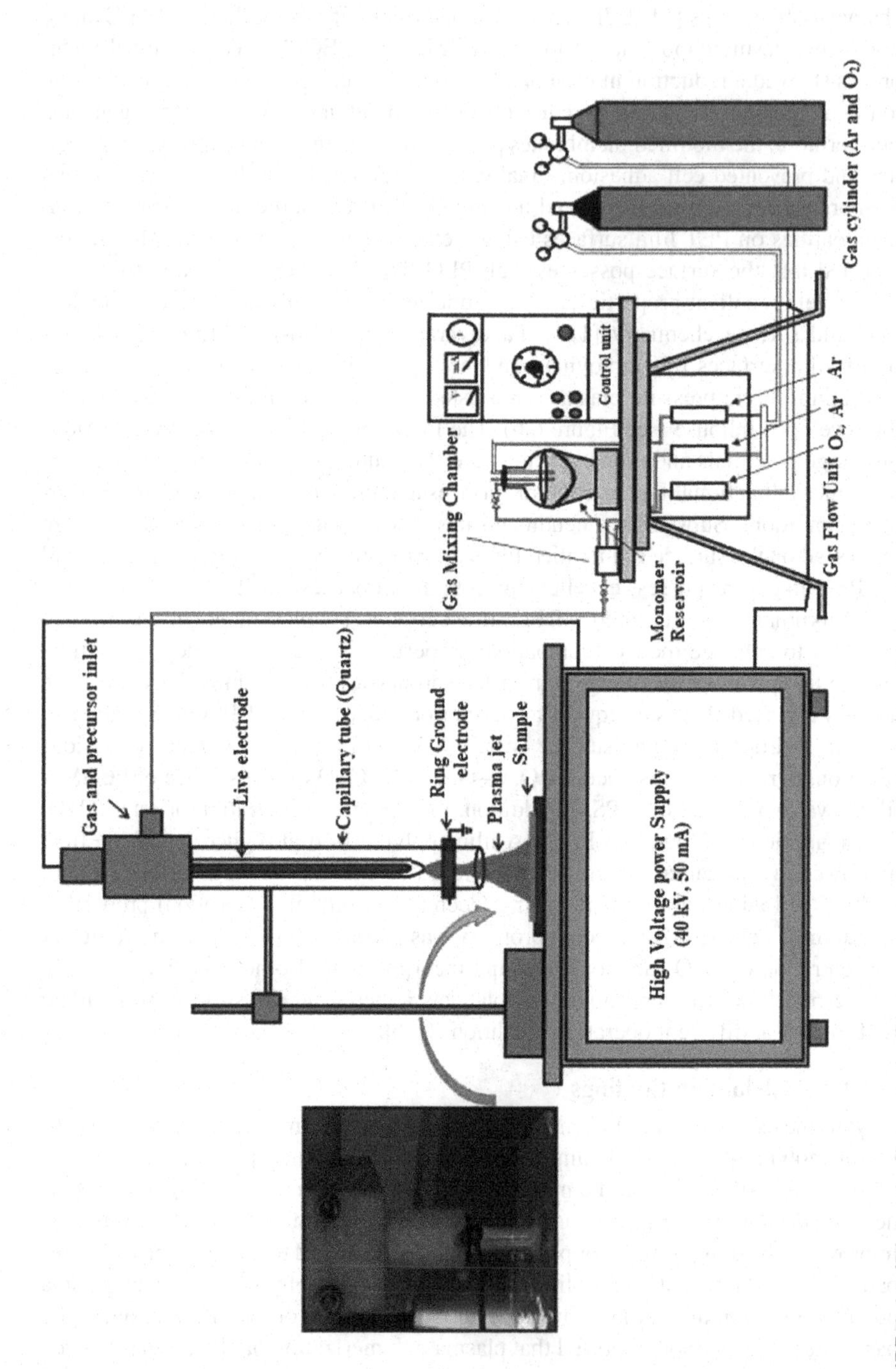

FIGURE 6.6 Schematic diagram of cold atmospheric pressure plasma jet-assisted polymerization reactor.

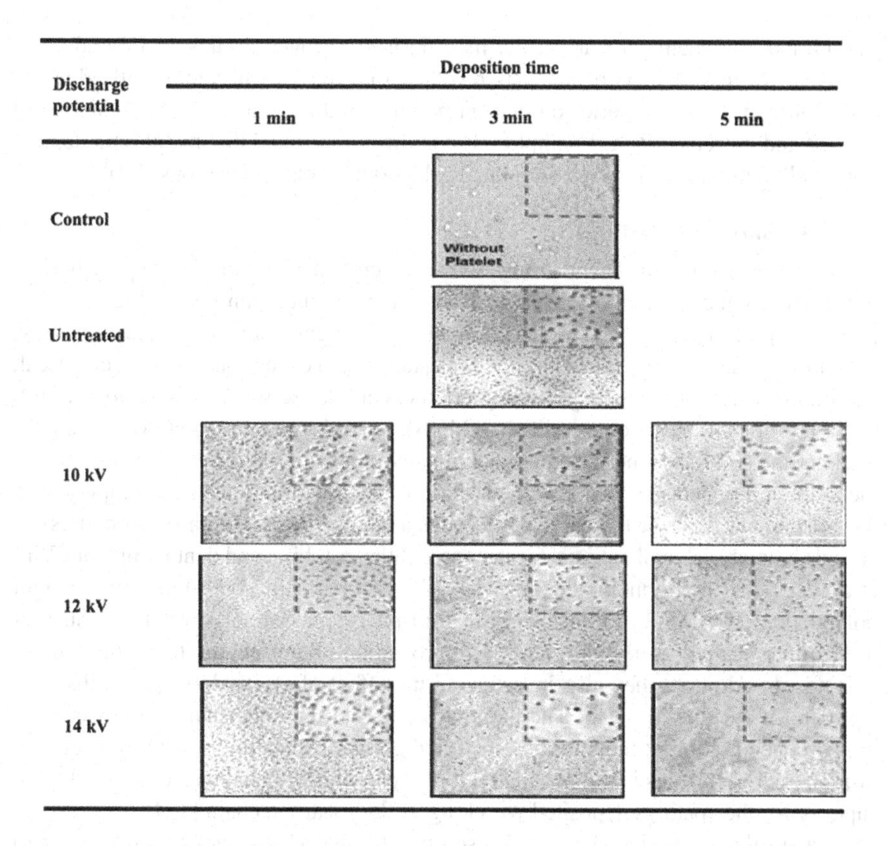

FIGURE 6.7 Morphology of platelets on PEGMA-deposited LDPE films at different operating conditions.

unveiled that the coatings substantially enhanced human foreskin fibroblast cell adhesion and proliferation [117]. Demirci et al. altered poly(vinyl chloride) (PVC) film surfaces using plasma polymerization of allylamine. Results showed a substantial reduction in water contact angle values and enhancement in surface energy and roughness on modified PVC film surfaces. FTIR results showed the presence of N-H peaks at anticipated wavelengths. Finally, they concluded that plasma polymerization of allylamine can be employed to increase the wettability of the PVC films [118].

In another study, Hamerli et al. employed microwave plasma-assisted polymerization of allylamine to enhance the cell adhesion properties of PET membranes and the coatings were deposited at different operating conditions. The nitrogen concentration was enhanced and oxygen was incorporated subsequent to plasma polymerization of allylamine. The contact angle was found to decrease with an increase in the concentration of amine functionalities. The deposited coating was found to be uniform as confirmed by SEM. *In vitro* biocompatibility analysis revealed that the modified PET membranes enhanced cell adhesion and viability of human skin fibroblast [119]. Barry et al. reported the deposition of allylamine on the surface of $P_{DL}LA$ surfaces by plasma grafting and polymerization to enhance its cell adhesion properties. The authors reported that the incorporation of nitrogen components was higher on the surface of $P_{DL}LA$ surfaces deposited by plasma polymerization compared to grafted allylamine.

In addition, the chemical state of the incorporated nitrogen groups was found to be different. Results of in vitro analysis revealed that the 3T3 fibroblast cell adhesion and proliferation was superior on plasma-polymerized allylamine films compared to grafted and untreated $P_{DL}LA$ samples. The authors concluded that the plasma deposition of allylamine is a suitable candidate for biocompatible applications [120].

6.3.3.4 Siloxane Coatings

Siloxane precursors offer a wide range of promising reactants for plasma polymerization reactions and are normally volatile at room temperature, non-flammable, and non-toxic [121]. Plasma polymers produced by means of organo-silicon precursors possess outstanding chemical resistance, thermal resistance, and exceptional optical, electrical, and biomedical properties. In general, monomers employed for the fabrication of SiO_2 or SiOx films are tetramethoxysilane (TMOS), tetraethoxysilane (TEOS), hexamethyldisilazane (HMDS), and hexamethyldisiloxane (HMDSO) [122, 123]. These monomers are components of various silicon-based materials that have been employed in the biomedical field for the past few decades as they possess desirable properties for various biomaterial applications, for instance, catheter tubing and dental implants. Yim et al. deposited hydrophobic coatings on the surface of ultra-high molecular weight polyethylene (UHMWPE) films using an atmospheric pressure plasma jet. The authors used different precursors with various fluorocarbon chain lengths to produce these coatings. In addition, they also investigated the effect of electrode-substrate distance and deposition time in detail. The authors found that the deposited coatings contain higher concentration of monomeric functional groups, indicating lower fragmentation of monomer molecules in the plasma region. Hydrophobic behaviour was found to be superior for the coatings deposited with longest fluorocarbon chain [124].

In a similar study, Dowling et al. used poly(dimethylsiloxane) as a precursor to study the influence of plasma modification and surface roughness on cell attachment and proliferation on the surface of polystyrene (PS) films. The authors obtained superhydrophobic (contact angle 155.8°) coatings by suitably varying the surface roughness and the deposition of the siloxane coatings on the surface of PS films. Cell adhesion outcome unveiled that the polymer films possessed higher surface roughness and showed enhanced MG63 osteosarcoma cell adhesion, whereas cell proliferation was decreased. Coatings that are extremely hydrophilic or hydrophobic resulted in a decreased level of cell adhesion [125]. Teshima et al. fabricated silica films on the surface of PET films using tetramethoxysilane with O_2. They reported that the growth on silica mainly relied on the existence of oxygen species in the plasma region. In addition, the oxygen species eliminates various contaminations substantially, resulting in the formation of dense silica film. Conclusively, the deposited films can be employed for food packing applications and also for semiconductor devices due to their excellent gas-barrier properties [126]. Schneider et al. deposited SiOx coatings on the surface of PET foils and PP using HMDSO and HMDSN via ECR plasma further to improve their barrier properties. Results indicated that the surface-modified films contained a lower amount of OH groups and carbon content, which confirms the enhancement in barrier properties of SiOx coatings [127].

Morent et al. modified the surface properties of polyethylene terephthalate (PET) via plasma polymerization of HMDSO using different carrier gas (argon and various argon/air mixtures). The authors reported that polymer films exhibited $[(CH_3)_2-Si-O]_n$ structure when argon was employed as a working gas. They observed that when increasing the air concentration, a steady transformation of organic coatings to inorganic coatings was observed. The thin films deposited using argon exhibited a hydrophobic property and in the case of air the films were found to be extremely hydrophilic. In addition, AFM results conclusively showed that increasing the air concentration resulted in a decrease in deposition rate and the coatings exhibited rougher morphology [128]. In another study, Morent et al. deposited polydimethylsiloxane (PDMS) on PET films using HMDSO. The effect of discharge power and monomer concentration on amendments in surface properties was studied in detail. The surface chemistry and deposition rate were greatly influenced by input energy per monomer molecule (W/FM). At higher W/FM values, a decrease in deposition rate was observed and the chemical structure was marginally oxidized PDMS [129]. Ward et al. fabricated polysiloxane-like films on the surface of polyethylene using tetramethylcyclotetrasiloxane and octamethylcyclotertasilozane monomers. Increasing the oxygen concentration resulted in formation of hydrophilic SiOx coatings with enhancement in gas barrier properties [130].

Pandiyaraj et al. tailored PP film surfaces via plasma polymerization of hexamethyldisiloxane (HMDSO) using $Ar+O_2$ discharge. The deposition was carried out at different plasma conditions (applied potential, 300, 400, and 500 V) and deposition time (5, 10, and 15 mins). XPS analysis showed the incorporation of higher oxidation state of Si^{4+} of Si and other functionalities such as C=O, C–O, and O–C=O on deposited SiOx film surfaces and the same was found to be higher for the films deposited at a higher deposition time of 15 mins and applied potential of 500 V. AFM analysis showed that the deposited SiOx coatings was found to be uniform and showed smooth topography. *In vitro* analysis showed improvement in cytocompatible properties, that is, the deposited SiOx films are non-toxic, enhancement in NIH-3T3 fibroblast cell adhesion, and proliferation was observed. The authors concluded that the enhancement in cytocompatible properties was mainly due to the incorporation of polar functionalities on modified PP film surfaces [131].

6.3.3.5 Titanium Coatings

Titanium dioxide (TiO_2) is a fascinating material and has gained interest among researchers owing to its exceptional physical and chemical properties [132–134]. It can be employed in an extensive range of technologies such as solar cells, photocatalysis, and water treatment and biomedical. Furthermore, TiO_2 is nontoxic and biocompatible and it can be used as an implant material and it is also used in numerous products, for example, paints, toothpaste, pharmaceuticals, and food additives. Kim et al. aimed to fabricate TiO_2 thin films on PP beads using TTIP. They reported that increasing the rotation speed of the reactor or monomer flow rate resulted in faster growth rate of TiO_2 film. The films deposited on PP beads were found to be homogenous and, finally, they concluded that the TiO_2 coated PP beads can be employed as photocatalysts for the elimination of organic pollutants from water and air [135]. In a similar work by the same group, Kim et al. examined the photocatalytic activity of TiO_2 by suitably depositing a thin layer of TiO_2 on PP beads and investigated the photodegradation of phenol in aqueous solution. The thickness of the TiO_2 thin film on PP beads varied, and was influenced by the operating conditions. The photodegradation of phenol was mainly influenced by an increase in the initial concentration of phenol or the quantity of TiO_2 film–deposited PP beads in aqueous solution. Conclusively, the modified PP beads can be employed for the elimination of pollutants in aqueous solution [136].

Pandiyaraj et al. deposited TiOx-based coatings on a PP film surface using a mixture of titanium tetrachloride and Ar + O_2 plasma at different discharge power. The Ti/C ratio and O/C ratio was increased with an increase in the discharge power, clearly showing the existence of O and Ti components on PP film surfaces. Extending the discharge power to higher values led to increased incorporation of Ti^{4+} components with simultaneous decrease in the concentration of Ti^{3+} components. In addition, incorporation of a higher amount of polar functionalities, such as O–C=O, C=O, and C–O, resulted in an outstanding hydrophilic property. The TiOx coatings showed nano-structured morphology on PP film surfaces as shown in Figure 6.8. In vitro analysis showed the TiOx coatings exhibited excellent cytocompatible properties, subjected to NIH-3T3 mouse embryonic fibroblast cells (see Figure 6.9) and excellent antibacterial activity against *S. aureus* and *Escherichia coli* (see Figure 6.10) [137].

Gancarz et al. achieved plasma polymerization of titanium isopropoxide on porous polypropylene membranes. The authors found that the existence of air in the

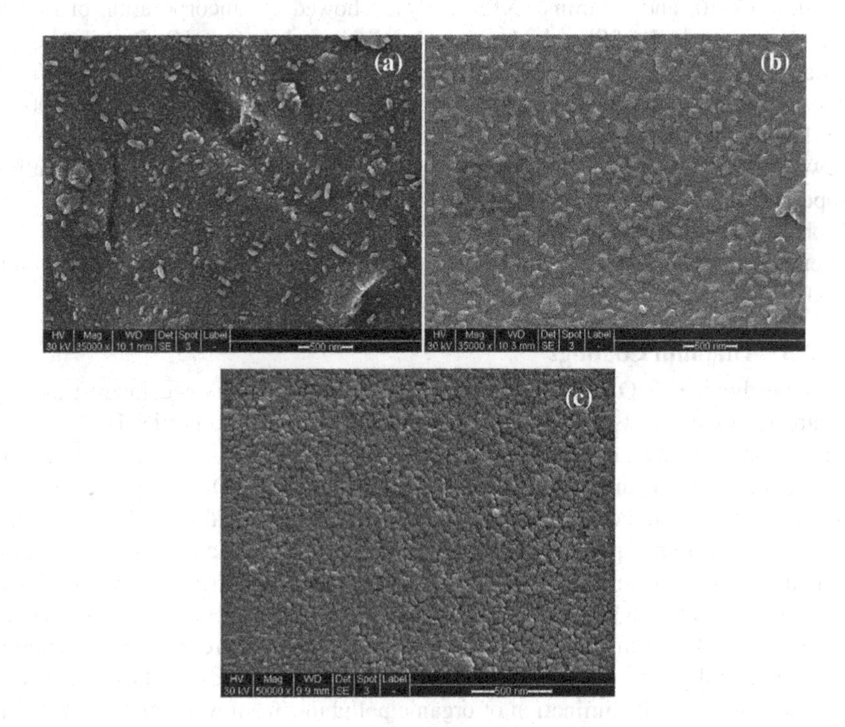

FIGURE 6.8 SEM images of TiOx coatings deposited on PP films at different applied power: (a) 100, (b) 200, and (c) 300W.

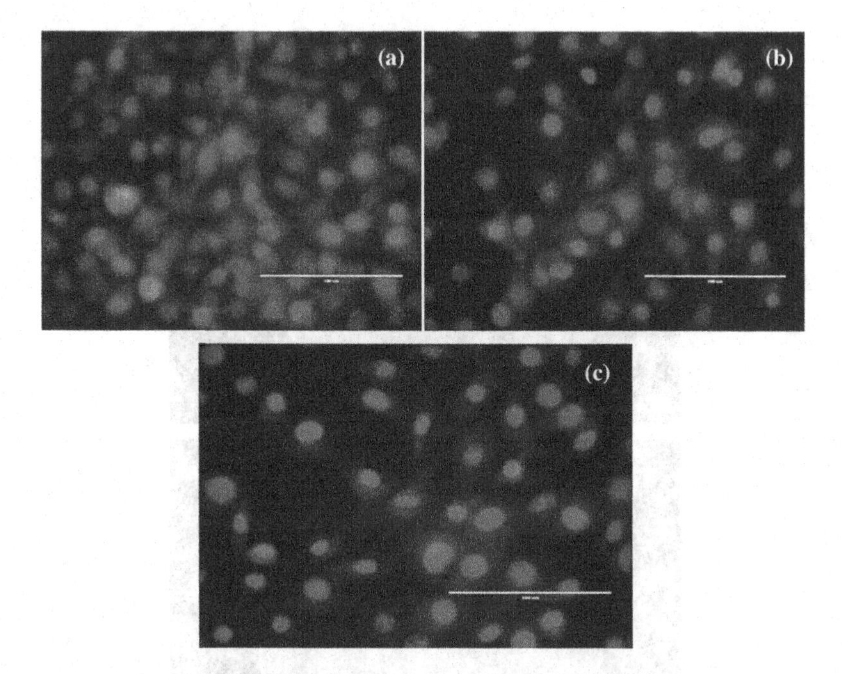

FIGURE 6.9 Fluorescence microscopic image of NIH3T3 cells on TiOx based coatings deposited on PP films surfaces: (a) TCP, (b) 200, and (c) 300 W.

plasma chamber enhanced the deposition of TiOx on polymer. Distance between the electrodes and the process time influenced the thickness of the deposition. The TiOx layer made the PP membrane more hydrophilic and resulted in the enhancement of photocleaning ability and reduction in adsorption of proteins [138].

6.4 CONCLUSION

Polymers have been employed extensively as implant material owing to their exceptional material properties. Conversely, their poor inherent surface properties restrict their use in biomedical applications. This can be overcome by tailoring the polymer surfaces with desirable properties. Surface modification of polymers has acquired great interest among researchers and has witnessed strong development in recent years. Various chemical surface modification methods are being used even now, but currently researchers are striving to prevent the use of harmful organic solvents, which could initiate difficulties with respect to cell viability. Therefore, solvent-free modification techniques, such as non-thermal plasma technology, have become an interesting area of research. Plasma technology is reliable, environmentally friendly, and offers specific control over the process. Plasma-based methods such as plasma treatment and polymerization are generally performed on polymers to enhance their surface properties and are explained in detail in this chapter. Plasma treatment can extensively improve the hydrophilic and biocompatible properties of polymers by incorporation of polar functionalities (i.e. oxygen-rich) with marked topographical changes. However, the polymers treated using plasma may undergo hydrophobic

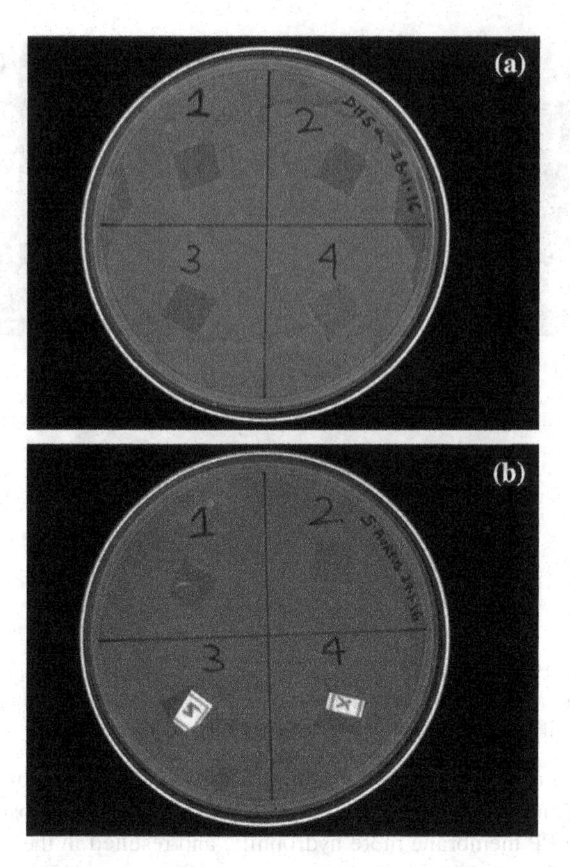

FIGURE 6.10 Photograph of petri plates for assessment of antibacterial effect of TiOx-based coatings on PP films at various discharge power against (a) *E. coli* DH5α and (b) *S. aureus* strains. (1) Uncoated, (2) 100W, (3) 200W, and (4) 300W.

recovery, which can be avoided by suitable choice of the coating material and polymer. Plasma polymerization produces thin films known as plasma polymers or coatings with exceptional physicochemical properties and are highly cross-linked and adherent to various substrates. Generally, acrylic coatings are used to improve the wettability of materials and to promote cell attachment and proliferation on surface of polymers. Poly(ethylene) glycol coatings showed substantial anti-fouling properties (i.e. decrease in adsorption of proteins and adhesion of platelets). Allylamine coating with a significant amount of amine functional groups incorporated, results in enhanced hydrophilicity and cytocompatibility. Siloxane coatings possess both hydrophilic and hydrophobic properties and exhibit substantial gas barrier properties. In addition, titanium oxide–based coatings are found to have antimicrobial activity against *E. coli* and *S. aureus*. From the literature review, it is obvious that when polymers are subjected to plasma-assisted surface modification, they acquire exceptional surface chemistry, roughness, hydrophilicity, and substantial cellular interaction; this can be employed for tissue engineering, cardiovascular, and orthopaedic applications.

ABBREVIATIONS

AFM: Atomic force microscopy
Ar: Argon
ATR: Attenuated total reflectance
BSA: Bovine Serum Albumin
CO_2: Carbon dioxide
DBD: Dielectric Barrier Discharge
ECR: erythrocyte sedimentation rate
FTIR: Fourier transform infrared
GMA: Glycidyl methacrylate
HDPE: High density polyethylene
HMDSO: Hexamethyldisiloxane
HUVEC: Human umbilical vein endothelial cell
LDPE: Low density polyethylene
LTE: Local thermodynamic equilibrium
N_2: Nitrogen
NH_3: Ammonia
O_2: Oxygen
O_3: Ozone
PAN: Poly aniline
PCL: Poly caprolactone
PCU: Poly carbonate urethane
PDMS: Poly dimethyl siloxane
PDMS: Poly (dimethylsiloxane)
PEGMA: Poly(ethylene glycol) methyl ether methacrylate
PEN: Poly (ethylene naphthalate)
PEO: Poly ethylene oxide
PES: Poly ether sulfonate
PET: Poly ethylene terephthalate
PHO: Poly(3-hydroxyoctanoate)
PLA: Poly lactic acid
PLGA: Poly(D,L-lactic acid-coglycolic aid)
PLLA: Poly(L-lactide)
PMMA: Poly methyl methacrylate
PPY: Polypyrrole
PP: Polypropylene
PS: Polystyrene
PSU: Polysulfone
PTFE: Poly-tetrafluoroethylene
PU: Polyurethane,
PVC: Poly(vinyl chloride)
RF: Radio-frequency
SEM: Scanning electron microscopy
TiO_2: Titanium dioxide
TEOS: Tetraethoxysilane

TMOS: Tetramethoxysilane
TTIP: Titanium Tetraisopropoxide
UHMWPE: Ultrahigh molecular weight polyethylene
UV: Ultraviolet
VOC: Volatile organic compounds
XPS: X-ray photoelectron spectroscopy
XRD: X-ray diffraction

REFERENCES

1. G. Lloyd, G. Friedman, G.S. Jafri, G. Schultz, A. Fridman, K. Harding, "Gas plasma: medical uses and developments in wound care", *Plasma Processes and Polymers*, 7 (2010) 194–211.
2. J. Lopez-Garcia, F. Bilek, J. Lehocky, I. Junkar, I. Mozeti, M. Sowe, "Enhanced printability of polyethylene through air plasma treatment", *Vacuum*, 95 (2013) 43–49.
3. S. Guruvenket, G.M. Rao, M. Komath, A.M. Raichur, "Plasma surface modification of polystyrene and polyethylene", *Applied Surface Science*, 236 (2004) 278–284.
4. K. Ishihara, Y. Iwasaki, S. Ebihara, Y. Shindo, N. Nakabayashi, "Photoinduced graft polymerization of 2-methacryloyloxyethyl phosphorylcholine on polyethylene membrane surface for obtaining blood cell adhesion resistance", *Colloids and Surfaces B: Biointerfaces*, 18 (2000) 325–335.
5. J.H. Liu, H.L. Jen, Y.C. Chung, "Surface modification of polyethylene membranes using phosphorylcholine derivatives and their platelet compatibility", *Journal of Applied Polymer Science*, 74 (1999) 2947–2954.
6. M. Ataeefard, S. Moradian, M. Mirabedini, M. Ebrahimi, S. Asiaban, "Investigating the effect of power/time in the wettability of Ar and O_2 gas plasma-treated low-density polyethylene", *Progress in Organic Coatings*, 64 (2009) 482–488.
7. F. Bilek, T. Krizova, M. Lehocky, "Preparation of active antibacterial LDPE surface through multistep physicochemical approach: I. Allylamine grafting, attachment of antibacterial agent and antibacterial activity assessment", *Colloids and Surfaces B: Biointerfaces*, 88 (2011) 440–447.
8. M. Pantoja, N. Encinas, J. Abenojar, M.A. Martinez, "Effect of tetraethoxysilane coating on the improvement of plasma treated polypropylene adhesion", *Applied Surface Science*, 280 (2013) 850–857.
9. R.A. Hoshi, R.V. Lith, M.C. Jen, J.B. Allen, K.A. Lapidos, G. Ameer, "The blood and vascular cell compatibility of heparin-modified ePTFE vascular grafts", *Biomaterials* 34 (2013) 30–41.
10. Y. Chen, Q. Gao, H. Wan, J. Yi, Y. Wei, P. Liu, "Surface modification and biocompatible improvement of polystyrene film by Ar, O_2 and Ar + O_2 plasma", *Applied Surface Science*, 265 (2013) 452–457.
11. S. Rimpelová, N.S. Kasálková, P. Slepička, H. Lemerová, V. Švorčík and T. Ruml, "Plasma treated polyethylene grafted with adhesive molecules for enhanced adhesion and growth of fibroblast", *Materials Science Engineering C*, 33 (2013) 1116–1124.
12. T. Jacobs, R. Morent, N. De Geyter, P. Dubruel, C. Leys, "Plasma surface modification of biomedical polymers: influence on cell-material interaction", *Plasma Chemistry Plasma Polymers*, 32 (2012) 1039–1073.
13. A.M.G. Borges, L.O. Benetoli, M.A. Licinio, V.C. Zoldan, M.C. Santos-Silva, J. Assreuy, A.A. Pasa, N.A. Debacher, V. Soldi, "Polymer films with surfaces unmodified and modified by non-thermal plasma as new substrates for cell adhesion", *Material Science and Engineering C*, 33 (3) (2013) 1315–1324.

14. K. Bazaka, M.V. Jacob, R.J. Crawford, E.P. Ivanova, "Plasma-assisted surface modification of organic biopolymers to prevent bacterial attachment", *Acta Biomaterialia*, 7 (5) (2011) 2015–2028.
15. K. Novotna, M. Bacakova, N.S. Kasalkova, P. Slepicka, V. Lisa, V. Svorcik, L. Bacakova, "Adhesion and growth of vascular smooth muscle cells on nanostructured and biofunctionalized polyethylene", *Materials* 6 (2013) 1632–1655.
16. P. Slepicka, N.S. Kasalkova, J. Siegel, Z. Kolska, L. Bacakova, V. Svorcik, "Nanostructured and functionalized surfaces for cytocompatibility improvement and bactericidal action", *Biotechnology Advances*, 33 (2015) 1120–1129.
17. P.K. Chu, J.Y. Chen, L.P. Wang, N. Huang, "Plasma-surface modification of biomaterials", *Material Science and Engineering R*, 36 (2002) 143–206.
18. H. Kaczmarek, J. Kowalonek, A. Szalla, A. Sionkowska, "Surface modification of thin polymeric films by air-plasma or UV-irradiation", *Surface Science*, 507–510 (2002) 883–888.
19. K. Walachova, V. Svorcík, L. Bacakova, V. Hnatowicz, "Colonization of ion-modified polyethylene with vascular smooth muscle cells in vitro", *Biomaterials*, 23 (14) (2002) 2989–2996.
20. R. Oosterom, T.J. Ahmed, J.A. Poulis, H.E.N. Bersee, "Adhesion performance of UHMWPE after different surface modification techniques", *Medical Engineering and Physics*, 28 (2006) 323–330.
21. H.N. Lim, H. Choi, T.M. Hwang, J.W. Kang, "Characterization of ozone decomposition in a soil slurry: kinetics and mechanism", *Water Research*, 36 (2002) 219–229.
22. S. Migita, K. Sakai, Z. Mori, H. Ota, R. Aoki, "Evaluation of ozone condensation system by thermal decomposition method", *Japanese Journal of Applied Physics*, 36 (1997) 94–97.
23. E.T. Kang, K.G. Neob, X. Zbang, K.L. Tan, D.J. Liaw, "Surface modification of electroactive polymer, films by ozone treatment", *Surface And Interface Analysis*, 24 (1996) 51–58.
24. K. Fujimoto, Y. Takebayashi, H. Inoue, Y. Ikada, "Ozone-induced graft polymerization onto polymer surface", *Journal of Polymer Science Part A: Polymer Chemistry*, 31 (1993) 1035–1043.
25. J.S. Kong, D.J. Lee, H.D. Kim, "Surface modification of low density polyethylene (LDPE) film and improvement of adhesion between evaporated copper metal film and LDPE", *Journal of Applied Polymer Science*, 82 (2001) 1677–1690.
26. D.S. Bag, V.P. Kumar, S. Maiti, "Chemical modification of LDPE film", *Journal of Applied Polymer Science*, 71 (1999) 1041–1048.
27. G.L. Tao, A.J. Gong, J.J. Lu, H.J. Sue, D.E. Bergbreiter, "Surface functionalized polypropylene: synthesis, characterization, and adhesion properties", *Macromolecules*, 34 (2001) 7672–7679.
28. J.M. Goddard, J.N. Talbert, J.H. Hotchkiss, "Covalent attachment of lactase to low density polyethylene films", *Journal of Food Science*, 72 (2007) E36–E41.
29. X.H. Liu, P.X. Ma, "Polymeric scaffolds for bone tissue engineering", *Annals of Biomedical Engineering*, 32 (2004) 477–486.
30. M. Herrero, R. Navarro, H. Reinecke, C. Mijangos, Y. Grohens, "Controlled wet-chemical modification and bacterial adhesion on PVC-surfaces", *Polymer Degradation and Stability*, 91 (2006) 1915–1918.
31. M. Gabriel, M. Dahm and C-F Vahl, "Wet-chemical approach for the cell-adhesive modification of polytetrafluoroethylene", *Biomedical Materials*, 6 (2011) 035007 (5pp).
32. D.E. Weibel, A.F. Michels, F. Horowitz, R.D. Cavalheiro, G.V. Mota, "Ultraviolet-induced surface modification of polyurethane films in the presence of oxygen or acrylic acid vapours", *Thin Solid Films*, 517 (2009) 5489–5495.

33. C. Decker, K. Zahouily, "Light-stabilization of polymeric materials by grafted UV-cured coatings", *Journal of Polymer Science Part A: Polymer Chemistry*, 36 (1998) 2571–2580.
34. B. Pan, K. Viswanathan, C.E. Hoyle, R.B. Moore, "Photoinitiated grafting of maleic anhydride onto polypropylene", *Journal of Polymer Science Part A: Polymer Chemistry*, 42 (2004) 1953–1962.
35. T. Kondo, M. Koyama, H. Kubota, R. Katakai, "Characteristics of acrylic acid and *N*-isopropyl acrylamide binary monomers-grafted polyethylene film synthesized by photo grafting", *Journal of Applied Polymer Science*, 67 (1998) 2057–2064.
36. D. Ruckert, G. Geuskens, "Surface modification of polymers-IV, Grafting of acrylamide via an unexpected mechanism using a water soluble photo-initiator", *European Polymer Journal*, 32 (1996) 201–208.
37. T-S. Hwang, J-W. Park, "UV-Induced graft polymerization of polypropylene-*g*-glycidyl methacrylate membrane in the vapor phase", *Macromolecular Research*, 11 (6) (2003) 495–500.
38. H. Salmi-Mani, G. Terreros, N. Barroca-Aubry, C. Aymes-Chodur, C. Regeard, P. Roger, "Poly (ethylene terephthalate) films modified by UV-induced surface graft polymerization of vanillin derived monomer for antibacterial activity", *European Polymer Journal* (2018), doi:10.1016/j.eurpolymj.2018.03.038.
39. H.W. Kim, C.W. Chung, Y.H. Rhee, "UV-induced graft copolymerization of monoacrylate-poly(ethylene glycol) onto poly(3-hydroxyoctanoate) to reduce protein adsorption and platelet adhesion", *International Journal of Biological Macromolecules* 35 (2005) 47–53.
40. Y-W. Song, H-S. Do, H-S. Joo, D-H. Lim, S. Kim, H-J. Kim, "Effect of grafting of acrylic acid onto PET film surfaces by UV irradiation on the adhesion of PSAs", *Journal of Adhesion Science and Technology*, 20 (12) (2006) 1357–1365.
41. Y. Feng, H. Zhao, M. Behl, A. Lendlein, J. Guo, D. Yang, "Grafting of poly(ethylene glycol) monoacrylates on polycarbonateurethane by UV initiated polymerization for improving hemocompatibility", *Journal of Material Science: Materials in Medicine*, 24 (2013) 61–70.
42. R. Ramanathan, K. Felipe, H. Pedro, de Leal Mour, "Surface modification of synthetic polymers using UV photochemistry in the presence of reactive vapours", *Macromolecular Symposium*, 299–300 (2011) 175–182.
43. M.F. Zaki, "Gamma-induced modification on optical band gap of CR-39 SSNTD", *Journal of Physics D: Applied Physics*, 41 (2008) 175404.
44. K. Gorna, S. Gogolewski, "The effect of gamma radiation on molecular stability and mechanical properties of biodegradable polyurethanes for medical applications", *Polymer Degradation and Stability*, 79 (2003) 465–474.
45. S. Siddhartha, K. Aarya, K. Dev, S.K. Raghuvanshi, J.B.M. Krishna, M.A. Wahab, "Effect of gamma radiation on the structural and optical properties of polyethyleneterephthalate (PET) polymer", *Radiation Physics and Chemistry*, 81 (2012) 458–462.
46. D. Sinha, K.L. Sahoo, U.B. Sinha, T. Swu, A. Chemseddine, D. Fink, "Gamma-induced modifications of polycarbonate polymer", *Radiation Effects and Defects in Solids*, 159 (2004) 587–595.
47. S. Jelvani, H. Pazokian, S. Moradi Farisar, "Effect of CO_2 pulsed laser irradiation on improving the biocompatibility of a polyethersulfone film", *Journal of Physics: Conference Series*, 414 (2013) 012021.
48. M. Dadsetan, H. Mirzadeh, N. Sharifi-Sanjani, P. Salehian, "*In vitro* studies of platelet adhesion on laser-treated polyethylene terephthalate surface", *Journal of Biomedical Material Research*, 54 (4) (2001) 540–546.

49. K.S. Tiaw, S.W. Goh, M. Hong, Z. Wang, B. Lan, S.H. Teoh, "Laser surface modification of poly(e-caprolactone) (PCL) membrane for tissue engineering applications", *Biomaterials* 26 (2005) 763–769.
50. M.T. Khorasani, H. Mirzadeh, P.G. Sammes, "Laser induced surface modification of polydimethylsiloxane as a super-hydrophobic material", *Radiation Physics and Chemistry*, 47 (1996) 881–888.
51. Z.K. Wang, H.Y. Zheng, C.P. Lim, Y.C. Lam, "Polymer surface wettability modification using femtosecond laser Irradiation", *SIMTech Technical Reports*, 11 (2010).
52. P. Rossini, P. Colpo, G. Ceccone, K.D. Jandt, F. Rossi, "Surfaces engineering of polymeric films for biomedical applications", *Materials Science and Engineering C*, 23 (3) (2003) 353–358.
53. P. Heyse, R. Dams, S. Paulussen, K. Houthoofd, K. Janssen, P.A. Jacobs, B.F. Sels, "Dielectric barrier discharge at atmospheric pressure as a tool to deposit versatile organic coatings at moderate power input", *Plasma Processes and Polymers*, 4 (2), (2007) 145–157.
54. S.E. Alexandrov, M.L. Hitchman, "Chemical vapor deposition enhanced by atmospheric pressure non-thermal non-equilibrium plasmas", *Chemical Vapor Deposition*, 11 (2005) 457–468.
55. M.H. Ho, L.T. Hou, C.Y. Tu, H.J. Hsieh, J.Y. Lai, W.J. Chen, D.M. Wang, "Promotion of cell affinity of porous PLLA scaffolds by immobilization of RGD peptides via plasma treatment", *Macromolecular Bioscience*, 6 (2006) 90–98.
56. H. Shen, X.X. Hu, F. Yang, J.Z. Bel, S.G. Wang, "Combining oxygen plasma treatment with anchorage of cationized gelatin for enhancing cell affinity of poly (lactide-co-glycolide)", *Biomaterials*, 28 (2007) 4219–4230.
57. J.J.A. Barry, M.M.C.G. Silva, K.M. Shakesheff, S.M. Howdle, M.R. Alexander, "Using plasma deposits to promote cell population of the porous interior of three-dimensional poly(D,L-Lactic Acid) tissue-engineering scaffolds", *Advanced Functional Materials*, 15 (2005) 1134–1140.
58. J.J.A. Barry, D. Howard, K.M. Shakesheff, S.M. Howdle, M.R. Alexander, "Using a core–sheath distribution of surface chemistry through 3D tissue engineering scaffolds to control cell ingress", *Advance Materials*, 18 (2006) 1406–1410.
59. N. Guerrouani, A. Baldo, A. Bouffin, C. Drakides, M.F. Guimon, A. Mas, "Allylamine plasma-polymerization on PLLA surface evaluation of the biodegradation", *Journal of Applied Polymer Science*, 105 (2007) 1978–1986.
60. M. Zelzer, D. Scurr, B. Abdullah, A.J. Urquhart, N. Gadegaard, J.W. Bradley, M.R. Alexander, "Influence of the plasma sheath on plasma polymer deposition in advance of a mask and down pores", *Journal of Physical Chemistry B*, 113 (2009) 8487–8494.
61. A. Fridman, (2008) "Plasma Chemistry", Cambridge University Press, New York.
62. P.C. Kong, E. Pfender, Plasma processes, in: A.W. Weimer (Ed.), "Carbide, nitride and boride materials—synthesis and processing, first ed.," Chapman & Hall, London, 1997, 359–383.
63. R. Matthes, I. Koban, C. Bender, K. Masur, E. Kindel, K.D. Weltmann, T. Kocher, A. Kramer, N.O. Hubner, "Antimicrobial efficacy of an atmospheric pressure plasma jet against biofilms of Pseudomonas aeruginosa and Staphylococcus epidermidis", *Plasma Processes and Polymers*, 10 (2013) 161–166.
64. J. Ehlbeck, U. Schnabel, M. Polak, J. Winter, T. Von Woedtke, R. Brandenburg, T. Von Dem Hagen, K.D. Weltmann, "Low temperature atmospheric pressure plasma sources for microbial decontamination", *Journal of Physics D: Applied Physics*, 44 (2011) 013002.
65. E. Gomez, D. Amutha Rani, C.R. Cheeseman, D. Deegan, M. Wise, A.R. Boccaccini, "Thermal plasma technology for the treatment of wastes: a critical review", *Journal of Hazardous Materials*, 161 (2009) 614–626.

66. R.R. Deshmukh, A.R. Shetty, "Surface characterization of polyethylene films modified by gaseous plasma", *Journal of Applied Polymer Science*, 104 (2007) 449–457.
67. G.A. Arolkar, S.M. Jacob, K.N. Pandiyaraj, V.R. Kelkar-Mane, R.R. Deshmukh, "Effect of TEOS plasma polymerization on corn starch/poly(ε-caprolactone) film: characterization, properties and biodegradation", *RSC Advances*, 620 (2016) 16779–16789.
68. R. Morent, N. De Geyter, L. Gengembre, C. Leys, E. Payen, S. Van Vlierberghe, E. Schacht, "Surface treatment of a polypropylene film with a nitrogen DBD at medium pressure", *European Physical Journal of Applied Physics*, 43 (2008) 289–294.
69. R. Morent, N. De Geyter, C. Leys, "Effects of operating parameters on plasma-induced PET surface treatment", *Nuclear Instruments and Methods in Physics Research Section B: Beam Interactions with Materials and Atoms*, 266 (2008) 3081–3085.
70. T. Desmet, R. Morent, N. De Geyter, C. Leys, E. Schacht, P. Dubruel, "Nonthermal plasma technology as a versatile strategy for polymeric biomaterials surface modification: a review", *Biomacromolecules*, 10 (2009), 2351–2378.
71. Z. Ding, J.N. Chen, S.Y. Gao, J.B. Chang, J.F. Zhang, E.T. Kang, "Immobilization of chitosan onto poly-L-lactic acid film surface by plasma graft polymerization to control the morphology of fibroblast and liver cells", *Biomaterials*, 25 (2004) 1059–1067.
72. R. Dorai, M.J. Kushner, "A model for plasma modification of polypropylene using atmospheric pressure discharges", *Journal of Physics D: Applied Physics*, 36 (2003) 666.
73. Y. Akishev, M. Grushin, N. Dyatko, I. Kochetov, A. Napartovich, N. Trushkin, T.M. Duc, S. Descours, "Studies on cold plasma polymer surface interaction by example of PP- and PET-films", *Journal of Physics D: Applied Physics*, 41 (2008) 235203.
74. M. Ataeefard, S. Moradian, M. Mirabedini, M. Ebrahimi, S. Asiaban, "Investigating the effect of power/time in the wettability of Ar and O2 gas plasma-treated low-density polyethylene", *Progress in Organic Coatings*, 64 (2009) 482–448.
75. E. Amanatides, D. Mataras, M. Katsikogianni, Y.F. Missirlis, "Plasma surface treatment of polyethylene terephthalate films for bacterial repellence", *Surface and Coatings Technology*, 200 (2006) 6331–6335.
76. G. Aziz, P. Cools, N. De Geyter, H. Declercq, R. Cornelissen, R. Morent, "Dielectric barrier discharge plasma treatment of ultrahigh molecular weight polyethylene in different discharge atmospheres at medium pressure: a cell-biomaterial interface study", *Biointerphases*, 10 (2) (2015) 029502.
77. P. Esena, C. Riccardi, S. Zanini, M. Tontini, G. Poletti, F. Orsini, "Surface modification of PET film by a DBD device at atmospheric pressure", *Surface and Coatings Technology*, 200 (2005) 664–667.
78. E. Gonzalez, M.D. Barankin, P.C. Guschl, R.F. Hicks, "Remote atmospheric-pressure plasma activation of the surfaces of polyethylene terephthalate and polyethylene naphthalate", *Langmuir*, 24 (21) (2008) 12636–12643.
79. P.A. Ramires, L. Mirenghi, A.R. Romano, F. Palumbo, G. Nicolardi, "Plasma-treated PET surfaces improve the biocompatibility of human endothelial cells", *Journal of Biomedical Materials Research*, 51 (3) (2000) 535–539.
80. M. Vishnuvarthanan, N. Rajeswari, "Effect of mechanical, barrier and adhesion properties on oxygen plasma surface modified PP", *Innovative Food Science and Emerging Technologies*, 30 (2015) 119–126.
81. A. Reznickova, Z. Novotna, Z. Kolska, N.S. Kasalkova, S. Rimpelova, V. Svorcik, "Enhanced adherence of mouse fibroblast and vascular cells to plasma modified polyethylene", *Materials Science and Engineering C*, 52 (2015) 259–266.
82. M. Nakagawa, F. Teraoka, S. Fujimoto, Y. Hamada, H. Kibayashi, J. Takahashi, "Improvement of cell adhesion on poly (L-lactide) by atmospheric plasma treatment", *Journal of Biomedical Material Research*, 77 (2006) 112–118.

83. M.T. Khorasani, H. Mirzadeh, S. Irani, "Plasma surface modification of poly (L-lactic acid) and poly (lactic-co-glycolic acid) films for improvement of nerve cells adhesion", *Radiation Physics and Chemistry*, 77 (2008) 280–287.
84. I. Junkar, C. Uros, L. Mariann, "Plasma treatment of biomedical materials", *Materials and Technology*, 45 (2011) 221–228.
85. P. Slepilka, N.S. Kasalkova, E. Stranska, L. Balakova, V. Vorlik, "Surface characterization of plasma treated polymers for applications as biocompatible carriers", *Express Polymers Letters*, 7 (2013) 535–545.
86. M.T. Khorasani, H. Mirzadeh, S. Irani, "Comparison of fibroblast and nerve cells response on plasma treated poly (L-lactide) surface", *Journal of Applied Polymer Science*, 112 (2009) 3429–3435.
87. A. Kuzminova, M. Vandrovcova, A. Shelemin, O. Kylian, A. Choukourov, J. Hanus, L. Bacakova, D. Slavinska, H. Biederman, "Treatment of poly(ethylene terephthalate) foils by atmospheric pressure air dielectric barrier discharge and its influence on cell growth", *Applied Surface Science*, 357 (2015) 689–695.
88. M. Jaganjac, A. Vesel, L. Milkovic, N. Recek, M. Kolar, N. Zarkovic, A. Latiff, K-S. Kleinschek, M. Mozetic, "Oxygen-rich coating promotes binding of proteins and endothelialization of polyethylene terephthalate polymers", *Journal of Biomedical Research Part A*, 102 (2014) 2305–2314.
89. N. De Geyter, R. Morent, T. Desmet, M. Trentesaux, L. Gengembre, P. Dubruel, C. Leys, E. Payen, "Plasma modification of polylactic acid in a medium pressure DBD", *Surface & Coatings Technology* 204 (2010) 3272–3279.
90. I. Junkar, A. Vesel, U. Cvelbar, M. Mozetic, S. Strnad, "Influence of oxygen and nitrogen plasma treatment on polyethylene terephthalate (PET) polymers", *Vacuum* 84 (2010) 83–85.
91. N. Gomathi, A. Sureshkumar, S. Neogi, "RF plasma-treated polymers for biomedical applications", *Current Science*, 94 (2008) 1478–1486.
92. R. Morent, N. De Geyter, S. Van Vlierberghe, P. Dubruel, C. Leys, L. Gengembre, E. Schacht, E. Payen, "Deposition of HMDSO-based coatings on PET substrates using an atmospheric pressure dielectric barrier discharge", *Progress in Organic Coatings*, 64 (2009) 304–310.
93. T. Desmet, R. Morent, N. De Geyter, C. Leys, E. Schacht, P. Dubruel, "Nonthermal plasma technology as a versatile strategy for polymeric biomaterials surface modification: a review", *Biomacromolecules*,10 (2009) 2351–2378.
94. J. Garcia-Torres, D. Sylla, L. Molina, E. Crespo, J. Mota, L. Bautista, "Surface modification of cellulosic substrates via atmospheric pressure plasma polymerization of acrylic acid: structure and properties", *Applied Surface Science*, 305 (2014) 292–300.
95. J.M. Bashir, W.B. Rees, Zimmerman, "Plasma polymerization in a micro capillary using an atmospheric pressure dielectric barrier discharge", *Surface and Coatings Technology*, 234 (2013) 82–91.
96. B. Gupta, C. Plummer, I. Bisson, P. Frey, J. Hilborn, "Plasma-induced graft polymerization of acrylic acid onto poly(ethylene terephthalate) films: characterization and human smooth muscle cell growth on grafted films", *Biomaterials*, 23 (2002) 863–871.
97. L. Detomaso, R. Gristina, G.S. Senesi, R. d'Agostino, P. Favia, "Stable plasma-deposited acrylic acid surfaces for cell culture applications", *Biomaterials*, 26 (2005) 3831–3841.
98. B.R. Pistillo, L. Detomaso, E. Sardella, P. Favia, R. d'Agostino, "RF-plasma deposition and surface characterization of stable (COOH)-rich thin films from cyclic L-lactide", *Plasma Processes and Polymers*, 4 (2007) S817–S820
99. K.S. Siow, L. Britcher, S. Kumar, H.J. Griesser, "Plasma methods for the generation ofchemically reactive surfaces for biomolecule immobilization and cell colonization: a review", *Plasma Processes and Polymers*, 3 (2006) 392–418.

100. F. Basarir, E.Y. Choi, S.H. Moon, K.C. Song, T.H. Yoon, "Electrochemical properties of PP membranes with plasma polymer coatings of acrylic acid", *Journal of Membrane Science*, 260 (2005) 66–74.
101. R. Jafari, M. Tatoulian, W. Morscheidt, F. Arefi-Khonsari, "Stable plasma polymerized acrylic acid coating deposited on polyethylene (PE) films in a low frequency discharge (70 kHz)", *Reactive and Functional Polymers*, 66 (2006) 1757–1765.
102. R. Morent, N. De Geyter, M. Trentesaux, L. Gengembre, P. Dubruel, C. Leys, E. Payen, "Stability study of polyacrylic acid films plasma-polymerized on polypropylene substrates at medium pressure", *Applied Surface Science*, 257 (2010) 372–380.
103. P. Cools, H. Declercq, N. De Geyter, R. Morent, "A stability study of plasma polymerized acrylic acid films", *Applied Surface Science*, 432 (2018) 214–223.
104. M.S. Kang, B. Chun, S.S. Kim, "Surface modification of polypropylene membrane by Low-temperature plasma treatment, *Journal of Applied Polymer Science*, 81 (6) (2001) 1555–1566.
105. I. Topala, N. Dumitrascu, G. Popa, "Properties of the acrylic acid polymers obtained by atmospheric pressure plasma polymerization", *Nuclear Instruments and Methods in Physics Research, Section B: Beam Interactions with Materials and Atoms*, 267 (2009) 442–445.
106. M.C. Ramkumar, K. NavaneethaPandiyaraj, A. ArunKumar, P.V.A. Padmanabhan, S. Uday Kumar, P. Gopinath, A. Bendavid, P. Cools, N. De Geyter, R. Morent, R.R. Deshmukh, "Evaluation of mechanism of cold atmospheric pressure plasma assisted polymerization of acrylic acid on low density polyethylene (LDPE) film surfaces: influence of various gaseous plasma pretreatment", Applied Surface Science, 439 (2018) 991–998.
107. M. Lindblad, M. Lestelius, A. Johansson, P. Tengvall, P. Tomsen, "Cell and soft tissue interactions with methyl and hydroxyl terminated alkane thiols on gold surfaces", *Biomaterials*, 18 (15) (1997) 1059–1068.
108. X.P. Zou, E.T. Kang, K.G. Neoh, "Plasma-induced graft polymerization of poly(ethylene glycol) methyl ether methacrylate on poly(tetrafluoroethylene) films for reduction in protein adsorption", *Surface and Coatings Technology*, 149 (2002) 119–128.
109. C. Choi, I. Hwang, Y.L. Cho, S.Y. Han, D.H. Jo, D. Jung, D.W. Moon, E.J. Kim, C.S. Jeon, J.H. Kim, T.D. Chung, T.G. Lee, "Fabrication and characterization of plasma-polymerized poly(ethylene glycol) film with superior biocompatibility", *ACS Applied Materials and Interfaces*, 5 (2013) 697–702.
110. B. Nisol, G. Oldenhove, N. Preyat, D. Monteyne, M. Moser, D. Perez-Morga, F. Reniers, "Atmospheric plasma synthesized PEG coatings: non-fouling biomaterials showing protein and cell repulsion", *Surface and Coatings Technology*, 252 (2014) 126–133.
111. R.A. D'Sa, B.J. Meenan, "Chemical grafting of poly(ethylene glycol) methyl ether methacrylate onto polymer surfaces by atmospheric pressure plasma processing", *Langmuir*, 26 (2013) 1894–1903.
112. D. Sakthi Kumar, M. Fujioka, K. Asano, A. Shoji, A. Jayakrishnan, Y. Yoshida, "Surface modification of poly(ethylene terephthalate) by plasma polymerization of poly(ethyleneglycol)", *Journal of Material Science: Materials in Medicine*, 18 (2007) 1831–1835.
113. J.F. Jhong, A. Venault, C.C. Hou, S.H. Chen, T.C. Wei, J. Zheng, J. Huang, Y. Chang, "Surface zwitterionization of expanded poly(tetrafluoroethylene) membranes via atmospheric plasma-induced polymerization for enhanced skin wound healing", *ACS Applied Materials and Interfaces*, 5 (2013) 6732–6742.
114. K.E. Bremmell, P. Kingshott, Z. Ademovic, B. Winther-Jensen, H.J. Griesser, "Colloid probe AFM investigation of interactions between fibrinogen and PEG-like plasma polymer surfaces", *Langmuir*, 22 (1) (2006) 313–318.

115. K.N. Pandiyaraj, A. Arun Kumar, M.C. RamKumar, P.V.A. Padmanabhan, A.M. Trimukhe, R.R. Deshmukh, P. Cools, R. Morent, N. De Geyter, V. Kumar, P. Gopinath, S.K. Jaganathan, "Influence of operating parameters on development of polyethylene oxide-like coatings on the surfaces of polypropylene films by atmospheric pressure cold plasma jet-assisted polymerization to enhance their antifouling properties", *Journal of Physics and Chemistry of Solids*, 123 (7) (2018) 76–86.
116. M.C. Ramkumar, K.N. Pandiyaraj, A. Arun Kumar, P.V.A. Padmanabhan, P. Cools, N. De Geyter, R. Morent, S. UdayKumar, V. Kumar, P. Gopinath, S.K. Jaganathan, R.R. Deshmukh, "Atmospheric pressure non-thermal plasma assisted polymerization of poly (ethylene glycol) methylether methacrylate (PEGMA) on low density polyethylene (LDPE) films for enhancement of biocompatibility", *Surface and Coatings Technology*, 329 (2017) 55–67.
117. G. Aziz, N. De Geyter, H. Declercq, R. Cornelissen, R. Morent, "Incorporation of amine moieties onto ultra-high molecular weight polyethylene (UHMWPE) surface via plasma and UV polymerization of allylamine", *Surface and Coatings Technology*, 271 (2015) 39–47.
118. N. Demirci, M. Demirel, N. Dilsiz, "Surface modification of PVC film with allylamine plasma polymers", *Advances in Polymer Technology*, 33 (4) (2014) 21435.
119. P. Hamerli, Th. Weigel, Th. Groth, D. Paul, "Surface properties of and cell adhesion onto allylamine-plasmacoated polyethylenterephtalat membranes", *Biomaterials*, 24 (2003) 3989–3999.
120. J.J.A. Barry, M.M.C.G. Silva, K.M. Shakesheff, S.M. Howdle, M.R. Alexander, "Using plasma deposits to promote cell population of the porous interior of three-dimensional poly(D,L-lactic Acid) tissue-engineering scaffolds", *Advanced Functional Materials*, 15 (2007) 1134–1140.
121. R.G. Meeks, "The dow corning siloxane research program: an overview and update", *Medical Device and Diagnostic Industry*, 21 (1999) 112.
122. H.G.P. Lewis, D.J. Edell, K.K. Gleason, "Pulsed-PECVD films from hexamethylcyclotrisiloxane for use as insulating biomaterials", *Chemistry of Materials*, 12 (2000) 3488–3494.
123. L. Dai, H.A.W. St. John, J. Bi, P. Zientek, R.C. Chatelier, H.J. Griesser, "Biomedical coatings by the covalent immobilization of polysaccharides onto gas-plasma-activated polymer surfaces", *Surface and Interface Analysis*, 29 (2000) 46–55.
124. J.H. Yim, V. Rodriguez-Santiago, A.A. Williams, T. Gougousi, D.D. Pappas, J.K. Hirvonen, "Atmospheric pressure plasma enhanced chemical vapor deposition of hydrophobic coatings using fluorine-based liquid precursors", *Surface & Coatings Technology* 234 (2013) 21–32.
125. D.P. Dowling, I.S. Miller, M. Ardhaoui, W.M. Gallagher, "Effect of surface wettability and topography on the adhesion of osteosarcoma cells on plasma modified polystyrene", *Journal of Biomedical Application*, 26 (2011) 327–347.
126. K. Teshimaa, Y. Inouec, H. Sugimuraa, O. Takaid, "Synthesis of silica films on apolymeric material by plasma-enhanced CVD using tetramethoxysilane", *Surface and Coatings Technology*, 169–170 (2003) 583–586.
127. J. Schneider, M.I. Akbar, J. Dutroncy, D. Kiesler, M. Leins, A. Schulz, M. Walker, U. Schumacher, U. Stroth, "Silicon oxide barrier coatings deposited on polymer materials for applications in food packaging industry", *Plasma Processes and Polymers*, 6 (2009) S700–S704.
128. R. Morent, N. De Geyter, S.V. Vlierberghe, P. Dubruel, C. Leys, E. Schacht, "Organic–inorganic behaviour of HMDSO films plasma-polymerized at atmospheric pressure", *Surface & Coatings Technology*, 203 (2009) 1366–1372 .

129. R. Morent, N. De Geyter, T. Jacobs, S.V. Vlierberghe, P. Dubruel, C. Leys, E. Schacht, "Plasma-polymerization of HMDSO using an atmospheric pressure dielectric barrier discharge", *Plasma Processes and Polymers*, 6 (2009) S537–S542.
130. L.J. Ward, W.C.E. Schofield, J.P.S. Badyal, "Atmospheric pressure glow discharge deposition of polysiloxane and SiOx films", *Langmuir*, 19 (2003) 2110–2114.
131. K. Navaneetha Pandiyaraj, A. Arun Kumar, M.C. Ramkumar, S. Uday Kumar, P. Gopinath, Pieter Cools, N. De Geyter, R. Morent, M. Bah, S. Ismat Shah, Pi-Guey Su, R.R. Deshmukh, "Effect of processing parameters on the deposition of SiOx-like coatings on the surface of polypropylene films using glow discharge plasma assisted polymerization for tissue engineering applications", *Vacuum*, 143 (2017) 412–422.
132. S. Girish Kumar, L. Gomathi Devi, "Review on modified TiO_2 photocatalysis under UV/visible light: selected results and related mechanisms on interfacial charge carrier transfer dynamics", *The Journal of Physical Chemistry A*, 115 (2011) 13211–13241.
133. Z. Abdin, M.A. Alim, R. Saidur, M.R. Islam,W. Rashmi, S. Mekhilef, A. Wadi, "Solar energy harvesting with the application of nanotechnology", *Renewable & Sustainable Energy Reviews*, 26 (2013) 837–852.
134. J. Maçaira, L. Andrade, A. Mendes, "Review on nanostructured photoelectrodes for next generation dye-sensitized solar cells", *Renewable & Sustainable Energy Reviews*, 27 (2013) 334–349.
135. D-J. Kim, H.C. Pham, K-S. Kim, "Application of PCVD process to uniform coating of Tio_2 thin films on polypropylene beads", *Surface Review and Letters*, 173 (2010) 329–335.
136. D-J. Kim, H-C. Pham, D-W. Park, K-S. Kim, "Preparation of TiO_2 thin films on polypropylene beads by a rotating PCVD process and its application to organic pollutant removal", *Chemical Engineering Journal*, 167 (2011) 308–313.
137. K.N. Pandiyaraj, A. ArunKumar, M.C. Ramkumar, A. Sachdev, P. Gopinath, Pieter Cools, N. DeGeyter, R. Morent, R.R. Deshmukh, P. Hegde, C. Han, M.N. Nadagouda, "Influence of non-thermal TiCl4/Ar + O2 plasma-assisted TiOx based coatings on the surface of polypropylene (PP) films for the tailoring of surface properties and cytocompatibility", *Materials Science and Engineering C* 62 (2016) 908–918.
138. I. Gancarz, M. Bryjak, J. Wolsk, A. Siekierk, W. Kujawski, "Membranes with a plasma deposited titanium isopropoxide layer", *Chemical Papers*, 70 (2016) 350–355.

7 Applications of Green Polymeric Nanocomposites

Mukesh Kumar Meher and Krishna Mohan Poluri

CONTENTS

7.1 INTRODUCTION

Though similarly named, "green polymers" is not a synonym for "biopolymers," and they do not represent the same materials. As their name suggests, bio-polymers are naturally derived or biologically produced by living organisms such as animals, plants and microorganisms. In a broad way, the biopolymers (collagen, cellulose, pectin, silk, etc.) are naturally occurring polymers that can be extracted or processed from raw material using various techniques. Even though the production of natural polymers is gradually increasing, the amount is still less than 5% of total plastics production. The highest quantity of plastic is that used in packaging, which also generates most non-biodegradable waste. Other than that, plastics are also widely used in textiles, automobiles, electrical products, industrial machinery, building and construction, which give rise to uncontrolled plastic waste generation. Currently, polymer technology emphasizes the production of green polymers, which may reduce the amount of environmental toxic waste generation. The growing demand of the green polymer market is driven mainly by the medical and pharmaceutical industries. Other industries are also focusing on, and trying to move towards, a "green" strategy. The expanding global waste generation problems have intensified by creating ecological imbalance (Singh et al. 2014, Zaman 2016).

In general, green chemistry is used for the production of green polymers. The concept of chemical products and their synthesis methods reducing the use and the production of toxic materials is called green chemistry (Sheldon et al. 2007). The renewable and sustainable properties of green polymeric nanocomposites will help to meet future needs for the development of novel nanomaterials. In the era of green polymer systems, numerous studies have been carried out for the development of green sustainable materials without any toxic effects (Mohanty et al. 2002). The growing toxic effect and non-biodegradability properties of waste materials have been emphasized and discussed in various scientific meetings which have led to technological transformation for greener materials. The design and production of green sustainable materials may suppress the production of traditionally used plastic materials. For a sustainable green chemistry, eight important subjects have been established: (1) use of a green catalyst for the reactions (e.g. bio-catalysts such as enzymes or complete living cells); (2) use of bio- or agriculture-based materials; (3) production of biodegradable polymers and minimal waste generation; (4) biological recycling of green polymer products and biocatalysts; (5) energy generation and its marginal usage; (6) optimal molecular design and activity; (7) use of benign solvents (e.g. water, ionic solutions and reactions without solvents); and (8) upgraded reaction processes and production (Cheng and Gross 2010, Anastas and Kirchhoff 2002).

Taking into consideration high demand and their biodegradable properties, green polymers were synthesized by assembling a large number of monomeric units into a single chain polymer or they were directly extracted from natural sources such as plants, animals and microorganisms. Polymers used earlier were less likely to be biodegradable and thus became an important source of global waste generation. Therefore, the production of green polymers was thought to be a better alternative to the traditionally used polymers. The principles of green polymer production include (1) high content of biomaterials, (2) a clean synthesis route, (3) absence of health and environmental toxicity-related issues, (4) low carbon footprint, (5) use of renewable resources, (6) no use of organic solvents during production, and (7) controlled product lifecycle and effective recycling (Dube and Salehpour 2014). The newly synthesized green polymers were developed with different excellent properties – for instance, hydrophobic or hydrophilic, biocompatible, biodegradable, water-soluble, pH- and temperature-responsive, heat resistant properties etc. – by changing the various parameters such as monomeric units, side-chain functional groups, polymerization process, extraction and processing of the polymers (Cheng et al. 2015).

The integration of nanotechnology into green polymer chemistry creates a new arena of study for the synthesis and application of sustainable green polymeric nanocomposites (GPNCs). The green polymers are fabricated in nanoscale dimensions, with materials having a size range of 1 to 1000 nm or incorporated into an organic/inorganic nanocomposite to exhibit their native functions. In a broad way, nanomaterials are usually available in the form of nanoparticles and nanofiber structures. Accordingly, the green polymers are directly fabricated or incorporated into nanoparticles or nanofibers. For example, Janes et al. prepared chitosan nanoparticles for the delivery applications of the anticancer drug doxorubicin (DOX), and in another study, Sugunan et al. synthesized chitosan-capped gold nanoparticles for heavy metal ion sensing (Janes et al. 2001, Sugunan et al. 2005). Similarly, single or multiple combinations of green polymers could be used for the fabrication of nanofibers by the direct electrospinning method. For example, Zhang et al. prepared gelatin nanofibers and Dhandayuthapani et al. fabricated chitosan-gelatin nanofibers for tissue engineering applications (Dhandayuthapani et al. 2010, Zhang et al. 2006). The unique physicochemical properties of nanomaterials, such as enhanced surface functionality due to higher surface-to-volume ratio, quantum effects and self-assembly, are inherited by the GPNCs (Kumar and Kumbhat 2016). In general, GPNCs are broadly divided into four categories on the basis of their nanoarchitecture: (1) polymeric nanoparticles, (2) polymeric nanofibers, (3) polymeric nanohydrogels, and (4) metal-polymer hybrid nanocomposites (see Figure 7.1). At present, GPNCs are widely studied, as plenty of research is needed for their innovation and their development from bulk green material. In this chapter we will review different types of green polymers, the mechanisms of developing various GPNCs and their applications.

7.2 GREEN POLYMERS

Green polymers are synthesized with an eco-friendly method which ensures the production of sustainable polymers with less toxic waste and prevents environmental pollution. Integrally, green chemistry, which is a strategy for designing the process and production using renewable resources by eliminating toxic by-products, is involved in the manufacture of green polymers (Cheng et al. 2015, Kobayashi 2017).

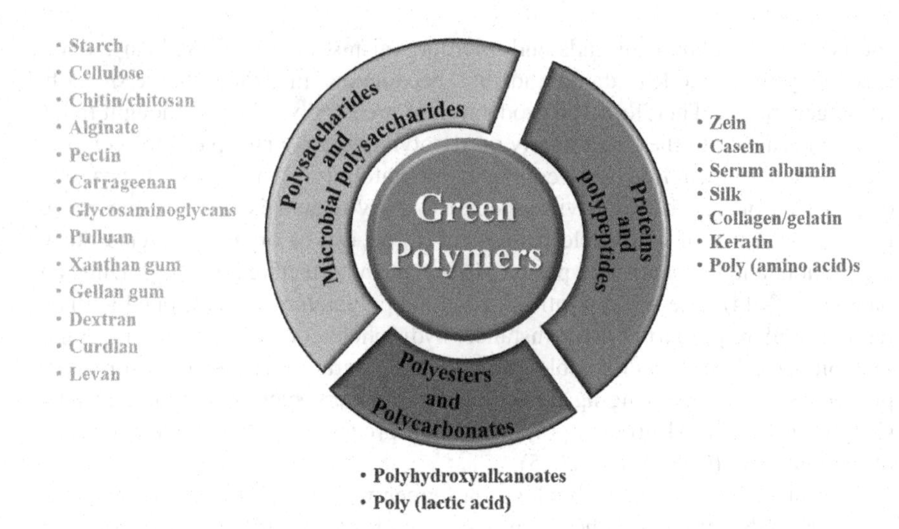

FIGURE 7.1 Schematic illustration of different types of green polymeric nanocomposites.

Green polymer chemistry also emphasizes energy minimization during the use, recycling and biodegradability of the material after application. Generally, green polymers are obtained from natural resources or produced by polymerizations by various green methods such as the use of enzymatic catalysis, of ionic liquids as solvents, light-source-induced and microwave-assisted polymerizations, benign atom transfer radical polymerizations (ATRP), reversible addition–fragmentation chain-transfer (RAFT) polymerization and bio-/chemical modifications (Kadokawa and Kobayashi 2010, Kobayashi and Makino 2009, Kubisa 2004, Dube and Salehpour 2014, Lalevée et al. 2010, Sivalingam et al. 2004, Guerrero-Sanchez et al. 2007, Tsarevsky and Matyjaszewski 2006, Semsarilar and Perrier 2010). The final product produced by these methods contains higher proportions of raw/starting materials, thus signifying the utilization of a maximum number of atoms (Cheng and Gross 2010). The biopolymers that are produced or derived from natural sources such as plants and animals are also considered to be green polymers due to their diverse set of functions, biocompatibility and biodegradability nature. The use of green polymers for various material applications is beneficial to nature due to good environmental compatibility with minimal waste burdens (Kaplan 1998). Additionally, the abundance of green polymers in renewable resources is setting a landmark in the green economy. Green polymers are extensively used in various applications such as coatings, membranes, packaging, solar energy systems, biomedical applications, adhesives, air and water filtrations units and in electronic systems (Jiang et al. 2017). These green polymers are largely divided into three major categories: (a) polypeptides, (b) polysaccharides and (c) polycarbonates and polyesters (see Figure 7.2).

7.2.1 Proteins and Polypeptides

Two or more amino acid units are linked together through peptide bonds to make a peptide chain and a chain of amino acids (number: 8–10) forms a polypeptide.

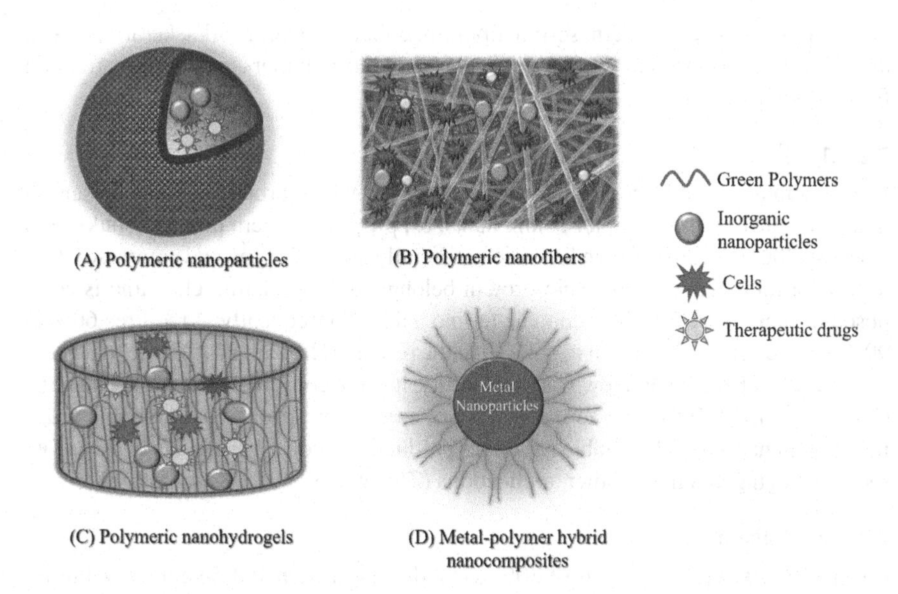

FIGURE 7.2 Schematic showing the classifications of different green polymers and their examples.

Each amino acid possesses at least two functional groups for polymerization and produces a functional structure. Similarly, two or more polypeptides are interconnected to make a long linear-chain polymer which is ultimately termed as a protein. Depending on the amino acids present in the chains, biopolymers can change their structure and properties or vice versa. According to the amino acids present in their chain, the interactions of the proteins vary and they behave uniquely according to their side-chain group charge (negative, positive or hydrophobic). The protein's structure/conformation, also determined by the amino acids and which change according to their spatial arrangements, range from primary to quaternary structures. In a primary structure, a protein exhibits linear amino acid chains; whereas in a secondary structure, it rearranges the backbone atoms for local rearrangements by producing three structural elements, that is, helices, sheets and turns. In a tertiary structure, the polypeptide chains undergo a complete side-chain interaction to form a compact three-dimensional structure. A quaternary structure comprises the interactions between two or more polypeptide chains (subunits) by forming homo- and heteromeric structures (Poluri and Gulati 2016).

In general, polypeptides and proteins are biopolymers, as they are derived from natural sources such as plants, animals and microorganisms. These intrinsic properties of natural origin enable polymers to degrade easily by naturally occurring enzymes. Proteins that are present abundantly within the cells and in the extracellular matrix (ECM), and which regulate multiple biological functions such as cellular transportations, cell structure, regulation of cellular pathways and act as a biocatalysts, are termed as "enzymes" (Gomes et al. 2008). Although polypeptides and proteins can denature in various conditions (i.e. temperature and pH) and by substrates (detergents), a large group of proteins and polypeptides have been used as green polymers such as zein, casein, serum albumin, collagen/gelatin, silks, keratin, elastin and polyamino acids. These naturally occurring bio-polymers are getting attention for nano-fabrication to exhibit various applications.

7.2.1.1 Zein

Zein is a major protein, accounting for 50% of the whole protein content present in maize. It consists of four major components: α-, β-, γ- and δ-zein protein (Shukla and Cheryan 2001). α-Zein protein is the commercially available type and comprises 70% to 85% of total zein protein. Zein protein belongs to the prolamin class and is composed of more than 50% hydrophobic amino acids. Consequently, it requires 60% to 90% aqueous ethanol solution to solubilize the protein (Hu et al. 2015). Due to its negative nitrogen balance and low water solubility, it is not consumed directly by humans (Pascoli et al. 2018). Hence, zein proteins are used to encapsulate hydrophobic drugs in colloidal nanoparticles, stable nanofiber productions, coatings, nano encapsulation, food packaging and many other applications (Zhang et al. 2016, Kasaai 2018).

7.2.1.2 Casein

Casein protein represents almost 80% of total milk protein and belongs to a family of phosphorylated proteins comprised of four members: α_{S1}- (199 amino acids), α_{S2}- (207 amino acids), β- (209 amino acids) and κ-caseins (169 amino acids) (Horne 2002, Otter 2003). Casein milk protein forms a micellar structure to transport large amount of minerals, that is, calcium and phosphate from mother to suckling offspring. Caseins have a higher number of surface-active molecules and a tendency for self-assembly. By reducing the milk pH to an acidic 4.5, casein proteins precipitate out to form sodium caseinate, a good stabilizer that is widely used in the food industry (De Kruif and Holt 2003, Schmidt 1982). Out of the four, three members of the group – α_{S1}-, α_{S2}-, β-caseins – precipitate out in the presence of ionic calcium, even in millimolar concentrations. On the other hand, κ-caseins inhibit the precipitations in a mixture of other caseins and produce stable micellar structures (Walstra and Jenness 1984). Casein micelles are generally spherical and have an average size in a range of 50 to 600 nm and molecular weight from 10^6 to 10^9 Da. Due to their cost effective, high nutritional value, bioactive/bio absorbable and nontoxic versatile properties, casein proteins are extensively used in the food product industry, as catalyst material and in biomedicines as a natural nanosphere for drug delivery systems (Rehan et al. 2019).

7.2.1.3 Serum Albumin

Serum albumin exists as a globular blood protein having an approximate molecular weight of 65 kDa. It is found in the blood of vertebrates, being especially abundant in mammal blood (Moman and Varacallo 2018). It consists of 585 amino acids which form three repetitive homologous domains and these are separated by two sub-domains in their globular structure. The most essential role of albumin is to maintain oncotic pressure in blood which is needed for proper body fluid distribution. Human and bovine serum albumins are the most studied albumins due to their obvious importance in metabolic, genetic and *in vitro/in vivo* clinical studies (Peters Jr 1985). Albumins are moderately soluble (up to 40% w/v) at physiological pH and perform as an excellent natural carrier for metabolic products such as steroid hormones, hemins, fatty acids, bilirubin, thyroxine and various drugs (van der Vusse 2009, Zilg et al. 1980). Albumin shows high stability in wide pH ranges from 4 to 9 and temperature up to 60°C for 10 hours of heating without any degradation/aggregation. Biodegradability and a high stability structure enable albumins to be employed as appropriate drug nano-carriers, and for encapsulating polymers in biomedical applications (Rai et al. 2017).

7.2.1.4 Silk

Silk is a natural fibrous protein with a unique molecular structure composed of large hydrophobic domains and relatively short hydrophilic regions. Silk protein obtained from the mulberry silkworm (*Bombyx mori*) comprises 5507 amino acids (Altman et al. 2003, Foo and Kaplan 2002). The amino acids are found in a repetitive order of six amino acid residues $(Gly–Ala–Gly–Ala–Gly–Ser)_n$ and most of the charged amino acids are found in hydrophilic regions, C-terminal and N-terminals (Elzoghby et al. 2015). Two proteins, fibroin and sericin, mainly comprise *B. mori* silk. Fibroin is the fibrous structured protein, while sericin is the glue that binds and successfully covers the fibroin proteins (Hardy et al. 2008). The fibroin proteins comprise compactly packed antiparallel β-sheets due to the higher number of hydrophobic domains. Silk protein exhibits notable properties such as high mechanical strength, biocompatibility, controlled proteolytic degradations, flexible structure, and environmental stability. Therefore, silk fibroin proteins are extensively studied to fabricate suitable nanomaterials for various uses (Wang et al. 2016, Cohen-Karni et al. 2012, Pina et al. 2015).

7.2.1.5 Collagen/Gelatin

Collagen is the most abundant structural protein in the body of mammals and comprises 25% of total protein concentration. Collagen proteins are found as tensile component fibrillary structures in tissue such as tendon, cartilage, bone and skin ECM to provide structural support to the cells. These proteins play a crucial role in connective tissue as well as in all organs for maintaining stability and structural integrity (Weiss 1993, Gelse et al. 2003). Therefore, collagen protein has numerous essential body functions in cell adhesion, cell migrations, tissue scaffolding, blood vessel formation, cancer, tissue repair and morphogenesis. A total of 29 different types of collagen have been discovered, which are composed of at least 46 different polypeptide chains (Shoulders and Raines 2009). Collagen exhibits a right-handed triple-helical structure composed of three parallel left-handed polypeptide strands and one amino acid residue staggered to the subsequent amino acids of proximate polypeptide chains. Every polypeptide chain contains a repeating unit of triple amino acids of Gly-X-Y, where X and Y are repeatedly found to be proline and 4-hydroxyproline, respectively. Moreover, each polypeptide chain contains integrin binding sites (RGD, Arg-Gly-Asp) for specific cell adhesion (Kadler et al. 2007). Gelatins are soluble protein forms achieved by partial hydrolysis/thermal denaturation of insoluble collagen. During conversion, the collagen triple-helix structure loses its integrity by breaking intermolecular covalent bonds and is converted into original α-helix conformations (Kozlov and Burdygina 1983, Gorgieva and Kokol 2011).

Gelatin characteristics are determined by the conversion (acidic or basic hydrolysis), type of collagen, and source of collagen protein (which is usually either bovine or porcine) (Aldana and Abraham 2017). Both collagen and gelatin proteins contain an RGD sequence which enables integrin mediated cell adhesion. Both proteins are non-immunogenic, have good biocompatibility and are biodegradable. Hence, collagen and gelatin are used widely in food production, tissue engineering, drug delivery and the pharmaceutical industry (Jalaja et al. 2016, Agheb et al. 2017, Olsen et al. 2003, Guillén et al. 2011).

7.2.1.6 Keratin

Keratin is the strongest protein among all the biopolymers owing to its high toughness and tensile strength. Keratin is the most abundant structural protein in the epidermal cell layer after collagen. By exhibiting its structural and functional robustness, keratin proteins produce various ultra-structures on the outer covering of vertebrates such as skin, hair, nail, fur, horn, wool, claw, shell and hoof (Wang et al. 2016). These structures impose highly efficient mechanical, structural and functional characteristics according to the physical requirements of the subject. Keratin biopolymers are composed of α-helix or β-sheet conformation polypeptide chains and contain a higher number of cysteine amino acid residues, which facilitate covalent disulfide bond formation within the polypeptide chains (McKittrick et al. 2012). Keratin proteins have a molecular weight in a range of 40 to 70 kDa and are categorized into two main classes (type I and type II) on the basis of their genomic similarity (Gu and Coulombe 2007). In addition, lysine and arginine are two amino acids present in the polypeptide chain that act as active sites for trypsin digestion. Due to their natural origin, keratin proteins are biocompatible as well as biodegradable. The diversified nature, abundance and strong bonding properties facilitate easy modification of keratin proteins for nanostructure preparations (Sun et al. 2017, Aluigi et al. 2008).

7.2.1.7 Elastin

Elastin is known as a connective tissue protein and is solely responsible for producing fibril-like structures that feature elasticity, flexibility and extensibility of the vertebrate tissues. Elastin proteins are the foremost insoluble proteins found in ECM that provide resilience, long range deformability and reflexive recoiling, which are essential properties of skin, heart, lungs and other connective tissues (Kielty et al. 2002). The elastin gene expresses 750 to 800 amino acid residues of tropoelastin protein in various cells and finally through a highly complex process of elastogenesis produces functional elastin fibers. The tropoelastin protein contains large hydrophobic domains which are dominated by aliphatic amino acids, mostly glycine (G), proline (P) and valine (V). Glycine has been found to be conserved during evolution. The structure–function relationship of elastin protein has been explained by various models and it was found that the most relaxed-state elastin proteins are highly hydrated, unstructured and have high entropy, which is precisely the opposite in an elongated state (Gray et al. 1973, Debelle and Tamburro 1999). In humans, elastin proteins are synthesized approximately before the age of 40 and after that their biosynthesis slows down, which can be directly correlated with wrinkles and the aging process. Therefore, elastins are overwhelmingly used and fabricated as various structural materials (i.e. cuspid valves, tubes, sheets and scaffolds) in various cosmetic industries, pharmaceutical industries and biomedical applications (Zhang et al. 2014, Despanie et al. 2016).

7.2.1.8 Poly(Amino Acid)s

Amino acids are the elementary units of peptides, homo-/polypeptides/proteins and poly(amino acid)s. To avoid confusion, poly(amino acid) chains are composed of a single amino acid, whereas peptides and polypeptides are made up of combinations of multiple amino acids (Khuphe and Thornton 2018). The polymeric assembly of amino acids through peptide bonds renders biodegradable and biocompatible polymers. Amino acids present in the polymer chains offer a site of interaction by exposing a series of side-chain functional group (thiol, amine, carboxyl and alcohol) moieties (Saxena et al. 2014). The initial amino acids of the polymer can be modified with various desirable functional groups or biomolecules for different biomedical applications, such as cell-specific targeting, crosslinking agents and fluoresce active molecules for bio-imaging (Lavasanifar et al. 2002, Khanna and Ullman 1982, Tian et al. 2017). Most poly(amino acid)s are manufactured by the N-carboxyanhydride ring-opening polymerization (NCA ROP) method where the desired amino acid cyclizes to form N-carboxyanhydride and undergoes ring-opening polymerization by the interaction of an initiator (Wibowo et al. 2014, Khuphe et al. 2016, Zhang and Li 2013). Furthermore, the NCA ROP method can also be used for the modification of poly(amino acid) polymers to achieve new functional properties such as amphiphilicity, pH-sensitivity, thermo-responsivity and oxidation/reduction sensitivity (Gyarmati et al. 2013, Takeuchi et al. 2006, Chen et al. 2011, Gyenes et al. 2008). Presently, poly(amino acid)s, especially poly(L-lysine), poly(glutamic acid) and poly(aspartic acid), are widely used as nanostructures in the form of micelles, vesicles, nanofibers and nanogels for biomedicine, as drug carriers and in personal-care product formulation (Park et al. 2017, Zhang et al. 2016, Lalatsa et al. 2012, Farrar et al. 2011, Xu et al. 2015).

7.2.2 Polysaccharides

Polysaccharides are the most commonly and abundantly found natural biopolymers. They are composed of monosaccharides as repeating units that are based on sugars. Monosaccharides are the smallest units and are composed of carbon (C), hydrogen (H) and oxygen (O) in a proportion of 1:2:1 ($C_nH_{2n}O_n$, $n \geq 3$). The repeating units of polysaccharides have many chiral carbon atoms which determine the actual saccharides, for example, five chiral carbon atoms in the case of hexose sugar ($C_6H_{12}O_6$). Monosaccharides are linked through a glycosidic bond where oxygen of carbon (1) interacts with the carbon (4) of another molecule by successive elimination of water molecules (Wang et al. 2013, Christian 2011). Polysaccharides are increasingly larger in size and have also been termed as complex carbohydrates. The chain length of the polysaccharides usually varies from one to several thousand repeating units. Polysaccharides are an important class of biopolymers and exhibit many essential roles in biological functioning. These macromolecules are found to be structurally diverse and fairly widespread in nature (Finkenstadt 2005,

Kadokawa 2016). Therefore, the polysaccharides are derived from natural sources such as plants, animals and microorganisms. The commonly used green polysaccharides in nanocomposite applications are starch, cellulose, alginate, pectin, carrageenan, glycosaminoglycans (GAGs), pullulan, elsinan, glucan, chitin/chitosan, levan, xanthan, polygalactosamine, gellan, dextran and polyhydroxybutyrate (PHB) (Mogoşanu and Grumezescu 2014, Wang et al. 2012, Wang et al. 2013).

7.2.2.1 Starch

Starch is one of the major sources, as well as a reserve, of carbohydrates in higher plants. It is mainly present in staple foods such as bread, rice, potato, etc., which are regularly consumed for essential nutrients and energy to sustain life (Carvalho 2008). Starch is a blend of two types of glucose polymers: amylopectin and amylose. Amylopectin is the highly branched larger polymer with an average molecular weight of around 10^8 Da, branched with α-1,4-linkage and interlinked with α-1,6-linkages. Amylose, the linear and small-sized polymer with a molecular weight of 10^5 Da, is branched with α-1,4-linkages (Le Corre and Angellier-Coussy 2014, Jackson 2003, Preiss 2010). Starch averagely comprises of 65% to 75% amylopectin and 25% to 35% amylose, although it greatly varies depending upon the plant source (Jackson 2003, Bertolini 2009). Due to its excellent qualities, such as being abundant in nature, inexpensive, nontoxic, biodegradable and naturally renewable, it has been extensively studied for fabricating a range of nanocomposites for various biomedical and industrial applications (Le Corre et al. 2010).

7.2.2.2 Cellulose

Cellulose is an excellent biopolymer due to its abundance and unique structural properties. Cellulose is a key structural element of plant cellular structure and is also produced by certain microorganisms such as bacteria, algae, fungi and some marine organisms (e.g. tunicate) (Rånby 2001). The cellulose proportion in plants varies according to the plant species and source such as 95% in cotton, 80% in flax, 60% to 70% in jute and 40% to 50% in wood. Cellulose is a hydrophilic and syndiotactic homopolymer of D-anhydroglucose ($C_6H_{11}O_5$) repeat units joined by β-1,4 linkages between the carbon atoms C (1) and C' (4) of neighboring glucose units. An individual repeat unit comprises three free hydroxyl groups at the carbon positions of 2, 3 and 6, which make the polymer more reactive (Rudnik 2012, Gupta et al. 2016, Festucci-Buselli et al. 2007). The crystalline packing of cellulose polymer provides strength and rigidity in plant structures. Cellulose cannot be digested by humans as they don't have an appropriate enzyme to break down the beta acetal linkages. Due to its outstanding properties, such as biodegradability, biocompatibility, low cost and abundance in nature, it is at the forefront of the nanomaterial production for different areas of application, such as medicine, textile, drug delivery and nano-sensor applications (Dahman 2017, Tian and He 2016, Mohan et al. 2015).

7.2.2.3 Chitin/Chitosan

Chitin or chitosan is a linear, naturally occurring biopolymer composed of randomly arranged N-acetyl-D-glucosamine (D-GlcNAc) and D-glucosamine (D-Glc) units linked by β (1→4) glycosidic bonds (Sánchez-Machado et al. 2019). Depending upon the higher amount of these monosaccharide units (D-GlcNAc/ D-Glc), the biopolymer is categorized as either chitin or chitosan. If the number of D-GlcNAc is higher than 50% in total polymer, then the polymer is termed as chitin. In contrast, if the number of D-Glc is higher than 50%, it is termed as chitosan (Rinaudo 2006). Commercially, these biopolymers are economically feasible, as they are obtained as food industry waste from crustacean shells such as those of crabs, shrimps, lobsters, krill and crayfish etc. (Dutta et al. 2004, Pillai et al. 2009). Naturally, chitin/chitosan is attached with proteins, minerals, lipids and pigments. Therefore, the chemical purifications of these polymers from crustacean shells require a two-step process followed by demineralization (by dilute HCl) and deprotonation (by NaOH or KOH) (Philibert et al. 2017). The most commercially available chitosan has an approximate molecular weight ranging from 10,000 to 1 million Da. Being a biodegradable, biocompatible and edible natural polymer, chitin/chitosan has been exploited as a nanomaterial for numerous applications in waste-water treatment, agriculture, the food industry, as an antimicrobial agent and in drug delivery systems (Suh and Matthew 2000, Zivanovic et al. 2015).

7.2.2.4 Alginate

Alginate or alginic acid is a negatively charged polysaccharide, abundantly present in the cell wall of brown seaweed algae including *Laminaria*, *Ascophyllum*, *Durvillaea*, *Ecklonia*, *Lessonia*, *Macrocystis*, *Sargassum* and *Macrocystis* species (Niekraszewicz and Niekraszewicz 2009, McHugh 2003, Qin 2008). It is a linear unbranched polysaccharide and 1–4' linked copolymer of β-D-mannuronic acid (M) and α-L-guluronic acid (G) residues (Lee and Mooney 2012, Draget 2009). Depending on the origin, the polymer chain differs by creating different arrangements of the residues such as MM, GG and MG block polymer. The arrangement of the block polymer and molecular size influence the various physiochemical properties such as viscosity and gelling properties. In the presence of divalent cations, that is, Mg^{2+}, Ca^{2+}, Ba^{2+}, Sr^{2+}, the alginate polymer crosslinks the carboxylate and glucouronate groups on its polymer backbone to form gels (Burdick and Stevens 2005). The mechanical properties of the alginate gel can be tailored by changing the monomer ratio, solvents, polymer concentrations and the molecular weight. Due to its remarkable gel-forming as well as biocompatible properties, alginate is used in various biomedical applications such as drug delivery, wound healing and tissue engineering (Lee and Mooney 2012, Roberts and Martens 2016).

7.2.2.5 Pectin

Pectin is a polysaccharide with a high molecular weight of more than 200 KDa, playing essential roles in almost all plant cell structures. The polymeric structure of pectin consists of galacturonic acids (at least 65% by weight) linked by α-1,4 glycosidic bonds, and the chain is partly esterified as methyl esters (Flutto 2003a). The pectin polymers have a high degree of polymerization ranging up to 1000 monomeric units in a chain. As a natural component, pectin is used as food as well as an excellent stabilizing agent for all kinds of packaged foods. Pectin is traditionally used for gelation properties in jam and jelly food products. The pectin gels can hold the food nutrient within the gel structure and maintain water levels between crosslinked pectin polymers (Flutto 2003b). Usually,

Pectin contains different amounts of neutral sugars such as D-glucose, D-xylose, L-rhamnose and L-arabinose in its backbone and sidechains. Different quantities and qualities of pectin are extracted from many fruits as a byproduct during the production of juice, oil and pulps (Endress and Christensen 2009). Due to the presence of galacturonic acid, pectin contains a high number of carboxylic acids as functional groups which facilitate amide bond formation by the interactions with amino acids. Thus, pectin polymers act as a good natural polymeric nano-host for drug delivery and other biomedical applications (Sharma et al. 2012, de Melo et al. 2012, Chittasupho et al. 2013).

7.2.2.6 Carrageenans

Carrageenans (CRG) or carrageenins are linear sulfated polysaccharides belonging to the family of glycans and are commonly isolated from red seaweed. The presence of sulfate groups in carrageenans renders the anionic properties of polymer (Sudha et al. 2014, Rioux and Turgeon 2015). Carrageenans are high-molecular-weight polysaccharides consisting of both sulfated and non-sulfated galactose units, that is, alternative repeats of β-1,4 and α-1,3 linked 3,6-anhydrogalactose and galactose-4-sulfate (Ahmed et al. 2014, Kariduraganavar et al. 2014). Due to its large structure and flexible helical properties, it assists the polymer in a curl-forming helical arrangement and facilitates the formation of different types of hydrogels in room temperature (Blakemore 2016). Considering the gel-forming and stabilizing properties, it is commonly used in the food industry for stabilizing meat and dairy products, as it has a high tendency to bind with proteins. Depending on the number of sulfations per disaccharide, commercially available carrageenans are divided into three different types, namely, *Kappa* (κ, one sulfate)-, *Iota* (ι, two sulfate)- and *Lambda* (λ, three sulfate)-carrageenan (Kariduraganavar et al. 2014). The degree of sulfation influences the gel-forming properties of carrageenans. κ-Carrageenans form strong and rigid gel in the presence of K^+ ions whereas ι-carrageenans form soft gel in the presence of Ca^{2+} ions and λ-carrageenans do not form gels (Tuvikene et al. 2007). Due to their high susceptibility to gel formation, biocompatibility and biodegradability, carrageenans are widely studied for their nanostructure functionality in different applications in food packing, drug delivery and the pharmaceutical industry (dos Santos and Grenha 2015, Goel et al. 2019).

7.2.2.7 Glycosaminoglycans

Glycosaminoglycans (GAGs) are long, unbranched and anionic hetero-polysaccharides. GAGs are generally present on cell surfaces, basement membrane and extracellular matrix (ECM). They exhibit various roles in cellular functions, such as cell signaling, tissue repair, wound healing and conferring mechanical strength to the connective tissue (Gulati and Poluri 2016, Kirker et al. 2002). Depending upon their chemical compositions, GAGs are divided into four groups: heparin or heparan sulfate (HS), chondroitin or dermatan sulfate (CS/DS), keratan sulfate (KS) and hyaluronan (HA). The GAGs polymeric chains are ideally composed of repeating disaccharide units containing hexouronic acid (D-glucuronic acid (D-GlcuA) or L-Iduronic acid (L-IdouA); with an exception of KS containing galactose instead of uronic acid) and hexosamine (N-acetyl-D-glucosamine (D-GlcNAc) or N-acetyl-D-galactosamine (D-GalNAc)) (Gulati et al. 2017). GAGs groups are classified according to the presence of the combinations of these different repeating disaccharide units. HS and HA contain glucosamine whereas CS and DS possess galactosamine; hence GAGs are further subdivided into glucosaminoglycans and galactosaminoglycans. GAGs with distinctive chemical structures, including the different sulfation patterns, decetylation and epimerizations, impart their ligand-specific interactions for particular biological activities (Antonio and Iozzo 2001, Hileman et al. 1998). Therefore, it is crucial to thoroughly understand the structure and functional properties of GAGs in nano-dimensions for various biomedical applications.

7.2.2.8 Microbial polysaccharides

A variety of microbial polysaccharides are known and are isolated from various microorganisms which are generally stored as exo-polysaccharides in their cells. These complex molecules function as energy reservoirs, structural components, and are involved in molecular communication processes. Due to their high diversity, microbial polysaccharides are receiving more attention than plant or animal produced polysaccharides (Linton et al. 1991). Although plants produce a large amount of polysaccharides, microbial polysaccharides are studied for their novel structure and functionality, chemical and physical properties, reproducibility and modifications. Some of the commonly used microbial polysaccharides are pullulan, xanthan, gellan, dextran, curdlan, scleroglucan, alternan, elsinan, levan, bacterial cellulose and bacterial alginate (Paul et al. 1986). According to their morphological localization, microbial polysaccharides are broadly divided into three different groups: intracellular, cell wall and extracellular polysaccharides. Evidently, microbial polysaccharides exhibit excellent biological activities such as antitumor, antibacterial, antioxidant, antiulcer and cholesterol-decreasing properties (Vibhuti et al. 2018, Pan and Mei 2010, Ramamoorthy et al. 2018, He et al. 2008, Giavasis 2014, Shu and Wen 2003). Additionally, microbial polysaccharides have also been discovered to be viscosifying, stabilizing, emulsifying or gelling agents in food and pharmaceutical products (Ahmad et al. 2015, Yangilar and Yildiz 2016).

7.2.2.8.1 Pullulan

Pullulan is a water soluble, neutral linear polysaccharide comprising of maltotriose trimer made up of α-(1→6)-linked (1→4)-α-d-triglucosides ($C_6H_{10}O_5$). Due to the presence of double linkage, pullulan exhibits high water solubility and flexibility. Pullulan comes under the extracellular polysaccharides group and is secreted by the fungus *Aureobasidium pullulans*. The molecular weights of pullulan range from 1 kDa to 2000 kDa. Pullulan is a biodegradable, non-hygroscopic, non-reducing, nontoxic, nonmutagenic, noncarcinogenic and edible polymer (Farris et al. 2014, Oğuzhan and Yangılar 2013, Kumar et al. 2012).

7.2.2.8.2 Xanthan Gum

Xanthan is a cold- and hot-water soluble extracellular polysaccharide, which is produced by the microorganism *Xanthomonas campestris*. But xanthan is commercially prepared by a fermentation process (BeMiller 2018, Sworn 2009b). The primary chain of the xanthan polymer is composed of cellulose-like linear backbone of (1→4)-linked β-D-glucose connected with a trisaccharide side chain on alternative glucose residue at C-3 position. The trisaccharide side chain consists of β-D-mannose

(1-4) linked β-D-acetyl glucose (1-2) linked α-D-mannose (McArdle and Hamill 2011, Prameela et al. 2018). Xanthan gums in cold water exhibit the rheology of being highly pseudo-plastic, which is described by the increased shear stress and reduced viscosity. Xanthan gels are thermoreversible in nature, which aids in storage of xanthan products without disturbing their textural characteristics (Fallourd and Viscione 2009). Xanthan is extensively used as a stabilizer and thickening agent in various food products (Cui et al. 2013).

7.2.2.8.3 Gellan Gum

Gellan gum is a long-chain extracellular polysaccharide, secreted by *Sphingomonas elodea* bacteria. Gellan biopolymer is an anionic linear heteropolysaccharide possessing repeating units of tetrasaccharide molecules such as α-L-rhamnose, β-D-glucose and β-D-glucuronic acid in the ratio of 1:2:1 with L-glycerate (Okiror and Jones 2012). In industry, gellan gum is manufactured by a fermentation process in an appropriate medium. Generally, gellan gum is synthesized in two acyl forms – high-acyl (HA) and low-acyl (LA) gellan gum. The gelation properties vary according to their acyl substitution form, as LA requires gel-promoting cations whereas HA produces gels by cooling down the hot gellan solutions (Sworn 2009a). Due to its gelling properties, gellan gums are widely used in the food and pharmaceutical industries (Krasaekoopt and Bhandari 2012, Morales and Ruiz 2016, Philp 2015).

7.2.2.8.4 Dextran

Dextran is a water-soluble homopolysaccharide composed of (1→6)-linked α-D-glucopyranose repeating units (Sidebotham 1974). Different varieties of dextran polysaccharides with different ranges of molecular weights and branching are produced by different microorganisms. Commercially available dextran is produced by the nonpathogenic bacteria *Leuconostoc mesenteroides* (Soetaert et al. 1995). The average molecular weight of dextran ranges from 1 kDa to 40,000 kDa. Dextrans have excellent volume expansion properties and inhibitory effects on thrombocyte aggregation and coagulation factors (Polifka and Habermann 2015, Reuvers 2001). Dextran has been used as an excellent coating polymer to improve biocompatibility (Estevanato et al. 2012). Therefore, dextran is extensively explored in biomedical applications and the pharmaceutical industry (Zahedi et al. 2017, Karandikar et al. 2017).

7.2.2.8.5 Curdlan

Curdlan is a neutral extracellular microbial polysaccharide composed of (1–3)-linked β-glucan units (Harada et al. 1993, Nishinari et al. 2009). Curdlan is produced by the bacteria *Alcaligenes faecalis* var. *myxogenes* during the fermentation process. The molecular weight of commercially available curdlan is more than 2000 kDa. It has excellent thermal stabilization and thermoreversible gelling properties when heated up to more than 80°C in aqueous solution. Curdlan particles get hydrated and soluble in water between 50°C to 60°C (Zhan et al. 2012). Curdlan produces solid and resilient gels which are considered as famous fat mimetics (Funami et al. 1998). Currently, most of the curdlans are used as food additives and stabilizers for processed foods (BeMiller 2018, Miwa et al. 1994).

7.2.2.8.6 Levan

Levan is a highly branched, high molecular weight, extracellularly produced non-structural microbial polysaccharide comprising of β(2→6)-linked fructans which are packed into a nano-sized spherical form, attaining greater stability than the linear polymer (Kasapis et al. 1994, Han 1989). Levan is biosynthesized by a variety of both gram-positive and gram-negative bacteria from fermentation of sucrose substrates through the activation of the levensucrase enzyme. Generally, levan is produced extracellularly by a wide range of bacteria: *Acetobacter xylinum, Bacillus subtilis, Bacillus lentus, Bacillus licheniformis, Halomonas smyrnensis, Halomonas smyrnensis, Paenibacillus polymyxa*, etc (Öner et al. 2016). In a study, levan polysaccharides were found to be effective hypoglycemic and antioxidant agents in diabetic conditions (Dahech et al. 2011). Due to its efficient biodegradability and biocompatibility properties, levan is used in personal care and medical applications (Öner et al. 2016).

7.2.3 Polycarbonates and Polyesters

Polycarbonates (PC) and polyesters (PE) are synthetic or naturally occurring oxygenated polymers and well-upgraded green polymers, which may contain different combinations of bio-derived cyclic monomeric units in their backbone. By green routes, PC and PE are generally obtained by the ring opening polymerization (ROP) and ring opening copolymerization (ROCOP) of monomers such as cyclic esters, cyclic carbonates, epoxides and anhydrides (Paul et al. 2015, Nakano et al. 2006). The unique structure and controlled synthesis process produce ecological biodegradable polymers with insignificant toxic by-products. Therefore, PC and PE have been explored for different applications including packaging, fibers, medical devices and engineering materials (Målberg et al. 2011, Auras et al. 2004, Wu et al. 2011).

7.2.3.1 Polyhydroxyalkanoates

Polyhydroxyalkanoates (PHAs) are naturally occurring biodegradable polymers produced by numerous bacteria and generally classified as aliphatic polyesters. PHAs are deposited in the bacterial cell cytoplasm in the form of water-soluble granules which serve as sources of energy and carbon for the microorganism (Reddy et al. 2003, Sudesh et al. 2000). PHAs have a wide spectrum of monomers that enable PHAs with distinctive properties. Depending on the carbon atoms present in the monomeric units, PHAs can be divided into two groups: long chain length (6–14 carbon atoms) PHA and short chain length (3–5 carbon atoms) PHA (Lee 1996). Poly(D-3-hydroxybutyrate) (P3HB) is the most common and intensively studied amongst the PHAs groups (Byun and Kim 2014). Depending on its monomeric units, the thermoplastic PHA polymer material shows extensive mechanical properties ranging from hard plastic material to elastic rubber type material (Verlinden et al. 2007, Ten et al. 2015, Raza et al. 2018).

7.2.3.2 Poly(Lactic Acid)

Poly(lactic acid) (PLA) has been known as a thermoplastic, high-strength and biodegradable aliphatic polyester. As the name suggests, PLA is a polymeric chain of lactic acids and is generally obtained by two main approaches: (1) the petrochemical route to produce D/L-lactic acids and (2) fermentation of sugars from different natural sources (such as sugarcane, corn starch and sugar beet) to produce L-lactic acids (Garlotta 2001, Wielgus et al. 2012). These monomeric units can be used to synthesize PLA polymers through three different processes: direct condensation polymerization, azeotropic dehydration condensation and ROP. The direct condensation and azeotropic dehydration processes are less followed as the yield of PLA is much less and produces fragile, ineffectual polymers. However, ROP yields large amounts of PLA and is suitable for commercial large-scale production due to its control in chemistry and resultant property modifications (Ross et al. 2017). The degradation of PLA is very effective as the ester bond has hydrolytically degradable nature (Kulkarni et al. 1971). Therefore, the green route production, biodegradability properties and high mechanical strength mean that this polymer features in the development of bio-nanocomposites for different potential applications (Jamshidian et al. 2010, Sin 2012).

7.3 APPLICATIONS OF GREEN POLYMERIC NANOCOMPOSITES (GPNCs)

GPNCs are no longer anomalies in the field of functional nanomaterials. GPNCs open up new avenues for the development of ecologically viable materials. In other words, enormous research has been done and for the past two decades various green synthesis methodologies have been implemented for the fabrication of GPNCs, which have overtaken traditional composites in research and production (Adeosun et al. 2012). The application of GPNCs requires a continuous development of eco-friendly nanomaterials from different green polymers. It is preferred that the synthesis process for the production of GPNCs is also green. Considering the increased concern for environmental sustainability, GPNCs offer a broad spectrum of applications in different sectors such as food packaging materials, nanofibers, construction, the automotive industry, development of electronic devices, air filters, water purification, dye degradation and absorption, tissue engineering implants and various biomedical devices production (see Figure 7.3) (Bhawani et al. 2018, Cheng et al. 2015). Some of the important applications are described concisely, with examples.

7.3.1 Tissue Engineering

Bioactive green polymer matrices, composites and scaffolds are essential components for the regeneration of tissue structure. Various natural and/or synthetic polymer combinations have been used to fabricate the tissue scaffold structures that are biocompatible, resorbable with excellent mechanical and tensile properties and less immunogenic (Zahedi et al. 2017, Aldana and Abraham 2017). The key objective of tissue engineering is to nurture and functionalize the cells over the polymer matrices. The cell adhesion can be regulated by the extracellular matrix proteins such as collagen, gelatin, laminin and fibronectin. Therefore, it is desirable to produce the engineered polymer scaffolds using these natural ECM proteins that will enhance cell growth and tissue repair (see Table 7.1) (Gulati et al. 2017, Gomes et al. 2008). In a study, Tambe et al. fabricated a collagen-immobilized PLA nanofiber scaffold to harness surface functionalization and better cellular adhesion in tissue engineering scaffolds. The collagen protein was

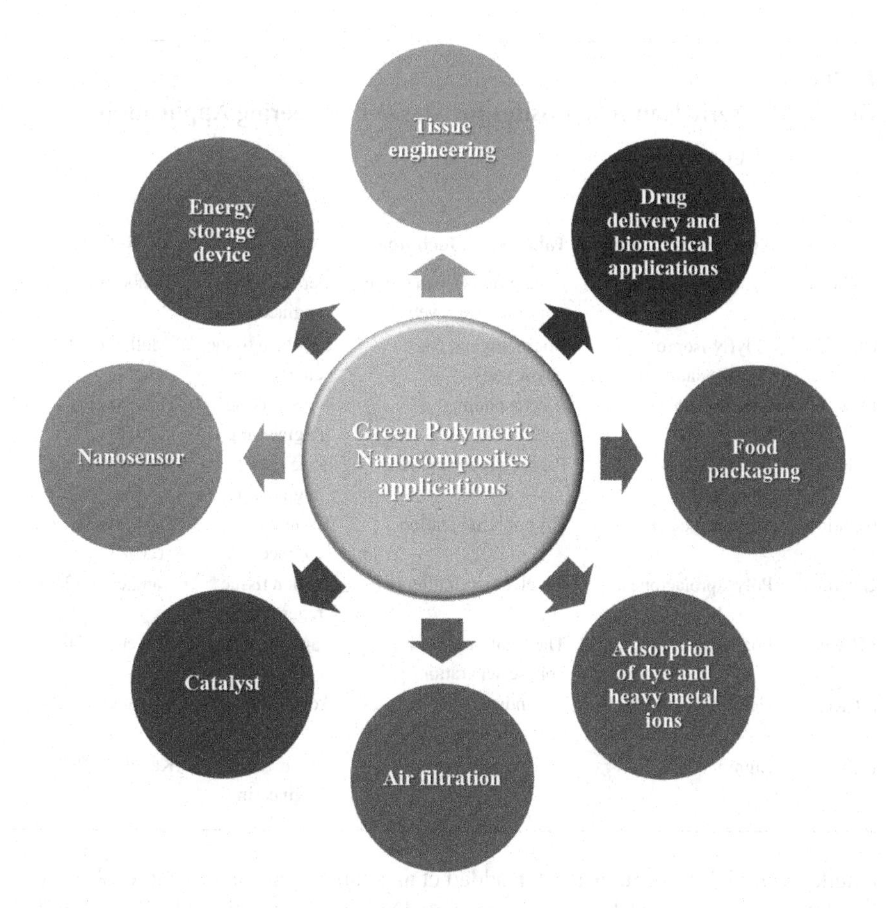

FIGURE 7.3 Schematic showing the versatile applications of green polymeric nanocomposites (GPNCs).

claimed to have better potency for the proliferation of human dermal fibroblast cells. Moreover, it was observed that cell growth is more efficient in genipin crosslinked collagen–PLA scaffold than the collagen–PLA scaffold without genipin. The incorporation of ECM protein, such as collagen, has improved the nonspecific binding of fibroblast cells to ECM and biocompatibility (Tambe et al. 2015).

In another study, Mellati et al. microfabricated thermoresponsive chitosan-grafted poly(N-isopropylacrylamide) (CS-g-PNIPAAm) 3D hydrogel for mimicking the zonal orientation of native articular cartilage which is crucial for appropriate tissue growth and function. The mesenchymal stem cells (MSCs) were harvested for their proliferation and differentiation over the 3D microengineered hydrogels. The CS-g-PNIPAAm hydrogels were allowed to produce different micromolds via photolithograghy to control the cellular alignment and elongation. In comparison with PNIPAAm hydrogels, the biochemical assays of CS-g-PNIPAAm hydrogels exhibited six- to seven-fold increases in the excretion of GAGs and collagen from MSCs within 28 days of incubation. These microengineered 3D cell-laden thermoresponsive hydrogels are thus testified to be a promising platform for the effective multi-zonal cartilage tissue engineering

TABLE 7.1
Green Polymeric Nanocomposites for Tissue Engineering Applications

Green Polymer	Co-Polymers/ Surface Modification/ Crosslinking Agents	Fabrication Technique	Applications	Reference
Collagen	PLA and genipin	Suface functionalization of nonwoven web	Dermal tissue engineering	Tambe et al. (2015)
Chitosan	Poly(N-isopro pylacrylamide)	Microengineered 3D hydrogels	Cartilage tissue engineering	Mellati et al. (2017)
PLA	Polyallylamine	Electospinning	Neural tissue engineering	Haddad et al. (2016)
Gelatin	Chitosan and bioglass	Freeze drying method	Bone tissue engineering	Maji et al. (2016)
Gelatin	Hyaluronic acid	Cryopolymerization	Tissue engineering	Rezaeeyazdi et al. (2018)
Gelatin	Polycaprolactone	Co-electrospinning	Tendon tissue regeneration	Yang et al. (2016)
PHA	PhaP, RGD	Thermally induced phase separation	Cartilage tissue engineering	You et al. (2011)
κ-CRG	Gelatin	UV-induced crosslinking	Adipose tissue	Tytgat et al. (2018)
PLA	Lignin	Electrospinning	Tissue engineering	Kai et al. (2016)

(Mellati et al. 2017). Furthermore, Haddad et al. synthesized electrospun PLA nanofiber 3D matrices for neural tissue engineering. Due to the suboptimal hydrophilicity of polyester surfaces for cellular adhesion, the PLA nanofibers were subjected to surface modification by aminolysis treatment using polyallylamine (PAAm) for the addition of amine groups in order to facilitate covalent binding of active biomolecules. The aminated PLA matrices were grafted with epidermal growth factor (EGF) and cultured with *in vitro* engineered neural stem-like cells (NSLC) derived from skin. The results showed that NSLCs were competent enough to proliferate and differentiate over the EGF-grafted PLA scaffolds and to exhibit viable growth with up to 14 days of incubation even in the absence of EGF in the medium (Haddad et al. 2016).

In another study, Maji et al. prepared a bioglass-infused gelatin–chitosan polymer nanocomposite scaffold for bone-tissue regeneration. The nanosized (20–30nm) bioactive bioglass was prepared by using the sol–gel method. By varying bioglass composition with 10% to 30% weight of chitosan and gelatin, a nanocomposite matrix was prepared using the freeze-drying method. The mechanical strength of the fabricated matrices was improved by crosslinking the matrices with glutaraldehyde to obtain interlinked 3D microstructures. Bioglass-infused gelatin-chitosan matrices exhibited noncytotoxic effects and allowed mesenchymal stem cell adhesion, proliferation and differentiation for regeneration of bone tissue (Maji et al. 2016). Rezaeeyazdi et al. have synthesized an injectable cryogel composed of gelatin and hyaluronic acid for tissue engineering applications. Formulation of these gelatin cryogels comply with the minimal invasive strategies, and prevented infections or damage to the adjacent cells. The addition of hyaluronic acid to gelatin matrices facilitates physical support for cellular adhesion and differentiation (Rezaeeyazdi et al. 2018).

On a similar note, Yang et al. fabricated a novel bio-nanocomposite scaffold by the co-electrospinning of polycaprolactone (PCL) and methacrylated gelatin (mGLT) for tendon tissue regeneration. Due to poor healing capacity of injured tendon and ligaments, it is necessary to regenerate the tissue over the 3D biomimetic scaffolds that impose the native features of tendon tissue. To mimic the tendon structure, electrospun PCL-mGLT nanofibers were interlocked with photocrosslinking, which improved the mechanical strength and allowed the development of a stack of scaffold sheets to produce a 3D scaffold. Human adipose-derived stem cells (hASCs) were impregnated in the PCL-mGLT scaffold and results showed excellent biocompatibility and cellular proliferation (Yang et al. 2016). Furthermore, You et al. have synthesized hydrophobic polyhydroxyalkanoate (PHA) scaffolds which are made of a copolyester of 3-hydroxybutyrate-cohydroxyhexanoate (PHBHHx), fused with PHA granule binding proteins (PhaP) for cartilage tissue engineering. The PhaPs were functionalized with RGD motifs for better adhesion, proliferation and homogeneous spread of cells. The PHA scaffolds were inoculated with human bone marrow mesenchymal stem cells (hBMSCs) for the formation of cartilage from chondrogenic differentiation. Over 14 days of incubation, enhanced extracellular matrix material and cartilage-specific extracellular substances such as GAGs and collagen were found in PhaP-RGD-coated scaffolds. The results showed a homogenously distributed cartilage-like matrice found over the PhaP-RGD-coated scaffold after 21 days of growth (You et al. 2011).

Tytgat et al. reported that the blend of κ-carrageenan and gelatin could produce potential hydrogel films for adipose tissue engineering. The functionalization of κ-carrageenan and gelatin has been carried out with methacrylate and methacrylamide moieties, respectively, which enables UV-induced crosslinking in the presence of a photo-initiator. The κ-carrageenan and gelatin blends have been evaluated for *in vitro* biocompatibility and cell viability of adipose tissue-derived mesenchymal stem cells. The result evidenced significant biocompatibility with excellent cell adhesion and proliferation (Tytgat et al. 2018). Similarly, Kai et al. synthesized electrospun PLA-lignin nanofiber composites with antioxidant activity for tissue engineering applications. To reduce the oxidative stress generated by PLA material in the cell, lignin was added to the PLA materials. The PLA–lignin was further blended with poly(L-lactic acid) (PLLA) and integrated into nanofiber matrices through electrospinning. The antioxidant assay of PLLA/PLA–lignin showed excellent radical scavenging capabilities over a period of 72 hours. In addition, these PLLA/PLA–lignin nanofibers were cultured with three different types of cells (PC12, human dermal fibroblasts and human mesenchymal stem cells). They showed good biocompatibility with balanced antioxidant activity (Kai et al. 2016).

7.3.2 Drug Delivery and Biomedical Applications

The application of these novel GPNC materials has generated a new research field of medical nanotechnology by providing solutions for new clinical problems such as disease diagnosis, therapeutics, systemic drug delivery, biosensing, bioimaging systems and surgical implants. Featuring lower cytotoxicity and excellent biocompatibility and biodegradability, these sustainable GPNC materials can be modified for distinct biological properties and functionalities under required physiological conditions. The design of GPNCs can be achieved by controlling their surface properties and size, which represents a strategy to achieve better response for specific biomedical applications (see Table 7.2) (Ramos et al. 2017, Jahangirian et al. 2017). For example, Salama et al. prepared a chitosan-grafted poly(3-sulfopropyl methacrylate) (CS-g-P(SPMA)) hydrogel nanocomposite through a free-radical polymerization method and subsequent calcium phosphate mineralization. The prepared hydrogel was evaluated for its drug release study and swelling behavior at different pH and saline conditions. A model protein bovine serum albumin (BSA) was loaded as a protein drug to the (CS-g-P(SPMA)) hydrogel nanocomposite and the results showed controlled drug release under physiological conditions (Salama 2018). Similarly, Dev et al. formulated PLA–CS nanoparticles (NPs) for controlled anti-HIV drug delivery applications. These PLA–CS NPs were synthesized by an emulsion method and loaded with a hydrophilic antiretroviral drug lamivudine. An *in vitro* drug release study displayed a higher drug-releasing capacity of 6%

TABLE 7.2
Green Polymeric Nanocomposites for Drug Delivery and Other Biomedical Applications

Green Polymer	Types of Nanocom posite Material	Synthesis Technique	Applications	Reference
Chitosan	Hydrogels	Free radical polymerization method	Controlled drug release	Salama (2018)
PLA and Chitosan	Nanoparticles	Emulsion method	Anti-HIV drug delivery	Dev et al. (2010)
Cellulose acetate	Nanofibers	Electrospinning	Absorbent material in feminine sanitary napkins	Yadav et al. (2016)
Chitosan	Nanoparticles	Solution phase synthesis	Antioxidant	Ahmad et al. (2018)
Hyaluronan and chitosan	Nanocoatings	Layer-by-layer self-assembly Coatings	Coatings for endovascular stents	Thierry et al. (2003)
Silk	Nanoparticles coated fibers	Exhaust method	Antibacterial sutures	Shubha et al. (2019)
Polylysine	Silver nanoparticles		Antibacterial sutures	Ho et al. (2013)
Albumin	Nanoparticles		Theranostic agent for cancer therapy	Wang et al. (2015

drug-loaded composition in alkaline pH compared to 3% drug-loaded composition in acidic pH conditions (Dev et al. 2010).

In another study, Yadav et al. reported the use of electrospun cellulose acetate (CA) nanofibers as a core absorbent material in feminine sanitary napkins. Feminine sanitary napkins are regularly used and are important disposable hygiene products. They contain superabsorbent polymers (SAPs) in their core in order to enhance the absorption capacity. Most of the SAPs used in sanitary napkins are not biodegradable and are linked with female health-related problems such as toxic shock syndrome. Therefore, Yadav et al. evaluated the absorption properties of CA nanofiber mats with/without SAPs under three conditions: free absorbency (deionized water), equilibrium absorbency (saline solution) and absorbency under load (synthetic urine). As a result, only CA electrospun nanofibers showed significantly high absorbency in all conditions, even in comparison with SAP-mixed CA nanofibers. The use of CA electrospun nanofibers in feminine sanitary napkins will thus enhance their biodegradability properties and reduce health-related problems (Yadav et al. 2016). Furthermore, Ahmad et al. synthesized chitosan–curcumin nanoparticles (CS–CC NPs) as antioxidants for the treatment of oxidative stress damage. Curcumin is a well-known antioxidant and a compound well-studied for its antioxidant properties. The CS–CC nanocomposites were prepared by using glutaraldehyde as a crosslinker in solution-phase synthesis. The CC-loaded CS NPs (alongside CS NPs and CC NPs) were evaluated for antioxidant properties in Swiss-strain adult male mice. The pre-treated (CS NPs, CC NPs and CS–CC NPs) mice were exposed to cadmium (Cd) at a rate of 10 mg/kg for 3 weeks. The results showed that CS–CC NPs appeared to be the best antioxidant in comparison to the other two nanoparticles (Ahmad et al. 2018).

According to Thierry et al., hyaluronan (HA) and CS can be used as bioactive nanocoatings for endovascular stents. For the coating experiment, a NiTi metal disc was coated primarily with polyethyleneimine (PEI), followed by layer-by-layer self-assembly coatings of HA and CS. The multilayered, coated NiTi disc exhibited enhanced antifouling properties and decreased platelet adhesion in an *in vitro* study. Due to the anti-inflammatory and wound-healing characteristics of HA and CS, this multilayered coated stent was reported as a potential solution for reducing the neointimal hyperplasia allied with stent grafting (Thierry et al. 2003). Most recently, Shubha et al. have prepared ZnO NPs coated silk fibers for high tensile strength antibacterial sutures. The ZnO NPs were synthesized with a bio-hydrothermal method and then coated over the degummed silk fibers. ZnO NPs coated silk fibers showed excellent tensile strength and antibacterial properties against gram-positive *Staphylococcus aureus* bacteria and proved to be a potential biomaterial for producing effective surgical sutures (Shubha et al. 2019). Similarly, Ho et al. synthesized hyperbranched polylysine-coated silver nanoparticles (HPL-Ag NPs) for the surface modification of poly(glycolic acid)-based surgical sutures. These HPL-Ag NP–modified sutures were evaluated for the release of silver ions into the surroundings. They showed constant release of silver ions over a period of 30 days, supporting their high antibacterial properties. They also tested as non-cytotoxic against L929 mouse fibroblast cells (Ho et al. 2013).

Furthermore, Wang et al. successfully prepared albumin-grafted, theranostic cyanine dye-linked gadolinium oxide nanocrystals (Alb-Cy-Gd_2O_3 NCs) for bioimaging and cancer therapy. Alb-Cy-Gd_2O_3 NCs have exhibited multiple imaging topographies including ultra-sensitivity and precise tumor localization. Due to their pH-responsive photothermal effect, the Alb-Cy-Gd_2O_3 NCs were preferable for cellular uptake, which, in turn, enhanced tumor accumulation. Therefore, these theranostic polymer nanocomposites represent promising materials for cancer therapy (Wang et al. 2015).

7.3.3 Food Packaging

Most food packaging materials are neither biodegradable nor environmentally friendly. These non-biodegradable packing materials are piled up in dump yards after use, creating a major source of soil pollution. So, these materials can be replaced by green polymers, which have better biodegradability and are renewable in nature. Due to poor mechanical strength and barrier properties, green polymers have limited usage in food packaging applications. To improve the mechanical properties, most of the green polymer composites were reinforced with nanomaterials through specific interactions during the synthesis process. Nanomaterials have a proportionally higher surface-area-to-volume ratio than their polymer counterparts, by which they facilitate the higher interaction of nanomaterials with the surfaces of the polymers, resulting in enhanced mechanical properties (see Table 7.3) (De Azeredo 2009). For example, Fortunati et al. prepared silver nanoparticles (Ag NPs) and cellulose nanocrystal (CNC)–infused PLA-based high-performance bionanocomposites for active food packaging applications. PLA/CNC/Ag NP films were developed by a solvent casting method. The effects of Ag NPs and CNCs were evaluated in terms of microstructure, barrier and nanoparticle migration properties, which showed that the use of a higher amount of CNCs (5%) and a lower amount of Ag NPs (1%) could produce an efficient barrier material with fewer migration properties (Fortunati et al. 2013).

TABLE 7.3
Green Polymeric Nanocomposites for Food Packaging Applications

Green Polymer	Nanocomposite Material	Synthesis Technique	Applications	Reference
PLA	Ag NP cellulose nano-crystals	Solvent casting method	Food packaging	Fortunati et al. (2013)
Guar gum	Cu–Ag NPs	Solvent casting	Food packaging method	Arfat et al. (2017)
HPMC	Ag NPs	Solvent casting	Food packaging	De Moura et al. (2012)
Pectin	Halloysite nanotubes	Solvent casting method	Food packaging	Makaremi et al. (2017)
PLA	ZnO NPs	Single screw extrusion	Food packaging	Marra et al. (2016)
κ-CRG	Ag NPs	Microwave green synthesis method	Food packaging	Goel et al. (2019)

In another similar study, Arfat et al. developed silver–copper alloy nanoparticle (Ag–Cu NP)–infused guar gum (GG)–based nanocomposites for active food packaging materials. The prepared film was investigated for mechanical strength and optical and thermal properties, NP loading capability and antimicrobial effect. The tensile test results showed that the incorporation of Ag–Cu NPs improved the mechanical strength of the film. In addition, it exhibited strong antibacterial effects against both gram-negative and gram-positive bacteria (Arfat et al. 2017). Furthermore, Moura et al. synthesized a Ag NP-incorporated hydroxypropyl methylcellulose (HPMC) matrix as an active food packaging material. Two different sizes (average size: 41 nm and 100 nm) of Ag NPs were used as active fillers in the polymer matrix. The mechanical strength and water-vapor-resistant barrier were found to be strong in the case of nanocomposites with a smaller size (41nm) of Ag NPs. Also, the bactericidal properties of Ag NPs/HPMC thin films were evaluated by the zone inhibition method and the results showed greater antibacterial properties for nanocomposites composed of 41 nm-sized Ag NPs (De Moura et al. 2012).

In another investigation, Makaremi et al. fabricated halloysite nanotube (HNT)–reinforced pectin bio nanocomposite films for active food packaging applications. For a comparison study, two types of HNT were used: (1) long and thin HNTs (200 to 30,000 nm length) and (2) short and stubby (50 to 3000 nm). Excellent thermal, mechanical and contact angle properties were observed for the films with long and thin HNTs. Furthermore, HNTs were encapsulated with salicylic acid as an active biocidal agent and studied for their controlled release kinetics. The results revealed greater antimicrobial activity against both gram-negative and gram-positive bacteria, indicating their potential for food packaging materials (Makaremi et al. 2017). Next, Marra et al. prepared zinc oxide nanoparticles (ZnO NPs) mixed PLA bio nanocomposite films for sustainable food packaging applications. The PLA films were fabricated with three different concentrations (1, 3 and 5wt%) of ZnO NPs and were examined for tensile strength, barrier and antibacterial properties against *Escherichia coli* gram-negative bacteria. The PLA bionanocomposite with 5% ZnO NPs showed better mechanical properties, reduced CO_2 and O_2 permeability and effective bactericidal properties (Marra et al. 2016). Most recently, Goel at al. formulated κ-CRG-Ag nanocomposites by a facile microwave green synthesis method for eradication of bacterial biofilms. κ-CRG was used as both reducing and stabilizing agents for the synthesis of κ-CRG-Ag NPs which were infused in KCL crosslinked κ-CRG solution to produce hydrogels. κ-CRG-Ag hydrogels showed excellent thermal stability as well as antimicrobial activity against *Staphylococcus aureus* and *Pseudomonas aeruginosa* biofilm (Goel et al. 2019).

7.3.4 Adsorption of Dye and Heavy Metal Ions

Environmental nanotechnology and polymer technology play a crucial role in the design and fabrication of green polymeric nanocomposites for detection, adsorption, degradation and exclusion of hazardous pollutants such as toxic dye and heavy metal ions. Due to higher surface interaction properties, the nanomaterials have a higher tendency for interaction with pollutants. The use of green polymers for nanomaterial-infused polymer matrices will enhance the pollutant retention properties as well as biodegradability of the nanocomposites. Hence, green polymeric nanocomposites offer sustainable, renewable, mechanically strong and highly catalytic materials for toxic dyes and heavy metal adsorption applications (see Table 7.4) (Chaturvedi et al. 2012). For instance, Haider et al. synthesized a copper nanoparticle (Cu NP)–loaded CS-Cellulose microfiber mat (CS-CMM) for dye absorption and degradation applications. Initially, CS polymers were adhered to the CMM and allowed for the adsorption of Cu^{2+} ions from a copper sulfate ($CuSO_4$) aqueous solution. Then, Cu^{2+} metal ions were reduced by sodium borohydride ($NaBH_4$) to produce Cu NPs in the microfiber matrices. The prepared Cu NP/CS-CMM bionanocomposites were evaluated for their catalytic activity against three dyes: 2-nitrophenol (2NP), 4-nitrophenol (4NP) and organic cresyl blue (CB) dye. Remarkably, the conversion rate constants for 2NP, 4NP and CB were observed as 1.2×10^{-3} s^{-1}, 2.1×10^{-3} s^{-1} and 1.3×10^{-3} s^{-1} respectively, which demonstrate higher catalytic activity against three dyes (Haider et al. 2016). Moreover, chitosan was also used by Hoa et al. who aimed to prepare Fe_3O_4-CS nanoparticle-filled graphene matrices (Fe_3O_4–CS@graphene) for textile dye adsorption applications. Fe_3O_4–CS@graphene nanocomposites were produced by a one-step facile solvothermal method. For the adsorption behavior evaluation of the prepared Fe_3O_4–CS@graphene nanocomposites, methylene blue (MB) dye was used as a model dye for study. The adsorption test results revealed rapid and high adsorption (98% within the contact time of 5 min at pH 9) of MB dye to the Fe_3O_4–CS@graphene nanocomposites (Van Hoa et al. 2016). Similarly, Min et al.

TABLE 7.4
Green Polymeric Nanocomposites for the Efficient Adsorption of Dye and Heavy Metal Ions

Green Polymer	Nanocomposite Material	Synthesis Technique	Dye and Heavy Metal Adsorption	Reference
Chitosan	CuNPs		2-Nitrophenol	Haider et al. (2016)
Cellulose			4-Nitrophenol cresyl blue	
Chitosan	Fe_3O_4 NPs	Solvothermal	Methylene blue	Van Hoa et al. (2016)
		Graphene	Method	
Chitosan Fe^{3+} ions	CS nanofibers	Electrospinning	Arsenate	Min et al. (2016)
κ-CRG		PVA	Freeze-thaw crystal violet	Hosseinzadeh et al. (2015)
		SM nanoclay	Technique	
Cellulose	CMC	Chemical	Methylene blue	Peng et al. (2016)
		Nanoclay	Crosslinking	
Gelatin		MNPs	Emulsification methylene blue	Saber-Samandari et al. (2017)
		CNT	Method direct red	
Zein	Electrospinning	Reactive black 5		Qureshi et al. (2017)

prepared iron (Fe^{3+}) functionalized electrospun CS nanofibers (Fe–CS NF) for trace arsenate (As(V)) removal from water. The fabricated Fe–CS NFs were found to be highly effective for As(V) adsorption (maximum 11.2 mg/g) at pH 7.2. Fe–CS NFs were capable of regeneration using 0.003M NaOH solution and the removal rate was found to be more than 98% up to 10 cycles of adsorption/desorption (Min et al. 2016).

In an independent investigation, Hosseinzadeh et al. prepared κ-carrageenan and poly(vinyl alcohol) (PVA) hydrogels incorporated with sodium montmorillonite (SM) nanoclay. The blend of polymers and SM nanoclay were crosslinked with the freeze–thaw technique and subsequent addition of K^+ ions to the mixture solution facilitated gel formation. The nanocomposite hydrogels were evaluated for the adsorption of cationic crystal violet dye by batch adsorption systems. In comparison with nanoclay free hydrogel composites, the nanocomposites exhibited maximum adsorption capacity of 151 mg/g (Hosseinzadeh et al. 2015). Next, Peng et al. developed superabsorbent cellulose–clay nanocomposite hydrogels for dye absorption from polluted water. Cellulose (CL) and carboxymethyl cellulose (CMC) were chemically crosslinked to prepare CL–CMC hydrogels and clay was intercalated by the treatment of NaOH/urea solution. CL–CMC/clay nanocomposites exhibited effective absorption and removal capacity of MB by 98% of total concentrations (Peng et al. 2016).

Saber-Samandari et al. synthesized magnetic Fe_3O_4 nanoparticle (MNP)–incorporated gelatin (Gel) hydrogels for effective removal of cationic and anionic dyes from aqueous solution. Gelatin hydrogels were entrapped with carboxylic acid functionalized multi-walled carbon nanotubes (CNTs) and embedded MNPs. The CNT–MNP Gel beads were evaluated for both anionic direct red 80 (DR) dye and cationic MB dye in aqueous solutions. Remarkably, the magnetic bionanocomposites showed excellent dye absorption capacity and removed 96.1% of DR and 76.3% MB from the test solutions (Saber-Samandari et al. 2017). Furthermore, Qureshi et al. have prepared electrospun zein nanofibers as a recyclable adsorbent for the removal of dye from aqueous solutions. Due to the higher surface area-to-volume ratio, zein nanofiber forms are more commonly studied for use in sustainable green nano-adsorbent materials compared with the powder and film forms. Twenty-five percent (w/v) zein powder was dissolved in 80% aqueous ethanol solution to make a proper blend for the electrospinning of zein nanofiber. The electrospun zein nanofibers were evaluated for reactive black 5 (RB5) dye removal properties and the results showed 97 ± 1.8% of RB5 dye adsorption and removal from aqueous solution at normal pH and room temperature within 20 minutes of interaction (Qureshi et al. 2017).

7.3.5 Air Filtration

The increasing rate of air contaminants/pollutants in the environment triggers various health-related issues, lung diseases and respiratory problems. For the betterment of human life, these pollutants need to be filtered out of the air. Recently, various machines and methods have been developed to eliminate pollutants from the atmosphere. However, these techniques are inadequate for the present circumstances and require high amounts of energy and long periods of time to purify the atmosphere. Hence, local filtration units, which can be provided by nanotechnology, may provide the best alternative for air filtration problems. Among various nanocomposites, nanofiber matrices can be used as air filtration material for the adsorption or elimination of pollutants. The major properties of nanofiber materials are high surface-to-volume ratio, high pore density, nanoscale fiber diameter, high-density pores and very low weight (Balamurugan et al. 2011).

Various green polymer compositions/blends were used for the fabrication of biodegradable, biocompatible and sustainable nanofiber matrices (see Table 7.5). For example, Nicosia et al. prepared electrospun cellulose acetate (CA) nanofibers on nylon substrates for air filtration applications. CA nanofibers were obtained by the electrospinning of 14% CA in acetone-DMSO-acetic acid solution and collected over the nylon substrate. The prepared CA nanofibers had average diameters of 300 nm and exhibited heat-resistant properties with high air filtration performance with a quality factor (QF) value of $0.080 \pm 0.050\ Pa^{-1}$ (Nicosia et al. 2016). Furthermore, Wang et al. synthesized Ag NP-infused silk nanofiber as a lightweight air filter. Silk fibroin (SF) nanofibers were fabricated by the electrospinning method and their air filtration properties were evaluated for $PM_{2.5}$ and submicron particles. The SF nanofibers were further improved by addition of AgNPs for potential antibacterial activity. The SF nanofibers showed enhanced filtration efficiency of 98.8% for $PM_{2.5}$ and 96.2% for 300 nm particles, thus representing a potential multifunctional green polymeric nanocomposite for air filtration applications (Wang et al. 2016).

Souzandeh et al. designed gelatin (GA) nanofibers for efficient air particulate/pollutant and toxic gas (e.g. HCHO and CO) filtration. GA powder (type A) of 18 wt% was dissolved in 80% acetic acid aqueous solution and the GA nanofibers were prepared by the electrospinning method with operating voltage of 18 to 20 kV. The synthesized GA nanofiber mats were found to have extremely high particulate removal efficiency of 99.3% for $PM_{0.3}$, 99.6% for $PM_{2.5}$ and toxic gas absorbing capacity of 80% for HCHO and 76% for CO (Souzandeh et al. 2016). Further, Wang et al. have prepared ultrafine CS nanofibers containing TiO_2 and/or Ag NPs for air filtration applications. The nanofiber production rate was increased and it reached up to 50 gh^{-1} by the needle-less electrospinning method which was much less nanofiber production by traditional electrospinning method. The CS nanofibers produced in this method were smooth and had uniform surface morphology with diameters ranging

TABLE 7.5
Green Polymeric Nanocomposites Used for Air Filtration

Green Polymer	Nanocomposite Material	Synthesis Technique	Application	Reference
Cellulose	Nanofibers	Electrospinning	Heat resistant air filter	Nicosia et al. (2016)
Silk	AgNPs-nanofibers	Electrospinning	$PM_{2.5}$ and 300nm particulate filter	Wang et al. (2016)
Gelatin	Nanofibers	Electrospinning	Pollutant and toxic gas filtration	Souzandeh et al. (2016)
Chitosan	Ag/TiO2 NPs nanofiber	Needleless electrospinning	Antibacterial air filter	Wang et al. (2016)

from 25 to 60 nm. The CS/PVA/Ag/TiO_2 hybrid nanofibers exhibited a substantial rise in aerosol filtration efficiency of more than 99%, with excellent antibacterial properties against both gram-positive and gram-negative bacteria (Wang et al. 2016).

7.3.6 Catalyst

Nanomaterials are very widely used in the chemical industry due to their high catalytic activity. Nanocatalysts have been used on a massive scale in the production of petrochemical and chemical products, the conversion and purification of crude oil and natural gases into fuels, dye degradations, environmental remediation, conversion of biomass and chemical conversion processes (Chaturvedi et al. 2012). The chemical processes require a large amount of catalyst that increases the production rate within a particular time period. Hence, the catalyst technology relies on nanomaterials due to the higher number of surface interactions with respect to volume and higher surface energy. The addition of nanocatalysts to the reaction improves the selectivity of the reaction and reduces energy consumption (Sharma et al. 2015). In order to increase the efficiency and environmental compatibility of the nanocatalyst, green polymers are supplemented to fabricate new bio nanocomposites for catalytic applications (see Table 7.6). For example, Kamal et al. prepared nickel nanoparticle-embedded and

TABLE 7.6
Green Polymeric Nanocomposites as a Catalyst for Dye Degradation and Heavy Metal Adsorption

Green Polymer	Nanocomposite Material	Synthesis Technique	Applications	Reference
Chitosan	Ni NPs	Chemical reduction	4-NP to 4-AP	Kamal et al. (2016)
Cellulose			2-NP to 2-AP reduction of methyl orange dye	
Chitosan	AgNPs	Chemical reaction	4-NP to 4-AP	Alshehri et al. (2016)
Chitosan	MWCNTs AgNPs Fe_3O_4 NPs	Multiple emulsion-Chemical crosslinking	4-NP to 4-AP	Xu et al. (2017)
κ-CRG	Fe_3O_4 NPs	Green synthesis	Synthesis and purification of rhodanines	Rostamnia et al. (2015)
Dextran	Ag NPs	Green synthesis	Reduction of MB and - antibacterial activity	Ma et al. (2009)
Pectin	ZSP	Sol–gel method	Reduction of MB Heavy metal adsorption antibacterial activity	Pathania et al. (2015)
Alginate	QC-Ag NPs	Emulsification technique	4-NP to 4-AP	You et al. (2014)

CS-coated CL filter paper for pollutant degradations. The CL filter paper was coated with 1% CS solution and further dipped into 0.2M $NiCl_2$ aqueous solution. The CS–CL filter paper was adsorbed with Ni^{2+} ions and finally it was reduced by 0.1M $NaBH_4$ aqueous solution to produce Ni nanoparticles on the surface of CS–CL filter papers. The prepared nanocomposite exhibited excellent catalytic properties by conversion of three toxic compounds: 4-nitrophenol (4-NP) to 4-aminophenol (4-AP), 2-nitrophenol (2-NP) to 2-aminophenol (2-AP) and reduction of methyl orange dye (Kamal et al. 2016). In a similar study, Alshehri et al. have achieved the catalytic reduction of 4-nitrophenol by Ag NPs and multi-walled carbon nanotubes (MWCNTs) decorated CS polymer matrix. Twenty percent w/v CS powder was dissolved in 2% acetic acid aqueous solution and further added with MWCNTs–COOH. The mixture solution was added with 25% glutaraldehyde and placed in a magnetic stirrer for 3 hours at 60°C temperature. The subsequent MWCNT–polymer matrix was washed with water and vacuum dried for 24 hours at 60°C. $AgNO_3$ was reduced in a MWCNT–polymer matrix by 1 mM trisodium citrate to produce a Ag NPs@MWCNTs–CS polymer nanocomposite. The developed nanocomposite showed effective catalytic activity against 4-NP in the presence of $NaBH_4$ aqueous solution (Alshehri et al. 2016).

In a similar context, Xu et al. also prepared CS bionanocomposites for catalytic reduction of 4-NP by designing magnetic separable CS microspheres. Fe_3O_4 magnetic nanoparticles were deposited in the core of CS microcapsules through multiple emulsion–chemical crosslinking methods and then coated with Ag NPs on the microsphere surface. The catalytic activity of Fe_3O_4/CS–Ag NPs was investigated for catalytic properties against 4-NP and showed 98% conversion of 4-NP to 4-AP within 15 minutes (Xu et al. 2017). In a different study, Rostamnia et al. synthesized magnetically separable κ-carrageenan-stabilized Fe_3O_4 nanoparticles as a catalyst for the green synthesis of rhodanines. The biodegradable Fe_3O_4/κ-CRG nanocomposites were employed for the green synthesis of rhodanines and subsequently it facilitated the purification/separation of the final product (Rostamnia et al. 2015). Ma et al. prepared AgNPs incorporated dextran (DX) hydrogel for catalytic as well as antibacterial applications. The DX hydrogel was obtained by the crosslinking of DX with N, N-methylenebisacrylamide in aqueous sodium hydroxide (NaOH) solution. The stable Ag NPs were prepared in the DX hydrogel network without using any reducing agent. The hybrid AgNP–DX nanocomposites exhibited excellent catalytic properties against MB in the presence of $NaBH_4$ and had potent antibacterial action against *Bacillus cereus* (Ma et al. 2009). Further, Pathania et al. have described a multipotent method for the photo-catalysis and reduction of MB, heavy metal adsorption and antimicrobial activity by pectin-linked zirconium (IV) silicophosphate nanocomposite (PC@ZSP). Initially, PC@ZSP was prepared as an ion exchanger by sol-gel method and further it was explored for the photo catalytic activity. The PC@ZSP was utilized for the degradation of MB dye in the presence of solar light and 97% degradation was found within 60 minutes of solar exposure (Pathania et al. 2015). In a different study, alginate (AL) was used by You et al. for the AgNP-deposited AL microspheres for catalytic application. AL microspheres were prepared by emulsification technique and the microsphere surface was coated with quaternized cellulose (QC)-Ag NPs. The prepared AL-QC-Ag NPs exhibited excellent catalytic activity for 4-NP conversion into 4-AP with a high reaction rate constant of 2.75 min^{-1} (You et al. 2014).

7.3.7 Nanosensor

Recent developments in nanotechnology have created the research area of bionanosensors. The essential properties of nanosensors need to be enhanced in terms of sensitivity, specificity, detection time, toxicity, affordable price range, and biodegradability. The incorporation of green polymers into nanosensors is thus expected to improve their biocompatibility and biodegradability, along with the other required properties (Sapountzi et al. 2017). The engineered nanomaterials have unique transducer surfaces, which facilitate sensing activity. The production of nanosensors is usually constituted by various conductive materials, metal/metal oxide nanoparticles, polymer nanofibers, carbon nanotubes and nanorods/nanowires (see Table 7.7) (Bogue 2008). The biosensing properties can be further enhanced by the addition of biomolecules or bioreceptors with higher specificity (Wujcik et al. 2014). Wu et al. developed a novel method for the biosensing application of enzymatic glucose. The nano biocomposite was prepared by the combine electrospinning of CS, PVA and Prussian blue. The principle of this prepared bionanosensor was based on the quantitative measurement of intermediate product hydrogen peroxide (H_2O_2). The obtained nanofibers exhibited food biocompatibility properties and had a porous structure, which enhanced the sustainability of Prussian blue in the nanofiber structure. Furthermore, for biosensing study, glucose was allowed to be oxidized by the enzyme glucose oxidase to produce H_2O_2. The fabricated nanofibers exhibited good sensing behavior for H_2O_2 by which glucose quantification could be done with a low detection limit of 3.61×10^{-7} M concentration (Wu and Yin 2013).

Saithongdee et al. fabricated CC-containing zein electrospun nanofibers for iron (III, Fe^{3+}) ion detection. The design of the nanosensor allowed the detection of the concentration of Fe^{3+} ions with the naked eye by dipping the CC-loaded zein nanofibers down into the sample. The presence of Fe^{3+} ions in the solution changed the color of the nanofiber from yellow to brown. The detection of Fe^{3+} ions was also investigated in solutions with different pH and other metal ions. It was observed that Fe^{3+} ion

TABLE 7.7
Green Polymeric Nanocomposites for Nanosensing Application

Green Polymer	Nanocomposite Material	Synthesis Technique	Applications	Reference
CS	PVA and Prussian blue loaded CS nanofiber	Electrospinning	Glucose quantification	Wu and Yin (2013)
Zein	CC-loaded zein nanofiber	Electrospinning	Fe^{3+} ions detection	Saithongdee et al. (2014)
Cellulose Acetate	CC-loaded CA nanofiber	Electrospinning	Pb^{2+} ions detection	Raj and Shankaran (2016)
Albumin	FITC	Electrospinning	pH change detection	Kowalczyk et al. (2008)
Starch CMC	CMC–starch–CuO NP nanocomposites	Casting method Sensor	Humidity	Hadi and Hashim (2017)

detection, by using CC loaded zein nanofibers, was highly efficient in a solution with pH 2 and could detect the limit up to 0.4 mg/L (Saithongdee et al. 2014). Similarly, Raj et al. have also developed CC based electrospun cellulose acetate nanofibers for lead (Pb^{2+}) detection with a visible detection limit of 20 μM (Raj and Shankaran 2016).

Further, Kowalczyk et al. electrospun bovine serum albumin (BSA) protein into a nanofiber structure and used it as local pH sensor. The BSA protein was blended with poly(ethylene oxide) (PEO) polymer as a supporting polymer to make the protein solution spinnable. Then, the BSA/PEO nanofibers were covalently linked with fluorescein isothiocyanate (FITC) to make the nanofibers fluorescent. The FTIC-labeled BSA/PEO nanocomposites exhibited changes in fluorescent intensity with change from neutral pH to alkaline pH 10, thus making detection possible without a sophisticated instrument (Kowalczyk et al. 2008). In a different investigation, Hadi et al. synthesized a novel nanocomposite material composed of CMC, starch and copper oxide nanoparticles (CuO NPs) for humidity-sensing applications. It was observed that the DC current conductivity of the CMC–starch–CuO blend was increased by the increase in concentration of CuO nanoparticles. As a result, the electric capacitance was measured in different samples with humidity ranging from 40 to 90 RH%. CMC-starch-CuONPs nano composites showed higher sensitivity in a humidity range of 60 to 90 RH% (Hadi and Hashim 2017).

7.3.8 Energy Storage Devices

Energy storage systems are widely used in portable electronic devices, with an endless demand for enhanced storage capacity. Considering the high demand for batteries and their recycling and toxicity properties, newly developed batteries are replacing the conventional acid battery with ecofriendly materials (Kafy et al. 2015). Currently, lithium-ion batteries (LIBs) are the most popularly used energy storage devices. In general, LIBs comprise four components: an anode, a cathode, a separating membrane and an electrolyte solution. The separating membrane of the battery plays a crucial role in insulating anode and cathode to prevent electrical short-circuit and simultaneously allows swift transfer of the charge carrier. Separators used in commercial energy storage devices are basically microporous polymer membranes (Li et al. 2015). These separators and electrolyte solutions have received ample attention aimed at increasing biodegradability and biocompatibility, which could be achieved by the use of green polymer technology. Various green polymers have been utilized and fabricated to produce biodegradable gel polymer electrolytes and polymeric separators (see Table 7.8). For instance, Kafy et al. prepared graphene oxide (GO) nanoplate–incorporated CL nanocomposites for environmentally stable energy and memory storage devices. A GO–CL nanocomposite was prepared by a solvent casting evaporation method and the dispersion of GO in the CL matrix was confirmed by Fourier transform infrared (FTIR) spectroscopy, X-ray diffraction (XRD) and scanning electron microscopy (SEM). The prepared GO–CL nanocomposites were evaluated for dielectric and ferroelectric properties with the variation of temperature and voltage. The results showed homogenous distributions of covalently bound GO nanoplates over a cellulose matrix, and ferroelectric properties were found to increase by the rise in applied electric field (Kafy et al. 2015).

TABLE 7.8
Green Polymeric Nanocomposites Used in Energy Storage Devices

Green Polymer	Nanocomposite Material	Synthesis Technique	Applications	Reference
Cellulose	Graphene oxide nanoplates	Solvent casting evaporation method	Energy and memory storage devices	Kafy et al. (2015)
Cellulose	CL nanofibers	Spinning process	LIB separators	Zhang et al. (2015)
Bacterial Cellulose	CL nanofibers	–	LIB separators	Jiang et al. (2015)
Bacterial Cellulose	SiO_2–CL nanofibers	–	LIB separators	Jiang et al. (2016)
Cellulose	CL nanocrystals	–	LIB separators	Ladhar et al. (2015)

Zhang et al. also used CL nanofiber materials as LIB separators in energy storage devices. Nonwoven polyethylene terephthalate (PET) fibers were fabricated by a wet-laid process and the CL nanofibers were coated over the PET surface by a papermaking process. The ecofriendly CL nanofibers have the advantage of a rapid electrolyte uptake of 250% while the traditionally used polypropylene (PP) can only uptake a maximum of 65%. Another finding showed that the CL nanofiber separator has heat resistant properties with no shrinkage while exposed to 180°C for 1 hour, whereas PP separator was shrunk by more than 50% (Zhang et al. 2015). In another study, Jiang et al. have prepared bacterial cellulose (B-CL) nanofibrous membrane as a thermally stable LIB separator. B-CL membrane separator exhibited excellent thermal stability up to 180°C, good ionic conductivity with efficient battery performance (Jiang et al. 2015). The same research group prepared core–shell nanofibrous B-CL membranes by covalently attaching SiO_2 nanoparticles over the surface. This novel SiO_2@B-CL nanocomposite showed enhanced electrolyte wettability, ionic conductivity and exerted higher coating strength than conventionally used ceramic coated separators (Jiang et al. 2016). Ladhar et al. fabricated a potential green battery separator using natural rubber (NR) and nanocellulose materials. The nanocellulose was extracted from the date palm tree in two forms: nanofibrillated cellulose (NFC) and CNCs. The fabricated CNC–NR material was found to be a good conductive material for battery separators (Ladhar et al. 2015).

7.4 CONCLUDING REMARKS

The integration of green polymers into nanoscience and nanotechnology bestows a more ecofriendly polymeric nanocomposite material for various routine applications. Along with environmentally friendly characteristics, nanocomposites exhibit their allied nanoscale properties along with the polymer's native functions. GPNC usage reduces the environmental, economic, and safety challenges of the product. One of the biggest advantages is the fact that most green polymers are low cost, easily accessible, biocompatible, biodegradable, renewable and have low energy consumption. Henceforth, green polymer technology is demystifying the development and systematic application of GPNCs. However, green polymers have some limitations, such as limited availability, immunogenicity, risk of pathogen transmission, lower synthesis rates and time-consuming purification processes that can, however, be improved with the aid of further studies. Other disadvantages include low mechanical/tensile strength requiring synthetic polymer substrates and water immiscibility requiring external coupling or crosslinking agents. In spite of these disadvantages, GPNCs are still considered to be promising and highly sustainable materials with tremendous development prospects. Hence, GPNCs are also considered to be effective alternatives for mitigating the rise in toxic and non-biodegradable nanomaterials in the environment.

ACKNOWLEDGMENTS

MKM acknowledges the MHRD fellowship from IIT-Roorkee for pursuing a PhD. KMP acknowledges the research support of DBT-IYBA fellowship, and research grants from NMCG-MoWR, SERB-DST, Govt. of India and BEST Pvt. Ltd.

REFERENCES

Adeosun, S. O., G. Lawal, S. A. Balogun, and E. I. Akpan. 2012. Review of Green Polymer Nanocomposites. *Journal of Minerals and Materials Characterization and Engineering* 11 (04): 385.

Agheb, M., M. Dinari, M. Rafienia, and H. Salehi. 2017. Novel Electrospun Nanofibers of Modified Gelatin-Tyrosine in Cartilage Tissue Engineering. *Materials Science Engineering: C* 71: 240–51.

Ahmad, M., G. M. A. Taweel, and S. Hidayathulla. 2018. Nano-Composites Chitosan-Curcumin Synergistically Inhibits the Oxidative Stress Induced by Toxic Metal Cadmium. *International Journal of Biological Macromolecules* 108: 591–97.

Ahmad, N. H., S. Mustafa, and Y. B. Che Man. 2015. Microbial Polysaccharides and Their Modification Approaches: A Review. *International Journal of Food Properties* 18 (2): 332–47.

Ahmed, A. B. A., M. Adel, P. Karimi, and M. Peidayesh. 2014. Pharmaceutical, Cosmeceutical, and Traditional Applications of Marine Carbohydrates. *Advances in Food and Nutrition Research*, 197–220: Elsevier.

Aldana, A. A., and G. A. Abraham. 2017. Current Advances in Electrospun Gelatin-Based Scaffolds for Tissue Engineering Applications. *International Journal of Pharmaceutics* 523 (2): 441–53.

Alshehri, S. M., T. Almuqati, N. Almuqati, E. Al-Farraj, N. Alhokbany, and T. Ahamad. 2016. Chitosan based polymer matrix with silver nanoparticles decorated multiwalled carbon nanotubes for catalytic reduction of 4-nitrophenol. *Carbohydrate Polymers* 151: 135–43.

Altman, G. H., F. Diaz, C. Jakuba, T. Calabro, R. L. Horan, J. Chen, H. Lu, J. Richmond, and D. L. Kaplan. 2003. Silk-Based Biomaterials. *Biomaterials* 24 (3): 401–16.

Aluigi, A., C. Vineis, A. Varesano, G. Mazzuchetti, F. Ferrero, and C. Tonin. 2008. Structure and Properties of Keratin/Peo Blend Nanofibres. *European Polymer Journal* 44 (8): 2465–75.

Anastas, P. T., and M. M. Kirchhoff. 2002. Origins, Current Status, and Future Challenges of Green Chemistry. *Accounts of Chemical Research* 35 (9): 686–94.

Antonio, J. D. S., and R. V. Iozzo. 2001. Glycosaminoglycans: Structure and Biological Functions. *eLS*. Wiley-Blackwell.

Arfat, Y. A., M. Ejaz, H. Jacob, and J. Ahmed. 2017. Deciphering the Potential of Guar Gum/Ag-Cu Nanocomposite Films as an Active Food Packaging Material. *Carbohydrate Polymers* 157: 65–71.

Auras, R., B. Harte, and S. Selke. 2004. An Overview of Polylactides as Packaging Materials. *Macromolecular Bioscience* 4 (9): 835–64.

Balamurugan, R., S. Sundarrajan, and S. Ramakrishna. 2011. Recent Trends in Nanofibrous Membranes and Their Suitability for Air and Water Filtrations. *Membranes* 1 (3): 232–48.

BeMiller, J. N. 2018. *Carbohydrate Chemistry for Food Scientists*. Elsevier.

Bertolini, A. 2009. *Starches: Characterization, Properties, and Applications*. CRC Press.

Bhawani, S., A. Bhat, F. Ahmad, and M. Ibrahim. 2018. Green Polymer Nanocomposites and Their Environmental Applications. *Polymer-Based Nanocomposites for Energy and Environmental Applications*, 2018: 617–33: Elsevier.

Blakemore, W. R. 2016. Polysaccharide Ingredients: Carrageenan. *Reference Module in Food Science*, 2016: 1–7. Elsevier.

Bogue, R. 2008. Nanosensors: A Review of Recent Progress. *Sensor Review* 28 (1): 12–17.

Burdick, J. A., and M. M. Stevens. 2005. Biomedical Hydrogels. *Biomaterials, Artificial Organs and Tissue Engineering*, 107–15. Elsevier.

Byun, Y., and Y. T. Kim. 2014. Bioplastics for Food Packaging: Chemistry and Physics. *Innovations in Food Packaging*, 353–68. Elsevier.

Carvalho, A. J. 2008. Starch: Major Sources, Properties and Applications as Thermoplastic Materials. *Monomers, Polymers and Composites from Renewable Resources*, 321–42: Elsevier.

Chaturvedi, S., P. N. Dave, and N. Shah. 2012. Applications of Nano-Catalyst in New Era. *Journal of Saudi Chemical Society* 16 (3): 307–25.

Chen, C., Z. Wang, and Z. Li. 2011. Thermoresponsive Polypeptides from Pegylated Poly-L-Glutamates. *Biomacromolecules* 12 (8): 2859–63.

Cheng, H., and R. A. Gross. 2010. *Green Polymer Chemistry: Biocatalysis and Biomaterials*. American Chemical Society Washington, DC.

Cheng, H., R. A. Gross, and P. B. Smith. 2015. Green Polymer Chemistry: Some Recent Developments and Examples. *Green Polymer Chemistry: Biobased Materials and Biocatalysis* 1191: 2–15.

Chittasupho, C., M. Jaturanpinyo, and S. Mangmool. 2013. Pectin Nanoparticle Enhances Cytotoxicity of Methotrexate against Hepg2 Cells. *Drug Delivery* 20 (1): 1–9.

Christian, P. 2011. Polymer Chemistry. *Electrospinning for Tissue Regeneration*, 34–50. Elsevier.

Cohen-Karni, T., K. J. Jeong, J. H. Tsui, G. Reznor, M. Mustata, M. Wanunu, A. Graham, *et al.* 2012. Nanocomposite Gold-Silk Nanofibers. *Nano Letters* 12 (10): 5403–06.

Cui, S., Y. Wu, and H. Ding. 2013. The Range of Dietary Fibre Ingredients and a Comparison of Their Technical Functionality. *Fibre-Rich and Wholegrain Foods: Improving Quality*, 96–119. Woodhead Publishing.

Dahech, I., K. S. Belghith, K. Hamden, A. Feki, H. Belghith, and H. Mejdoub. 2011. Antidiabetic Activity of Levan Polysaccharide in Alloxan-Induced Diabetic Rats. *International Journal of Biological Macromolecules* 49 (4): 742–46.

Dahman, Y. 2017. *Nanotechnology and Functional Materials for Engineers*. Elsevier.

De Azeredo, H. M. 2009. Nanocomposites for Food Packaging Applications. *Food Research International* 42 (9): 1240–53.

De Kruif, C., and C. Holt. 2003. Casein Micelle Structure, Functions and Interactions. *Advanced Dairy Chemistry—1 Proteins*, 233–76. Springer.

de Melo, L. S., A. S. Gomes, S. Saska, K. Nigoghossian, Y. Messaddeq, S. J. Ribeiro, and R. E. de Araujo. 2012. Singlet Oxygen Generation Enhanced by Silver-Pectin Nanoparticles. *Journal of Fluorescence* 22 (6): 1633–38.

De Moura, M. R., L. H. Mattoso, and V. Zucolotto. 2012. Development of Cellulose-Based Bactericidal Nanocomposites Containing Silver Nanoparticles and Their Use as Active Food Packaging. *Journal of Food Engineering* 109 (3): 520–24.

Debelle, L., and A. Tamburro. 1999. Elastin: Molecular Description and Function. *The International Journal of Biochemistry & Cell Biology* 31 (2): 261–72.

Despanie, J., J. P. Dhandhukia, S. F. Hamm-Alvarez, and J. A. J. J. o. C. R. MacKay. 2016. Elastin-Like Polypeptides: Therapeutic Applications for an Emerging Class of Nanomedicines. *Journal of Controlled Release* 240: 93–108.

Dev, A., N. Binulal, A. Anitha, S. Nair, T. Furuike, H. Tamura, and R. Jayakumar. 2010. Preparation of Poly (Lactic Acid)/Chitosan Nanoparticles for Anti-Hiv Drug Delivery Applications. *Carbohydrate Polymers* 80 (3): 833–38.

Dhandayuthapani, B., U. M. Krishnan, and S. Sethuraman. 2010. Fabrication and Characterization of Chitosan-Gelatin Blend Nanofibers for Skin Tissue Engineering. *Journal of Biomedical Materials Research Part B: Applied Biomaterials* 94 (1): 264–72.

dos Santos, M. A., and A. Grenha. 2015. Polysaccharide Nanoparticles for Protein and Peptide Delivery: Exploring Less-Known Materials. *Advances in Protein Chemistry and Structural Biology*, 223–61. Elsevier.

Draget, K. I. 2009. Alginates. *Handbook of Hydrocolloids*, 807–28. Elsevier.

Dube, M. A., and S. Salehpour. 2014. Applying the Principles of Green Chemistry to Polymer Production Technology. *Macromolecular Reaction Engineering* 8 (1): 7–28.

Dutta, P. K., J. Dutta, and V. Tripathi. 2004. Chitin and Chitosan: Chemistry, Properties and Applications. *Journal of Scientific & Industrial Research* (63): 20-23.

Elzoghby, A. O., M. M. Elgohary, and N. M. Kamel. 2015. Implications of Protein-and Peptide-Based Nanoparticles as Potential Vehicles for Anticancer Drugs. *Advances in Protein Chemistry and Structural Biology*, 169–221: Elsevier.

Endress, H.-U., and S. Christensen. 2009. Pectins. *Handbook of Hydrocolloids*, 274–97: Elsevier.

Estevanato, L. L., L. M. Lacava, L. C. Carvalho, R. B. Azevedo, O. Silva, F. Pelegrini, S. N. Báo, P. C. Morais, and Z. G. Lacava. 2012. Long-Term Biodistribution and Biocompatibility Investigation of Dextran-Coated Magnetite Nanoparticle Using Mice as the Animal Model. *Journal of Biomedical Nanotechnology* 8 (2): 301–08.

Fallourd, M., and L. Viscione. 2009. Ingredient Selection for Stabilisation and Texture Optimisation of Functional Beverages and the Inclusion of Dietary Fibre. *Functional and Speciality Beverage Technology*, 3–38. Elsevier.

Farrar, D., K. Ren, D. Cheng, S. Kim, W. Moon, W. L. Wilson, J. E. West, and S. M. Yu. 2011. Permanent Polarity and Piezoelectricity of Electrospun A-Helical Poly (A-Amino Acid) Fibers. *Advanced Materials* 23 (34): 3954–58.

Farris, S., I. U. Unalan, L. Introzzi, J. M. Fuentes-Alventosa, and C. A. Cozzolino. 2014. Pullulan-Based Films and Coatings for Food Packaging: Present Applications, Emerging Opportunities, and Future Challenges. *Journal of Applied Polymer Science* 131 (13): 40539.

Festucci-Buselli, R. A., W. C. Otoni, and C. P. Joshi. 2007. Structure, Organization, and Functions of Cellulose Synthase Complexes in Higher Plants. *Brazilian Journal of Plant Physiology* 19 (1): 1–13.

Finkenstadt, V. L. 2005. Natural Polysaccharides as Electroactive Polymers. *Applied Microbiology Biotechnology* 67 (6): 735–45.

Flutto, L. 2003a. Pectinl Food Use, *Encyclopedia of Food Sciences and Nutrition, 2nd ed.*, 4449–4456: Elsevier.

Flutto, L. 2003b. Pectinl Properties and Determination, *Encyclopedia of Food Sciences and Nutrition, 2nd ed.*, 4440–9: Elsevier.

Foo, C. W. P., and D. L. Kaplan. 2002. Genetic Engineering of Fibrous Proteins: Spider Dragline Silk and Collagen. *Advanced Drug Delivery Reviews* 54 (8): 1131–43.

Fortunati, E., M. Peltzer, I. Armentano, A. Jiménez, and J. M. Kenny. 2013. Combined Effects of Cellulose Nanocrystals and Silver Nanoparticles on the Barrier and Migration Properties of Pla Nano-Biocomposites. *Journal of Food Engineering* 118 (1): 117–24.

Funami, T., H. Yada, and Y. Nakao. 1998. Curdlan Properties for Application in Fat Mimetics for Meat Products. *Journal of Food Science* 63 (2): 283–87.

Garlotta, D. 2001. A Literature Review of Poly (Lactic Acid). *Journal of Polymers and the Environment* 9 (2): 63–84.

Gelse, K., E. Pöschl, and T. Aigner. 2003. Collagens—Structure, Function, and Biosynthesis. *Advanced Drug Delivery Reviews* 55 (12): 1531–46.

Giavasis, I. 2014. Bioactive Fungal Polysaccharides as Potential Functional Ingredients in Food and Nutraceuticals. *Current Opinion in Biotechnology* 26: 162–73.

Goel, A., M. K. Meher, P. Gupta, K. Gulati, V. Pruthi, and K. M. Poluri. 2019. Microwave Assisted K-Carrageenan Capped Silver Nanocomposites for Eradication of Bacterial Biofilms. *Carbohydrate Polymers* 206: 854–62.

Gomes, M., H. Azevedo, P. Malafaya, S. Silva, J. Oliveira, G. Silva, R. Sousa, J. Mano, and R. Reis. 2008. Natural Polymers in Tissue Engineering Applications. *Tissue Engineering*, 145–92: Elsevier.

Gorgieva, S., and V. Kokol. 2011. Collagen-Vs. Gelatine-Based Biomaterials and Their Biocompatibility: Review and Perspectives. *Biomaterials Applications for Nanomedicine*, 17–52. IntechOpen.

Gray, W. R., L. B. Sandberg, and J. A. Foster. 1973. Molecular Model for Elastin Structure and Function. *Nature* 246 (5434): 461.

Gu, L.-H., and P. A. Coulombe. 2007. Keratin Function in Skin Epithelia: A Broadening Palette with Surprising Shades. *Current Opinion in Cell Biology* 19 (1): 13–23.

Guerrero-Sanchez, C., M. Lobert, R. Hoogenboom, and U. S. Schubert. 2007. Microwave-Assisted Homogeneous Polymerizations in Water-Soluble Ionic Liquids: An Alternative and Green Approach for Polymer Synthesis. *Macromolecular Rapid Communications* 28 (4): 456–64.

Guillén, G., B. Giménez, M. López Caballero, and P. Montero García. 2011. Functional and Bioactive Properties of Collagen and Gelatin from Alternative Sources: A Review. *Food Hydrocolloids*25 (8): 1813–27.

Gulati, K., M. K. Meher, and K. M. Poluri. 2017. Glycosaminoglycan-Based Resorbable Polymer Composites in Tissue Refurbishment. *Regenerative Medicine* 12 (4): 431–57.

Gulati, K., and K. M. Poluri. 2016. Mechanistic and Therapeutic Overview of Glycosaminoglycans: The Unsung Heroes of Biomolecular Signaling. *Glycoconjugate Journal* 33 (1): 1–17.

Gupta, V., P. Carrott, R. Singh, M. Chaudhary, and S. Kushwaha. 2016. Cellulose: A Review as Natural, Modified and Activated Carbon Adsorbent. *Bioresource Technology* 216: 1066–76.

Gyarmati, B., B. Vajna, Á. Némethy, K. László, and A. Szilágyi. 2013. Redox-and Ph-Responsive Cysteamine-Modified Poly (Aspartic Acid) Showing a Reversible Sol–Gel Transition. *Macromolecular Bioscience* 13 (5): 633–40.

Gyenes, T., V. Torma, B. Gyarmati, and M. Zrínyi. 2008. Synthesis and Swelling Properties of Novel Ph-Sensitive Poly (Aspartic Acid) Gels. *Acta Biomaterialia* 4 (3): 733–44.

Haddad, T., S. Noel, B. Liberelle, R. El Ayoubi, A. Ajji, and G. De Crescenzo. 2016. Fabrication and Surface Modification of Poly Lactic Acid (Pla) Scaffolds with Epidermal Growth Factor for Neural Tissue Engineering. *Biomatter* 6 (1): 1231276.

Hadi, A., and A. Hashim. 2017. Development of a New Humidity Sensor Based on (Carboxymethyl Cellulose–Starch) Blend with Copper Oxide Nanoparticles. *Ukrainian Journal of Physics* 62 (12): 1044–44.

Haider, S., T. Kamal, S. B. Khan, M. Omer, A. Haider, F. U. Khan, and A. M. Asiri. 2016. Natural Polymers Supported Copper Nanoparticles for Pollutants Degradation. *Applied Surface Science* 387: 1154–61.

Han, Y. W. 1989. Levan Production Bybacillus Polymyxa. *Journal of Industrial Microbiology* 4 (6): 447–51.

Harada, T., M. Terasaki, and A. Harada. 1993. Curdlan. *Industrial Gums*, 427–45: Elsevier.

Hardy, J. G., L. M. Römer, and T. R. Scheibel. 2008. Polymeric Materials Based on Silk Proteins. *Polymer* 49 (20): 4309–27.

He, F., Y. Yang, G. Yang, and L. Yu. 2008. Components and Antioxidant Activity of the Polysaccharide from Streptomyces Virginia H03. *Zeitschrift für Naturforschung C* 63 (3–4): 181–88.

Hileman, R. E., J. R. Fromm, J. M. Weiler, and R. J. Linhardt. 1998. Glycosaminoglycan-Protein Interactions: Definition of Consensus Sites in Glycosaminoglycan Binding Proteins. *Bioessays* 20 (2): 156–67.

Ho, C. H., E. K. Odermatt, I. Berndt, and J. C. Tiller. 2013. Long-Term Active Antimicrobial Coatings for Surgical Sutures Based on Silver Nanoparticles and Hyperbranched Polylysine. *Journal of Biomaterials Science, Polymer Edition* 24 (13): 1589–600.

Horne, D. S. 2002. Casein Structure, Self-Assembly and Gelation. *Current Opinion in Colloid Interface Science* 7 (5–6): 456–61.

Hosseinzadeh, H., S. Zoroufi, and G. R. Mahdavinia. 2015. Study on Adsorption of Cationic Dye on Novel Kappa-Carrageenan/Poly (Vinyl Alcohol)/Montmorillonite Nanocomposite Hydrogels. *Polymer Bulletin* 72 (6): 1339–63.

Hu, K., X. Huang, Y. Gao, X. Huang, H. Xiao, and D. J. McClements. 2015. Core–Shell Biopolymer Nanoparticle Delivery Systems: Synthesis and Characterization of Curcumin Fortified Zein–Pectin Nanoparticles. *Food Chemistry* 182: 275–81.

Jackson, D. 2003. Starchl Structure, Properties, and Determination, *Encyclopedia of Food Sciences and Nutrition, 2nd ed.*, 5561–7: Elsevier.

Jahangirian, H., E. G. Lemraski, T. J. Webster, R. Rafiee-Moghaddam, and Y. Abdollahi. 2017. A Review of Drug Delivery Systems Based on Nanotechnology and Green Chemistry: Green Nanomedicine. *International Journal of Nanomedicine* 12: 2957.

Jalaja, K., D. Naskar, S. C. Kundu, and N. R. James. 2016. Potential of Electrospun Core–Shell Structured Gelatin–Chitosan Nanofibers for Biomedical Applications. *Carbohydrate Polymers* 136: 1098–107.

Jamshidian, M., E. A. Tehrany, M. Imran, M. Jacquot, and S. Desobry. 2010. Poly-Lactic Acid: Production, Applications, Nanocomposites, and Release Studies. *Comprehensive Reviews in Food Science and Food Safety* 9 (5): 552–71.

Janes, K. A., M. P. Fresneau, A. Marazuela, A. Fabra, and M. a. J. Alonso. 2001. Chitosan Nanoparticles as Delivery Systems for Doxorubicin. *Journal of Controlled Release* 73 (2–3): 255–67.

Jiang, F., L. Yin, Q. Yu, C. Zhong, and J. Zhang. 2015. Bacterial Cellulose Nanofibrous Membrane as Thermal Stable Separator for Lithium-Ion Batteries. *Journal of Power Sources* 279: 21–27.

Jiang, F., Y. Nie, L. Yin, Y. Feng, Q. Yu, and C. Zhong. 2016. Core–Shell-Structured Nanofibrous Membrane as Advanced Separator for Lithium-Ion Batteries. *Journal of Membrane Science* 510: 1–9.

Jiang, G., D. J. Hill, M. M. Kowalczuk, and I. Radecka. 2017. *Green Polymer Composites Technology.* CRC Press Tylor & Francis Group.

Kadler, K. E., C. Baldock, J. Bella, and R. P. Boot-Handford. 2007. Collagens at a Glance. *Journal of Cell Science* 120 (12): 1955–58.

Kadokawa, J.-i., and S. Kobayashi. 2010. Polymer Synthesis by Enzymatic Catalysis. *Current Opinion in Chemical Biology* 14 (2): 145–53.

Kadokawa, J. 2016. Chemical Synthesis of Well-Defined Polysaccharides. *Reference Module in Materials Science and Materials Engineering*. Elsevier.

Kafy, A., K. K. Sadasivuni, H.-C. Kim, A. Akther, and J. Kim. 2015. Designing Flexible Energy and Memory Storage Materials Using Cellulose Modified Graphene Oxide Nanocomposites. *Physical Chemistry Chemical Physics* 17 (8): 5923–31.

Kai, D., W. Ren, L. Tian, P. L. Chee, Y. Liu, S. Ramakrishna, and X. J. Loh. 2016. Engineering Poly (Lactide)–Lignin Nanofibers with Antioxidant Activity for Biomedical Application. *ACS Sustainable Chemistry & Engineering* 4 (10): 5268–76.

Kamal, T., S. B. Khan, and A. M. Asiri. 2016. Nickel Nanoparticles-Chitosan Composite Coated Cellulose Filter Paper: An Efficient and Easily Recoverable Dip-Catalyst for Pollutants Degradation. *Environmental Pollution* 218: 625–33.

Kaplan, D. L. 1998. Introduction to Biopolymers from Renewable Resources. *Biopolymers from Renewable Resources*, 1–29: Springer.

Karandikar, S., A. Mirani, V. Waybhase, V. B. Patravale, and S. Patankar. 2017. Nanovaccines for Oral Delivery-Formulation Strategies and Challenges. *Nanostructures for Oral Medicine*, 263–93: Elsevier.

Kariduraganavar, M. Y., A. A. Kittur, and R. R. Kamble. 2014. Polymer Synthesis and Processing. *Natural and Synthetic Biomedical Polymers*, 1–31: Elsevier.

Kasaai, M. R. 2018. Zein and Zein-Based Nano-Materials for Food and Nutrition Applications: A Review. *Trends in Food Science Technology* 79: 184–97.

Kasapis, S., E. R. Morris, M. Gross, and K. Rudolph. 1994. Solution Properties of Levan Polysaccharide from Pseudomonas Syringae Pv. Phaseolicola, and Its Possible Primary Role as a Blocker of Recognition During Pathogenesis. *Carbohydrate Polymers* 23 (1): 55–64.

Khanna, P., and E. F. Ullman. 1982. Novel Ether Substituted Fluorescein Polyamino Acid Compounds as Fluorescers and Quenchers. Google Patents.

Khuphe, M., C. S. Mahon, and P. D. Thornton. 2016. Glucose-Bearing Biodegradable Poly (Amino Acid) and Poly (Amino Acid)-Poly (Ester) Conjugates for Controlled Payload Release. *Biomaterials Science* 4 (12): 1792–801.

Khuphe, M., and P. D. Thornton. 2018. Poly (Amino Acids). *Engineering of Biomaterials for Drug Delivery Systems*, 199–228: Elsevier.

Kielty, C. M., M. J. Sherratt, and C. A. Shuttleworth. 2002. Elastic Fibres. *Journal of Cell Science* 115 (14): 2817–28.

Kirker, K. R., Y. Luo, J. H. Nielson, J. Shelby, and G. D. Prestwich. 2002. Glycosaminoglycan Hydrogel Films as Bio-Interactive Dressings for Wound Healing. *Biomaterials* 23 (17): 3661–71.

Kobayashi, S. 2017. Green Polymer Chemistry: New Methods of Polymer Synthesis Using Renewable Starting Materials. *Structural Chemistry* 28 (2): 461–74.

Kobayashi, S., and A. Makino. 2009. Enzymatic Polymer Synthesis: An Opportunity for Green Polymer Chemistry. *Chemical Reviews* 109 (11): 5288–353.

Kowalczyk, T., A. Nowicka, D. Elbaum, and T. A. Kowalewski. 2008. Electrospinning of Bovine Serum Albumin. Optimization and the Use for Production of Biosensors. *Biomacromolecules* 9 (7): 2087–90.

Kozlov, P., and G. Burdygina. 1983. The Structure and Properties of Solid Gelatin and the Principles of Their Modification. *Polymer* 24 (6): 651–66.

Krasaekoopt, W., and B. Bhandari. 2012. Properties and Applications of Different Probiotic Delivery Systems. *Encapsulation Technologies and Delivery Systems for Food Ingredients and Nutraceuticals*, 541–94. Elsevier.

Kubisa, P. 2004. Application of Ionic Liquids as Solvents for Polymerization Processes. *Progress in Polymer Science* 29 (1): 3–12.

Kulkarni, R. K., E. Moore, A. Hegyeli, and F. Leonard. 1971. Biodegradable Poly (Lactic Acid) Polymers. *Journal of Biomedical Materials Research* 5 (3): 169–81.

Kumar, D., N. Saini, V. Pandit, and S. Ali. 2012. An Insight to Pullulan: A Biopolymer in Pharmaceutical Approaches. *International Journal of Basic and Applied Sciences* 1 (3): 202–19.

Kumar, N., and S. Kumbhat. 2016. *Essentials in Nanoscience and Nanotechnology*. John Wiley & Sons.

Ladhar, A., M. Arous, H. Kaddami, M. Raihane, A. Kallel, M. Graça, and L. Costa. 2015. Ionic Hopping Conductivity in Potential Batteries Separator Based on Natural Rubber–Nanocellulose Green Nanocomposites. *Journal of Molecular Liquids* 211: 792–802.

Lalatsa, A., A. G. Schätzlein, M. Mazza, T. B. H. Le, and I. F. Uchegbu. 2012. Amphiphilic Poly (L-Amino Acids)—New Materials for Drug Delivery. *Journal of Controlled Release* 161 (2): 523–36.

Lalevée, J., N. Blanchard, M.-A. Tehfe, F. Morlet-Savary, and J. P. Fouassier. 2010. Green Bulb Light Source Induced Epoxy Cationic Polymerization under Air Using Tris (2, 2′-Bipyridine) Ruthenium (Ii) and Silyl Radicals. *Macromolecules* 43 (24): 10191–95.

Lavasanifar, A., J. Samuel, and G. S. Kwon. 2002. Poly (Ethylene Oxide)-Block-Poly (L-Amino Acid) Micelles for Drug Delivery. *Advanced Drug Delivery Reviews* 54 (2): 169–90.

Le Corre, D., and H. Angellier-Coussy. 2014. Preparation and Application of Starch Nanoparticles for Nanocomposites: A Review. *Reactive Functional Polymers* 85: 97–120.

Le Corre, D., J. Bras, and A. Dufresne. 2010. Starch Nanoparticles: A Review. *Biomacromolecules* 11 (5): 1139–53.

Lee, K. Y., and D. J. Mooney. 2012. Alginate: Properties and Biomedical Applications. *Progress in Polymer Science* 37 (1): 106–26.

Lee, S. Y. 1996. Bacterial Polyhydroxyalkanoates. *Biotechnology and Bioengineering* 49 (1): 1–14.

Li, M., X. Wang, Y. Yang, Z. Chang, Y. Wu, and R. Holze. 2015. A Dense Cellulose-Based Membrane as a Renewable Host for Gel Polymer Electrolyte of Lithium Ion Batteries. *Journal of Membrane Science* 476: 112–18.

Linton, J., S. Ash, and L. Huybrechts. 1991. Microbial Polysaccharides. *Biomaterials*, 215–61. Springer.

Ma, Y.-Q., J.-Z. Yi, and L.-M. Zhang. 2009. A Facile Approach to Incorporate Silver Nanoparticles into Dextran-Based Hydrogels for Antibacterial and Catalytical Application. *Journal of Macromolecular Science®, Part A: Pure and Applied Chemistry* 46 (6): 643–48.

Maji, K., S. Dasgupta, K. Pramanik, and A. Bissoyi. 2016. Preparation and Evaluation of Gelatin-Chitosan-Nanobioglass 3d Porous Scaffold for Bone Tissue Engineering. *International Journal of Biomaterials* 2016: 9825659.

Makaremi, M., P. Pasbakhsh, G. Cavallaro, G. Lazzara, Y. K. Aw, S. M. Lee, and S. Milioto. 2017. Effect of Morphology and Size of Halloysite Nanotubes on Functional Pectin Bionanocomposites for Food Packaging Applications. *ACS Applied Materials & Interfaces* 9 (20): 17476–88.

Målberg, S., A. Höglund, and A.-C. Albertsson. 2011. Macromolecular Design of Aliphatic Polyesters with Maintained Mechanical Properties and a Rapid, Customized Degradation Profile. *Biomacromolecules* 12 (6): 2382–88.

Marra, A., C. Silvestre, D. Duraccio, and S. Cimmino. 2016. Polylactic Acid/Zinc Oxide Biocomposite Films for Food Packaging Application. *International Journal of Biological Macromolecules* 88: 254–62.

McArdle, R., and R. Hamill. 2011. Utilisation of Hydrocolloids in Processed Meat Systems. *Processed Meats*, 243–69: Elsevier.

McHugh, D. 2003. A Guide to the Seaweed Industry Fao Fisheries Technical Paper 441. Food Agriculture Organization of the United Nations, Rome.

McKittrick, J., P.-Y. Chen, S. Bodde, W. Yang, E. Novitskaya, and M. Meyers. 2012. The Structure, Functions, and Mechanical Properties of Keratin. *Jom* 64 (4): 449–68.

Mellati, A., C. M. Fan, A. Tamayol, N. Annabi, S. Dai, J. Bi, B. Jin, *et al.* 2017. Microengineered 3d Cell-Laden Thermoresponsive Hydrogels for Mimicking Cell Morphology and Orientation in Cartilage Tissue Engineering. *Biotechnology and Bioengineering* 114 (1): 217–31.

Min, L.-L., L.-B. Zhong, Y.-M. Zheng, Q. Liu, Z.-H. Yuan, and L.-M. Yang. 2016. Functionalized Chitosan Electrospun Nanofiber for Effective Removal of Trace Arsenate from Water. *Scientific Reports* 6: 32480.

Miwa, M., Y. Nakao, and K. Nara. 1994. Food Applications of Curdlan. *Food Hydrocolloids*, 119–24: Springer.

Mogoşanu, G. D., and A. M. Grumezescu. 2014. Natural and Synthetic Polymers for Wounds and Burns Dressing. *International Journal of Pharmaceutics* 463 (2): 127–36.

Mohan, T., S. Hribernik, R. Kargl, and K. Stana-Kleinschek. 2015. Nanocellulosic Materials in Tissue Engineering Applications. *Cellulose-Fundamental Aspects and Current Trends*, 251–73. IntechOpen.

Mohanty, A. K., M. Misra, and L. Drzal. 2002. Sustainable Bio-Composites from Renewable Resources: Opportunities and Challenges in the Green Materials World. *Journal of Polymers and the Environment* 10 (1–2): 19–26.

Moman, R. N., and M. Varacallo. 2018. Physiology, Albumin. *Statpearls*, StatPearls Publishing.

Morales, M., and M. Ruiz. 2016. Microencapsulation of Probiotic Cells: Applications in Nutraceutic and Food Industry. *Nutraceuticals*; Grumezescu, AM, Ed.; Academic Press: Cambridge, MA: 627–68.

Nakano, K., T. Kamada, and K. Nozaki. 2006. Selective Formation of Polycarbonate over Cyclic Carbonate: Copolymerization of Epoxides with Carbon Dioxide Catalyzed by a Cobalt (Iii) Complex with a Piperidinium End-Capping Arm. *Angewandte Chemie International Edition* 45 (43): 7274–77.

Nicosia, A., T. Keppler, F. Müller, B. Vazquez, F. Ravegnani, P. Monticelli, and F. Belosi. 2016. Cellulose Acetate Nanofiber Electrospun on Nylon Substrate as Novel Composite Matrix for Efficient, Heat-Resistant, Air Filters. *Chemical Engineering Science* 153: 284–94.

Niekraszewicz, B., and A. Niekraszewicz. 2009. The Structure of Alginate, Chitin and Chitosan Fibres. *Handbook of Textile Fibre Structure*, 266–304: Elsevier.

Nishinari, K., H. Zhang, and T. Funami. 2009. Curdlan. *Handbook of Hydrocolloids*, 567–91: Elsevier.

Oğuzhan, P., and F. Yangılar. 2013. Pullulan: Production and Usage in Foodındustry. *African Journal of Food Science and Technology* 4 (3): 2141–5455.

Okiror, G. P., and C. L. Jones. 2012. Effect of Temperature on the Dielectric Properties of Low Acyl Gellan Gel. *Journal of Food Engineering* 113 (1): 151–55.

Olsen, D., C. Yang, M. Bodo, R. Chang, S. Leigh, J. Baez, D. Carmichael, *et al.* 2003. Recombinant Collagen and Gelatin for Drug Delivery. *Advanced Drug Delivery Reviews* 55 (12): 1547–67.

Öner, E. T., L. Hernández, and J. Combie. 2016. Review of Levan Polysaccharide: From a Century of Past Experiences to Future Prospects. *Biotechnology Advances* 34 (5): 827–44.

Otter, D. 2003. Milkl Physical and Chemical Properties, *Encyclopedia of Food Sciences and Nutrition, 2nd ed.*, 395763: Elsevier.

Pan, D., and X. Mei. 2010. Antioxidant Activity of an Exopolysaccharide Purified from Lactococcus Lactis Subsp. Lactis 12. *Carbohydrate Polymers* 80 (3): 908–14.

Park, C. W., H.-M. Yang, M.-A. Woo, K. S. Lee, and J.-D. Kim. 2017. Completely Disintegrable Redox-Responsive Poly (Amino Acid) Nanogels for Intracellular Drug Delivery. *Journal of Industrial Engineering Chemistry* 45: 182–88.

Pascoli, M., R. de Lima, and L. F. Fraceto. 2018. Zein Nanoparticles and Strategies to Improve Colloidal Stability: A Mini-Review. *Frontiers in Chemistry* 6: 6.

Pathania, D., G. Sharma, and R. Thakur. 2015. Pectin@ Zirconium (Iv) Silicophosphate Nanocomposite Ion Exchanger: Photo Catalysis, Heavy Metal Separation and Antibacterial Activity. *Chemical Engineering Journal* 267: 235–44.

Paul, F., A. Morin, and P. Monsan. 1986. Microbial Polysaccharides with Actual Potential Industrial Applications. *Biotechnology Advances* 4 (2): 245–59.

Paul, S., Y. Zhu, C. Romain, R. Brooks, P. K. Saini, and C. K. Williams. 2015. Ring-Opening Copolymerization (Rocop): Synthesis and Properties of Polyesters and Polycarbonates. *Chemical Communications* 51 (30): 6459–79.

Peng, N., D. Hu, J. Zeng, Y. Li, L. Liang, and C. Chang. 2016. Superabsorbent Cellulose–Clay Nanocomposite Hydrogels for Highly Efficient Removal of Dye in Water. *ACS Sustainable Chemistry & Engineering* 4 (12): 7217–24.

Peters Jr, T. 1985. Serum Albumin. *Advances in Protein Chemistry*, 161–245: Elsevier.

Philibert, T., B. H. Lee, and N. Fabien. 2017. Current Status and New Perspectives on Chitin and Chitosan as Functional Biopolymers. *Applied Biochemistry and Biotechnology* 181 (4): 1314–37.

Philp, K. 2015. Polysaccharide Ingredients. *Reference Module in Food Science*. Elsevier.

Pillai, C., W. Paul, and C. P. Sharma. 2009. Chitin and Chitosan Polymers: Chemistry, Solubility and Fiber Formation. *Progress in Polymer Science* 34 (7): 641–78.

Pina, S., J. M. Oliveira, and R. L. Reis. 2015. Natural-Based Nanocomposites for Bone Tissue Engineering and Regenerative Medicine: A Review. *Advanced Materials* 27 (7): 1143–69.

Polifka, J. E., and J. Habermann. 2015. Anticoagulants, Thrombocyte Aggregation Inhibitors, Fibrinolytics and Volume Replacement Agents. *Drugs During Pregnancy and Lactation*, 225–49: Elsevier.

Poluri, K. M., and K. Gulati. 2016. *Protein Engineering Techniques: Gateways to Synthetic Protein Universe*. Springer.

Prameela, K., C. M. Mohan, and C. Ramakrishna. 2018. Biopolymers for Food Design: Consumer-Friendly Natural Ingredients. *Biopolymers for Food Design*, 1–32: Elsevier.

Preiss, J. 2010. Biochemistry and Molecular Biology of Glycogen Synthesis in Bacteria and Mammals and Starch Synthesis in Plants.

Qin, Y. 2008. Alginate Fibres: An Overview of the Production Processes and Applications in Wound Management. *Polymer International* 57 (2): 171–80.

Qureshi, U. A., Z. Khatri, F. Ahmed, M. Khatri, and I.-S. Kim. 2017. Electrospun Zein Nanofiber as a Green and Recyclable Adsorbent for the Removal of Reactive Black 5 from the Aqueous Phase. *ACS Sustainable Chemistry & Engineering* 5 (5): 4340–51.

Rai, A., J. Jenifer, and R. T. P. Upputuri. 2017. Nanoparticles in Therapeutic Applications and Role of Albumin and Casein Nanoparticles in Cancer Therapy. *Asian Biomedicine* 11 (1): 3–20.

Raj, S., and D. R. Shankaran. 2016. Curcumin Based Biocompatible Nanofibers for Lead Ion Detection. *Sensors and Actuators B: Chemical* 226: 318–25.

Ramamoorthy, S., A. Gnanakan, S. S. Lakshmana, M. Meivelu, and A. Jeganathan. 2018. Structural Characterization and Anticancer Activity of Extracellular Polysaccharides from Ascidian Symbiotic Bacterium Bacillus Thuringiensis. *Carbohydrate Polymers* 190: 113–20.

Ramos, A. P., M. A. Cruz, C. B. Tovani, and P. Ciancaglini. 2017. Biomedical Applications of Nanotechnology. *Biophysical Reviews* 9 (2): 79–89.

Rånby, B. 2001. Natural Cellulose Fibers and Membranes: Biosynthesis. *Reference Module in Materials Science and Materials Engineering*, 5938–43. Elsevier.

Raza, Z. A., S. Abid, and I. M. Banat. 2018. Polyhydroxyalkanoates: Characteristics, Production, Recent Developments and Applications. *International Biodeterioration & Biodegradation* 126: 45–56.

Reddy, C., R. Ghai, and V. C. Kalia. 2003. Polyhydroxyalkanoates: An Overview. *Bioresource Technology* 87 (2): 137–46.

Rehan, F., N. Ahemad, and M. Gupta. 2019. Casein Nanomicelle as an Emerging Biomaterial–a Comprehensive Review. *Colloids Surfaces B: Biointerfaces.*

Reuvers, M. 2001. Anticoagulant and Fibrinolytic Drugs. *Drugs During Pregnancy and Lactation: Handbook of Prescription Drugs and Comparative Risk Assessment*, 85. Gulf Professional Publishing.

Rezaeeyazdi, M., T. Colombani, A. Memic, and S. Bencherif. 2018. Injectable Hyaluronic Acid-Co-Gelatin Cryogels for Tissue-Engineering Applications. *Materials* 11 (8): 1374.

Rinaudo, M. 2006. Chitin and Chitosan: Properties and Applications. *Progress in Polymer Science* 31 (7): 603–32.

Rioux, L.-E., and S. L. Turgeon. 2015. Seaweed Carbohydrates. *Seaweed Sustainability*, 141–92: Elsevier.

Roberts, J., and P. Martens. 2016. Engineering Biosynthetic Cell Encapsulation Systems. *Biosynthetic Polymers for Medical Applications*, 205–39: Elsevier.

Ross, G., S. Ross, and B. J. Tighe. 2017. Bioplastics: New Routes, New Products. *Brydson's Plastics Materials*, 631–52: Elsevier.

Rostamnia, S., B. Zeynizadeh, E. Doustkhah, A. Baghban, and K. O. Aghbash. 2015. The Use of K-Carrageenan/Fe3o4 Nanocomposite as a Nanomagnetic Catalyst for Clean Synthesis of Rhodanines. *Catalysis Communications* 68: 77–83.

Rudnik, E. 2012. *Compostable Polymer Materials* . Elsevier.

Saber-Samandari, S., S. Saber-Samandari, H. Joneidi-Yekta, and M. Mohseni. 2017. Adsorption of Anionic and Cationic Dyes from Aqueous Solution Using Gelatin-Based Magnetic Nanocomposite Beads Comprising Carboxylic Acid Functionalized Carbon Nanotube. *Chemical Engineering Journal* 308: 1133–44.

Saithongdee, A., N. Praphairaksit, and A. Imyim. 2014. Electrospun Curcumin-Loaded Zein Membrane for Iron (Iii) Ions Sensing. *Sensors and Actuators B: Chemical* 202: 935–40.

Salama, A. 2018. Chitosan Based Hydrogel Assisted Spongelike Calcium Phosphate Mineralization for in-Vitro Bsa Release. *International Journal of Biological Macromolecules* 108: 471–76.

Sánchez-Machado, D. I., J. López-Cervantes, M. A. Correa-Murrieta, R. G. Sánchez-Duarte, P. Cruz-Flores, and G. S. de la Mora-López. 2019. Chitosan. *Nonvitamin and Nonmineral Nutritional Supplements*, 485–93: Elsevier.

Sapountzi, E., M. Braiek, J.-F. Chateaux, N. Jaffrezic-Renault, and F. Lagarde. 2017. Recent Advances in Electrospun Nanofiber Interfaces for Biosensing Devices. *Sensors* 17 (8): 1887.

Saxena, T., L. Karumbaiah, and C. M. Valmikinathan. 2014. Proteins and Poly (Amino Acids). *Natural and Synthetic Biomedical Polymers*, 43–65. Elsevier.

Schmidt, D. 1982. Association of Caseins and Casein Micelle Structure. *Developments in Dairy Chemistry 1—Protetins*. Elsevier.

Semsarilar, M., and S. Perrier. 2010. 'Green'reversible Addition-Fragmentation Chain-Transfer (Raft) Polymerization. *Nature Chemistry* 2 (10): 811.

Sharma, N., H. Ojha, A. Bharadwaj, D. P. Pathak, and R. K. Sharma. 2015. Preparation and Catalytic Applications of Nanomaterials: A Review. *Rsc Advances* 5 (66): 53381–403.

Sharma, R., M. Ahuja, and H. Kaur. 2012. Thiolated Pectin Nanoparticles: Preparation, Characterization and Ex Vivo Corneal Permeation Study. *Carbohydrate Polymers* 87 (2): 1606–10.

Sheldon, R. A., I. Arends, and U. Hanefeld. 2007. *Green Chemistry and Catalysis.* John Wiley & Sons.

Shoulders, M. D., and R. T. Raines. 2009. Collagen Structure and Stability. *Annual Review of Biochemistry* 78: 929–58.

Shu, C.-H., and B.-J. Wen. 2003. Enhanced Shear Protection and Increased Production of an Anti-Tumor Polysaccharide by Agaricus Blazei in Xanthan-Supplemented Cultures. *Biotechnology Letters* 25 (11): 873–76.

Shubha, P., M. L. Gowda, K. Namratha, S. Shyamsunder, H. Manjunatha, and K. Byrappa. 2019. Ex-Situ Fabrication of Zno Nanoparticles Coated Silk Fiber for Surgical Applications. *Materials Chemistry and Physics* 231: 21–26.

Shukla, R., and M. Cheryan. 2001. Zein: The Industrial Protein from Corn. *Industrial Crops Products* 13 (3): 171–92.

Sidebotham, R. L. 1974. Dextrans. *Advances in Carbohydrate Chemistry and Biochemistry*, 371–444: Elsevier.

Sin, L. T. 2012. *Polylactic Acid: Pla Biopolymer Technology and Applications*. William Andrew.

Singh, J., R. Laurenti, R. Sinha, and B. Frostell. 2014. Progress and Challenges to the Global Waste Management System. *Waste Management & Research* 32 (9): 800–12.

Sivalingam, G., N. Agarwal, and G. Madras. 2004. Kinetics of Microwave-Assisted Polymerization of ε-Caprolactone. *Journal of Applied Polymer Science* 91 (3): 1450–56.

Soetaert, W., D. Schwengers, K. Buchholz, and E. Vandamme. 1995. A Wide Range of Carbohydrate Modifications by a Single Micro-Organism: Leuconostoc Mesenteroides. *Progress in Biotechnology*, 351–58: Elsevier.

Souzandeh, H., Y. Wang, and W.-H. Zhong. 2016. "Green" Nano-Filters: Fine Nanofibers of Natural Protein for High Efficiency Filtration of Particulate Pollutants and Toxic Gases. *Rsc Advances* 6 (107): 105948–56.

Sudesh, K., H. Abe, and Y. Doi. 2000. Synthesis, Structure and Properties of Polyhydroxyalkanoates: Biological Polyesters. *Progress in Polymer Science* 25 (10): 1503–55.

Sudha, P. N., T. Gomathi, P. A. Vinodhini, and K. Nasreen. 2014. Marine Carbohydrates of Wastewater Treatment. *Advances in Food and Nutrition Research*, 103–43: Elsevier.

Sugunan, A., C. Thanachayanont, J. Dutta, and J. Hilborn. 2005. Heavy-Metal Ion Sensors Using Chitosan-Capped Gold Nanoparticles. *Science and Technology of Advanced Materials* 6 (3–4): 335.

Suh, J.-K. F., and H. W. Matthew. 2000. Application of Chitosan-Based Polysaccharide Biomaterials in Cartilage Tissue Engineering: A Review. *Biomaterials* 21 (24): 2589–98.

Sun, Z., Z. Yi, H. Zhang, X. Ma, W. Su, X. Sun, and X. Li. 2017. Bio-Responsive Alginate-Keratin Composite Nanogels with Enhanced Drug Loading Efficiency for Cancer Therapy. *Carbohydrate Polymers* 175: 159–69.

Sworn, G. 2009a. Gellan Gum. *Handbook of Hydrocolloids*, 204–27: Elsevier.

Sworn, G. 2009b. Xanthan Gum. *Handbook of Hydrocolloids*, 186–203: Elsevier.

Takeuchi, Y., H. Uyama, N. Tomoshige, E. Watanabe, Y. Tachibana, and S. Kobayashi. 2006. Injectable Thermoreversible Hydrogels Based on Amphiphilic Poly (Amino Acid) S. *Journal of Polymer Science Part A: Polymer Chemistry* 44 (1): 671–75.

Tambe, N., J. Di, Z. Zhang, S. Bernacki, A. El-Shafei, and M. W. King. 2015. Novel Genipin-Collagen Immobilization of Polylactic Acid (Pla) Fibers for Use as Tissue Engineering Scaffolds. *Journal of Biomedical Materials Research Part B: Applied Biomaterials* 103 (6): 1188–97.

Ten, E., L. Jiang, J. Zhang, and M. P. Wolcott. 2015. Mechanical Performance of Polyhydroxyalkanoate (Pha)-Based Biocomposites. *Biocomposites*, 39–52. Elsevier.

Thierry, B., F. M. Winnik, Y. Merhi, J. Silver, and M. Tabrizian. 2003. Bioactive Coatings of Endovascular Stents Based on Polyelectrolyte Multilayers. *Biomacromolecules* 4 (6): 1564–71.

Tian, H., and J. He. 2016. Cellulose as a Scaffold for Self-Assembly: From Basic Research to Real Applications. *Langmuir* 32 (47): 12269–82.

Tian, J., R. Jiang, P. Gao, D. Xu, L. Mao, G. Zeng, M. Liu, *et al.* 2017. Synthesis and Cell Imaging Applications of Amphiphilic Aie-Active Poly (Amino Acid) S. *Materials Science Engineering: C* 79: 563–69.

Tsarevsky, N. V., and K. Matyjaszewski. 2006. Environmentally Benign Atom Transfer Radical Polymerization: Towards "Green" Processes and Materials. *Journal of Polymer Science Part A: Polymer Chemistry* 44 (17): 5098–112.

Tuvikene, R., K. Truus, A. Kollist, O. Volobujeva, E. Mellikov, and T. Pehk. 2007 Gel-Forming Structures and Stages of Red Algal Galactans of Different Sulfation Levels. Paper presented at the Nineteenth International Seaweed Symposium.

Tytgat, L., M. Vagenende, H. Declercq, J. Martins, H. Thienpont, H. Ottevaere, P. Dubruel, and S. Van Vlierberghe. 2018. Synergistic Effect of K-Carrageenan and Gelatin Blends Towards Adipose Tissue Engineering. *Carbohydrate Polymers* 189: 1–9.

van der VUSSE, G. J. 2009. Albumin as Fatty Acid Transporter. *Drug Metabolism Pharmacokinetics* 24 (4): 300–07.

Van Hoa, N., T. T. Khong, T. T. H. Quyen, and T. S. Trung. 2016. One-Step Facile Synthesis of Mesoporous Graphene/Fe3o4/Chitosan Nanocomposite and Its Adsorption Capacity for a Textile Dye. *Journal of Water Process Engineering* 9: 170–78.

Verlinden, R. A., D. J. Hill, M. Kenward, C. D. Williams, and I. Radecka. 2007. Bacterial Synthesis of Biodegradable Polyhydroxyalkanoates. *Journal of Applied Microbiology* 102 (6): 1437–49.

Vibhuti, R. K., R. Shukla, R. Gupta, and J. K. Saini. 2018. Food Grade Microorganisms for Nutraceutical Production for Industrial Applications. *Nutraceuticals and Innovative Food Products for Healthy Living and Preventive Care*, 342–67. IGI Global.

Walstra, P., and R. Jenness. 1984. *Dairy Chemistry & Physics*. John Wiley & Sons.

Wang, B., W. Yang, J. McKittrick, and M. A. Meyers. 2016. Keratin: Structure, Mechanical Properties, Occurrence in Biological Organisms, and Efforts at Bioinspiration. *Progress in Materials Science* 76: 229–318.

Wang, C., S. Wu, M. Jian, J. Xie, L. Xu, X. Yang, Q. Zheng, and Y. Zhang. 2016. Silk Nanofibers as High Efficient and Lightweight Air Filter. *Nano Research* 9 (9): 2590–97.

Wang, H., Y. Liu, Z. Qi, S. Wang, S. Liu, X. Li, H. Wang, and X. Xia. 2013. An Overview on Natural Polysaccharides with Antioxidant Properties. *Current Medicinal Chemistry* 20 (23): 2899–913.

Wang, L., C. Zhang, F. Gao, and G. Pan. 2016. Needleless Electrospinning for Scaled-up Production of Ultrafine Chitosan Hybrid Nanofibers Used for Air Filtration. *RSC Advances* 6 (107): 105988–95.

Wang, W., S.-X. Wang, and H.-S. Guan. 2012. The Antiviral Activities and Mechanisms of Marine Polysaccharides: An Overview. *Marine Drugs* 10 (12): 2795–816.

Wang, Y., T. Yang, H. Ke, A. Zhu, Y. Wang, J. Wang, J. Shen, *et al.* 2015. Smart Albumin-Biomineralized Nanocomposites for Multimodal Imaging and Photothermal Tumor Ablation. *Advanced Materials* 27 (26): 3874–82.

Weiss, J. 1993. *The Collagens: Biochemistry and Pathophysiology*. Wiley Online Library.

Wibowo, S. H., A. Sulistio, E. H. Wong, A. Blencowe, and G. G. Qiao. 2014. Polypeptide Films Via N-Carboxyanhydride Ring-Opening Polymerization (Nca-Rop): Past, Present and Future. *Chemical Communications* 50 (39): 4971–88.

Wielgus, K., K. Grajek, M. Szalata, and R. Słomski. 2012. Bioengineered Natural Textile Fibres. *Handbook of Natural Fibres*, 291–313: Elsevier.

Wu, J., and F. Yin. 2013. Sensitive Enzymatic Glucose Biosensor Fabricated by Electrospinning Composite Nanofibers and Electrodepositing Prussian Blue Film. *Journal of Electroanalytical Chemistry* 694: 1–5.

Wu, R., J.-F. Zhang, Y. Fan, D. Stoute, T. Lallier, and X. Xu. 2011. Reactive Electrospinning and Biodegradation of Cross-Linked Methacrylated Polycarbonate Nanofibers. *Biomedical Materials* 6 (3): 035004.

Wujcik, E., H. Wei, X. Zhang, J. Guo, X. Yan, N. Sutrave, S. Wei, and Z. Guo. 2014. Antibody Nanosensors: A Detailed Review. *RSC Advances* 4 (82): 43725–45.

Xu, H., Q. Yao, C. Cai, J. Gou, Y. Zhang, H. Zhong, and X. Tang. 2015. Amphiphilic Poly (Amino Acid) Based Micelles Applied to Drug Delivery: The in Vitro and in Vivo Challenges and the Corresponding Potential Strategies. *Journal of Controlled Release* 199: 84–97.

Xu, P., X. Liang, N. Chen, J. Tang, W. Shao, Q. Gao, and Z. Teng. 2017. Magnetic Separable Chitosan Microcapsules Decorated with Silver Nanoparticles for Catalytic Reduction of 4-Nitrophenol. *Journal of Colloid and Interface Science* 507: 353–59.

Yadav, S., M. P. Illa, T. Rastogi, and C. S. Sharma. 2016. High Absorbency Cellulose Acetate Electrospun Nanofibers for Feminine Hygiene Application. *Applied Materials Today* 4: 62–70.

Yang, G., H. Lin, B. B. Rothrauff, S. Yu, and R. S. Tuan. 2016. Multilayered Polycaprolactone/Gelatin Fiber-Hydrogel Composite for Tendon Tissue Engineering. *Acta Biomaterialia* 35: 68–76.

Yangilar, F., and P. O. Yildiz. 2016. Microbial Polysaccharides and the Applications in Food Industry. *Journal of Biotechnology* (231): S38.

You, J., C. Zhao, J. Cao, J. Zhou, and L. Zhang. 2014. Fabrication of High-Density Silver Nanoparticles on the Surface of Alginate Microspheres for Application in Catalytic Reaction. *Journal of Materials Chemistry A* 2 (22): 8491–99.

You, M., G. Peng, J. Li, P. Ma, Z. Wang, W. Shu, S. Peng, and G.-Q. Chen. 2011. Chondrogenic Differentiation of Human Bone Marrow Mesenchymal Stem Cells on Polyhydroxyalkanoate (Pha) Scaffolds Coated with Pha Granule Binding Protein Phap Fused with Rgd Peptide. *Biomaterials* 32 (9): 2305–13.

Zahedi, E., S. Ansari, B. M. Wu, S. Bencharit, and A. Moshaverinia. 2017. Hydrogels in Craniofacial Tissue Engineering. *Biomaterials for Oral and Dental Tissue Engineering*, 47–64. Elsevier.

Zaman, A. U. 2016. A Comprehensive Study of the Environmental and Economic Benefits of Resource Recovery from Global Waste Management Systems. *Journal of Cleaner Production* 124: 41–50.

Zhan, X.-B., C.-C. Lin, and H.-T. Zhang. 2012. Recent Advances in Curdlan Biosynthesis, Biotechnological Production, and Applications. *Applied Microbiology and Biotechnology* 93 (2): 525–31.

Zhang, H., X. Wang, and Y. Liang. 2015. Preparation and Characterization of a Lithium-Ion Battery Separator from Cellulose Nanofibers. *Heliyon* 1 (2): e00032.

Zhang, S., and Z. Li. 2013. Stimuli-Responsive Polypeptide Materials Prepared by Ring-Opening Polymerization of A-Amino Acid N-Carboxyanhydrides. *Journal of Polymer Science Part B: Polymer Physics* 51 (7): 546–55.

Zhang, Y.-X., Y.-F. Chen, X.-Y. Shen, J.-J. Hu, and J.-S. Jan. 2016. Reduction-and Ph-Sensitive Lipoic Acid-Modified Poly (L-Lysine) and Polypeptide/Silica Hybrid Hydrogels/Nanogels. *Polymer* 86: 32–41.

Zhang, Y., L. Cui, F. Li, N. Shi, C. Li, X. Yu, Y. Chen, and W. Kong. 2016. Design, Fabrication and Biomedical Applications of Zein-Based Nano/Micro-Carrier Systems. *International Journal of Pharmaceutics* 513 (1–2): 191–210.

Zhang, Y., J. Venugopal, Z.-M. Huang, C. T. Lim, and S. Ramakrishna. 2006. Crosslinking of the Electrospun Gelatin Nanofibers. *Polymer* 47 (8): 2911–17.

Zhang, Z., O. Ortiz, R. Goyal, and J. Kohn. 2014. Biodegradable Polymers. *Handbook of Polymer Applications in Medicine and Medical Devices*, 303–35: Elsevier.

Zilg, H., H. Schneider, and F. Seiler. 1980. Molecular Aspects of Albumin Functions: Indications for Its Use in Plasma Substitution. *Developments in Biological Standardization* 48: 31–42.

Zivanovic, S., R. Davis, and D. Golden. 2015. Chitosan as an Antimicrobial in Food Products. *Handbook of Natural Antimicrobials for Food Safety and Quality*, 153–82. Elsevier.

8 Bionanocomposites
Green Materials for Orthopedic Applications

Archita Gupta, Padmini Padmanabhan, and Sneha Singh

CONTENTS

8.1 INTRODUCTION

In bone diseases, fractures account for the most adverse effects. These include osteoporosis, Paget's disease, osteomalacia, rickets and osteogenesis imperfecta which severely affect bone strength, density, and structure leading to fractures. Osteoporosis, the leading cause of fractures, affects about 1.5 million individuals annually of all age groups worldwide (Office of the Surgeon General 2004). These statistics are estimated to double by the year 2020 as a consequence of an increasing aging population, obesity, genetic abnormalities and accidents (Amini et al. 2012). Due to these diseases or abnormalities, orthopedic implants are most often used to repair or replace damaged bone tissue. Since bone diseases are the most prevalent, orthopedic implants and tissue transplantation employed clinically are increasing drastically every year. However, the direct cost, related to surgery and therapies, as well as the indirect cost comprising lost productivity of patients, creates a burden for billions worldwide (Office of the Surgeon General 2004 and Listl et al. 2015). Also, these implants last for a maximum of 15 years, thereby necessitating revision surgeries (Webster et al. 2003). Therefore, to reduce patient discomfort and the associated cost, research in the area of designing next-generation orthopedic prostheses with increased efficacy and lifetimes is the prime requirement. Thus, repairing bone injuries has become the leading area of research in the field of tissue engineering and regenerative medicine. Since conventional grafts (autograft, allograft, and xenograft) have limitations, of foreign body rejection, donor site morbidity, surgical site infections, and disease transmission, development of synthetic biomaterials is currently a booming area of interest.

In biological systems, different physicochemical, biological, and mechanical properties are of great importance for the development of biomaterials. Nowadays, synthetic biomaterials, such as polymers, bioglasses, and their composites, are widely used for the development of hard tissues. In the last 25 years, researchers have been devoted to utilizing composites for biomedical applications, as living tissues themselves are composed of a different hierarchy of composites (Liu et al. 2007). Many organic/inorganic biocomposites are capable of attaining great mechanical strength and biological activity, but optimization is required before utilization in biomedical applications. Nowadays, nanotechnology-based approaches have gathered great attention for the generation of orthopedic scaffolds, as they mimic the natural structure of bone. They can also overcome the limitations of poor mechanical properties, impaired cell proliferation, and inadequate differentiation (Walmsley et al. 2015). Hence, currently, bionanocomposites (BnCs) have provided improvement in the efficacy of orthopedic prostheses (Liu et al. 2007). Not only the components but the synthesis procedures also influence the properties of any BnC. Thus, complete knowledge of the constituents and the fabrication techniques are required for successful biomedical applications.

Therefore, this chapter mainly focuses on BnCs used for treating bone diseases through repairing and/or replacing the damaged tissues by the emphasis of nanotechnological interventions during scaffold design which can enhance the physiochemical, mechanical, and biological properties of BnCs. This chapter mainly deals with three areas; first, it defines the natural bone, its remodeling, and requirements for a scaffold used in bone repair. Second, it introduces BnCs and elaborates their properties, classification, and composition. Third, the chapter highlights the manufacturing techniques and recent applications of BnCs for bone tissue engineering.

8.2 BONE STRUCTURE AND REMODELING

Bone is a connective tissue comprising fibers, cells, and the ground substances, forming a matrix. The bone skeleton weighs around 4 kg with a total tissue volume of 1.75

L and average bone calcium content of approximately 1 kg (Parfitt 1983). The matrix component of bone tissues is mineralized providing high strength and rigidity to the bone, unlike other soft tissues. This hard tissue performs several important functions, some of which are locomotion, internal support, storage for calcium and phosphorus, an attachment site for muscles and tendons, and protection of soft organs from injuries (Weatherholt et al. 2012). Bones are macroscopically divided into two types, that is, cortical (80% of the skeleton) and cancellous (20% of the skeleton) bone, which dominate the long bones of extremities and the vertebrae and pelvis, respectively (Eriksen et al. 1993). In cross-section, the ends of the bone represent cortical (or compact) bone surrounding the inner cancellous (or spongy) bone. Cortical bone is found at the diaphysis of long bone while the cancellous bone is present at metaphyses and epiphyses of both cuboidal and long bones. Out of the two, cancellous bone is more metabolically active and can be remodeled more easily than cortical bone.

At the microstructural level, bone can be structurally classified as woven or lamellar. Woven bone is immature bone that remodels to form matured lamellae (Liu et al. 2007). Natural bone is itself a composite material comprising 30% (w/v) organic material, mainly collagen fibrils and 70% (w/v) inorganic minerals, such as calcium phosphate (main hydroxyapatite/HA) nanocrystals (Kargozar et al. 2019), although composition does vary based on species, age, dietary habits, anatomical location, and health status. This composite is arranged in different length scales, concurrently defining the architecture of the bone as represented in Figure 8.1. There are mainly three levels of hierarchical structure: nanostructure, such as non-collagenous organic proteins, collagen fibrils, and infiltrating mineral crystals; microstructure, such as osteons, lamellae, and Haversian systems; and macrostructure, such as cortical and cancellous bone. Therefore it is important to understand the nanostructured components of the bone prominently consisting of collagen fibers reinforced with minerals. The collagen tissues are approximately 3 nm wide and 200 nm long, which are arranged to a fibrillar structure undergoing aggregation to form fibrils of about 500 nm. Their aggregation provides small gaps which act as the platform for the deposition of mineral crystals ranging from 10 to 20 nm in length and 2 to 3 nm in width (Rho et al. 1998). The organic phase consists of approximately 90% type I collagen and 10% non-collagenous proteins. The collagen provides an architecture to the bone while hydroxyapatite crystals are present in and/or between the collagen fibers.

The non-collagenous proteins, such as bone inductive proteins (osteopontin, osteonectin, and osteocalcin), growth factors and cytokines (insulin, such as growth factor and osteogenic proteins), and proteins of extracellular matrix (bone sialoprotein and proteoglycans), provide a major contribution to the biological function of the bone (Liu et al. 2007). While the collagen makes fibers and minerals deposit in between them, simultaneously the ground substances (proteins, polysaccharide, and mucopolysaccharide) fill the space between fibers and minerals. Since the release of ions from the bone regulates the cell function, the presence of doping elements such as magnesium, strontium, potassium, and sodium can affect the bone cell-mediated functions. These organic and inorganic components play an essential role in bone metabolism, but the major function is actually regulated by the bone cells. Mainly four different cell types are found in the osseous tissue as represented in Figure 8.2, which play a major role in remodeling, namely osteoprogenitor cells, osteocytes,

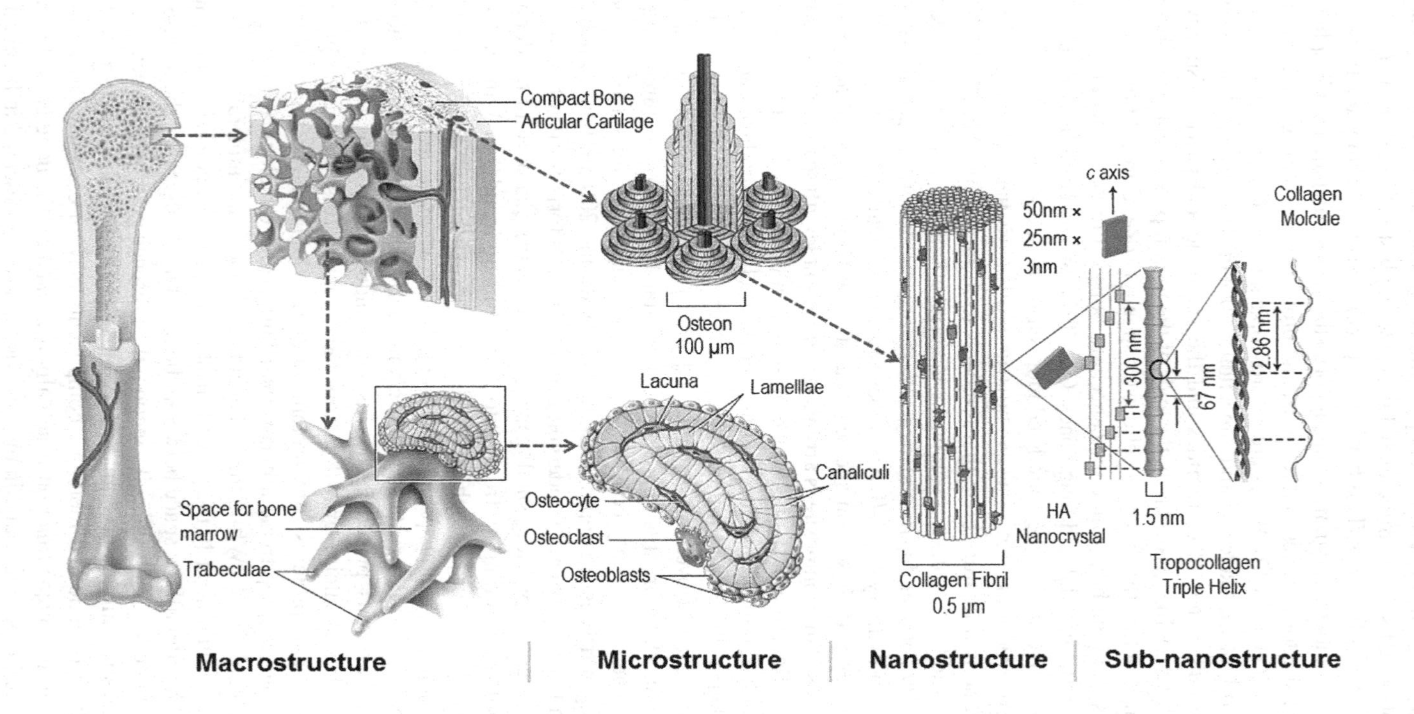

FIGURE 8.1 The hierarchical arrangement of bone at a different scale (macro to nano). (Reprinted with permission from Wang et al. "Topological design and additive manufacturing of porous metals for bone scaffolds and orthopaedic implants: A review". *Biomaterials* 83 (2016): 127–141.)

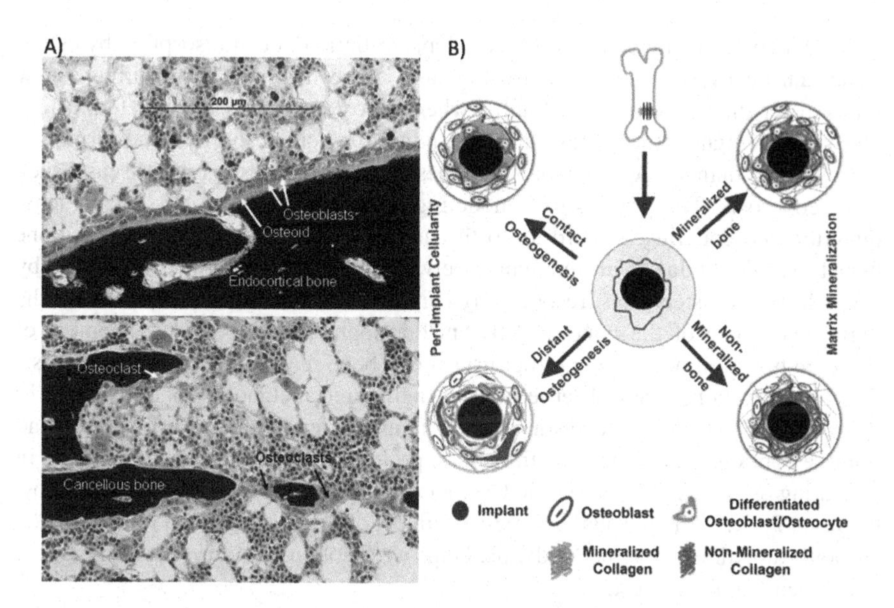

FIGURE 8.2 Representation of A) bone cells through von Kossa stained sections of bone B) mineralization in vicinity of bone cells at implant site. (Reprinted with permission from A) Iwaniec et al., Histological analysis of bone. [In *Alcohol*, Humana Press, 2008] pp. 325–341). B) Ragamouni et al. "Histological analysis of cells and matrix mineralization of new bone tissue induced in rabbit femur bones by Mg–Zr based biodegradable implants". *Acta Histochemica* 115, no. 7 (2013): 748–756.)

osteoblasts, and osteoclasts (Kargozar et al. 2019). The osteoprogenitor cells are the undifferentiated ones present at the deep layers of bone marrow, periosteum, and endosteum that develop into osteoblasts. Osteocytes are the mature cells involved in maintaining the mineral concentration of the bone matrix. Osteoblasts, the differentiated bone cells, are present in the growing areas of the bone (periosteum and endosteum) and secrete collagen fibers during osteogenesis. Finally, the osteoclasts are the multinucleated cells formed by the fusion of precursor cells. They are present at the bone surface and injury site, thereby assisting the bone remodeling process by breaking down the matrix and releasing calcium.

During adult life, bone undergoes a continual process of repair, renewal, and remodeling. Bone tissues are dynamic in nature and remodel by intercommunication between the aforementioned bone cells. Bone modeling is associated with bone growth during childhood and there is no relationship between resorption and formation of the bone, unlike bone remodeling, which is a cyclical process. Bone has an inherent ability to regenerate and remodel after damage. It can remodel into different sizes, shapes, and properties to respond appropriately during any physical or mechanical stress. The same can be activated by the bone remodeling unit (BRU) comprising mainly three types of bone cells – osteoblasts, osteocytes, and osteoclasts, presented in Figure 8.2, specifying bone formation, maintenance, and resorption respectively. A number of local factors, such as growth factors and cytokines as well as systemic factors such as estrogen and calcitonin, control the process of remodeling.

Remodeling mainly comprises three steps: initiation of cell resorption by osteoclasts during hypocalcemia; then osteoblasts come into play by forming a new matrix over the resorbed bone; finally, when the osteogenesis is completed, osteoblasts differentiate into osteocytes and maintain the bone resorption and formation in a systemic manner. Any imbalance in these processes leads to bone disease, such as osteoporosis. The process of remodeling is mainly based on two phenomena: quantum and coupling. According to the quantum concept, the variation in bone density is mainly due to an imbalance between bone resorption and formation by osteoclasts and osteoblasts, respectively (Frost 1963 and Parfitt 1979). BRU is the main constituent of the bone at different developmental stages. Arrangement of these units and minerals define the turnover of bone matrix and become the basis for change in bone mass observed during aging and metabolic bone diseases. In this concept, first the osteoclastic precursors are differentiated to osteoclasts, and bone resorption starts. The growth factors, proteins, and cytokines are involved in activating osteoclast for resorption. Osteoclasts express tartrate-resistant acid phosphatase (TRAP) which leads to a decrease in pH due to hydrogen ion release by the carbonic anhydrase system. Acidic pH helps in dissolving HA crystals and organic components of the matrix.

The rate of BRU formation is called activation frequency and determines the tissue turnover along with the function rate of the BRU (Parfitt 1979; Eriksen 1986 and Charles et al. 1987). This activation frequency is very significant in regulating the bone mass and turnover as it varies by 50% to 100% with every change of 10% to 20% of resorption depth. The coupling phenomenon explains the formation of bone after bone resorption termination within its periphery through the coupling procedure (Frost 1963 and Parfitt 1982). The osteoblasts are activated by the growth factors such as insulin-like growth factor that is secreted by osteoclasts. Osteocytes modulate osteoblast differentiation from non-calcium secretory to calcium secretory for regulating bone formation through insulin-like growth factor and tissue growth factor β (Trippel 1998). Osteoblasts proliferate and express genes for vitronectin, type I collagen, and fibronectin. This process makes sure that the amount of bone resorbed is to be laid down by new bone formation. It determines that the bone formation is distinctly followed after the resorption (temporal coupling) and the amount of bone resorbed is equilibrated by the bone formed (quantitative coupling). Mechanical stimulation of the bone cells is also important for the coupling process. After proliferation, development and maturation of the extracellular matrix take place along with a ten-fold increase in alkaline phosphatase activity and mRNA expression of osteopontin and collagenase protein. Osteocalcin, a calcium-binding protein secreted by osteoblasts, interacts with HA, mediating the coupling process. Osteocytes, present around osteons concentrically and between the lamellae, helps in maintaining the balance between the two processes through its connective channels (canaliculi) (Liu et al. 2007).

Implant failures may result from either limited remodeling leading to malnourished collocated bone or too much remodeling leading to intemperate bone resorption. Therefore, knowledge about bone remodeling is important, as the degree of remodeling occurring at the implant site determines the life of the prostheses.

8.3 REQUIREMENTS OF BONE SCAFFOLDS

Bone scaffolds are widely used for regenerating bone in cases of fractures, trauma, infection, disease, joint replacement, and bone alignment. Approximately 2.2 million grafts are used annually worldwide under these conditions (Neighbour 2008). Conventional grafts, such as autografts, allografts, xenografts, and metallic implants, have long been used for repairing bone fractures but they have their limitations. The tissue regeneration capabilities of these grafts include their osteoinductive, osteoconductive, and osteogenic potential. Osteoinductivity includes the activation of mesenchymal stem cells to differentiate into pre-osteoblasts with the help of inductive signals such as growth factors. Osteogenic property is defined by the osteoblasts, osteocytes, and the mesenchymal stem cells. Osteoconductivity includes the scaffold matrix that stimulates the growth of the bone cells. Autografts, which are considered the gold standard, are also associated with pain, infections, donor non-availability, and site morbidity (Liu et al. 2007 and Saltzman 2004). Allografts and xenografts involve transplantation from members of different genera and species, leading to high risk of disease transmission and immunological responses (Nather 2001). Metallic implants do not exhibit the physicomechanical and dynamic characteristics of bone, so therefore cannot mimic, remodel, or self-repair with time. Ever since bone grafts have been widely used, their shortcomings have prompted researchers to work on different bone substitutes.

Bone tissue engineering provides a gateway for natural bone regeneration by incorporating bone cells with the scaffolds representing BnC as shown in Figure 8.3. Tissue engineering is a discipline that covers the application of engineering principles in life sciences and the design of efficient biological substitutes for improvising tissue functions (Skalak et al. 1988). Tissue substitutes are fabricated *in silico* by combining living cells with the biomaterial and are then implanted into subjects to repair the defect. The scaffold should be biodegradable, provide three-dimensional matrixes for cell attachment and proliferation, be resorbable, allow diffusion of oxygen, and provide for nutrients to reach the cells and for the removal of waste material. The main challenges faced by the researchers in designing scaffolds are improper adhesion, proliferation, and growth of the cells in 3D orientation, and cell function on these scaffolds (Sachlos et al. 2003). For the scaffold design to be successful all the properties at the macro and micro level should be considered and also interaction with the tissues at the nano level.

In the last 50 years, numerous advances have been made in the field of biomaterials wherein the materials under consideration for biomedical applications have been categorized into four different generations (Navarro et al. 2008). The first generation represents the bioinert materials; the second generation constitutes biodegradable and bioactive materials; the third generation incorporates the materials basically designed to generate specific responses at the molecular level; and the fourth generation includes materials relevant to bioelectric cues and cell communication (Ning et al. 2016).

Recently, advances in bone tissue engineering have led to the development of materials mimicking the 3D structure of bone. Scaffold must be biocompatible, non-toxic, cells should adhere, function properly, proliferate, differentiate, generate new

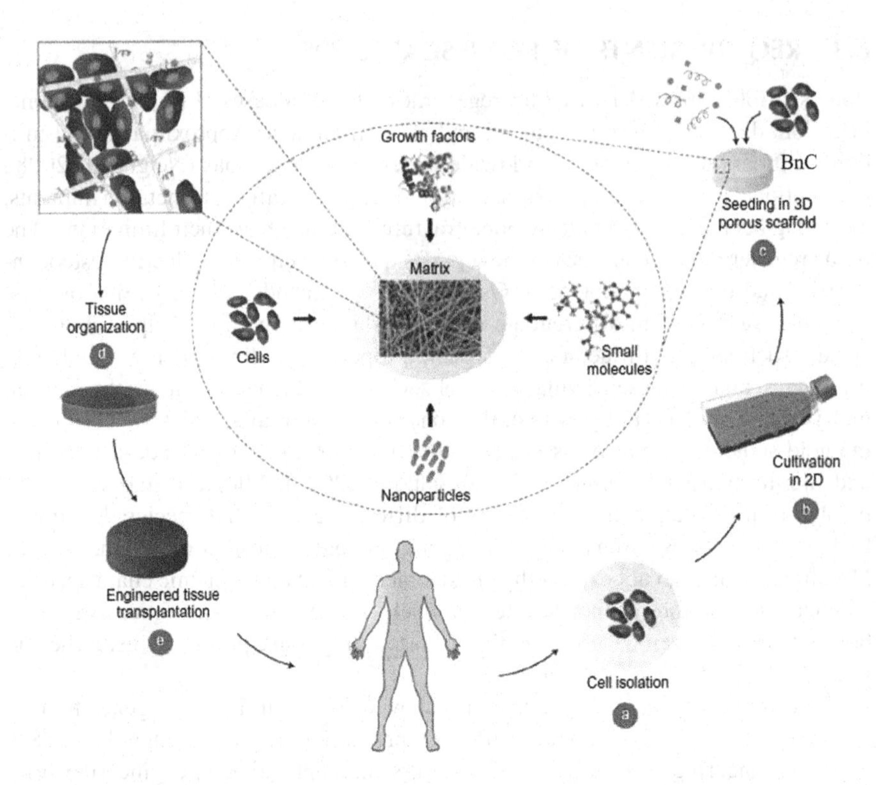

FIGURE 8.3 Schematic representation of regenerative strategy used for bone regeneration using BnC. (Reprinted with permission from Chee et al., Electrospun hydrogels composites for bone tissue engineering. [In *Applications of Nanocomposite Materials in Orthopedics*, Woodhead Publishing, 2019] pp. 39–70.)

matrix, be non-hemolytic, non-coagulative, and should not elicit immune response and infection (Bouët et al. 2014; Henke et al. 2013; Kumbar et al. 2014 and Salgado et al. 2004). In addition, the scaffold should be degradable or resorbable with time as bone growth progresses. The biocompatibility of the scaffold is dependent on the fabrication techniques. The chemicals involved in the synthesis of the biomaterial, such as organic solvents, stabilizers, crosslinkers, etc., may leach out of the scaffolds. So, the degradation products should not be toxic to the body and should be eliminated by the body without any reactivity. Acidic byproducts can cause tissue necrosis or inflammation by a sudden decrease in pH (Goldberg and Caplan 2004). The resorption rate of the scaffold and the rate of bone formation should be balanced so that by the time the injury site is regenerated, the scaffolds should be totally degraded. The scaffold composition, structure, molecular weight of components, porosity, the surface-to-volume ratio, and pore size determine the resorption rate.

Next, the scaffold should be biocompatible, that is, it should not be capable of inducing an immunological response in the host and hence be non-immunogenic biomaterials. Also, scaffolds should interact with the bone cells, and help in cell migration, differentiation, and communication, which are essential for neo-bone formation (Chen et al. 2008 and Hofmann et al. 2011). The scaffold surface should be such that it facilitates cell attachment and further differentiation. The surface properties of the scaffold should be efficient enough to be bioactive as well as osteoconductive. Chemical properties imply the ability of the protein and its adsorption onto the scaffold for cell adherence. Topographical properties make them osteoconductive, as they must help in the cell movement. There have been reports that rough surfaces facilitate the cell movement through fibrin clot (Davies 1996 and Albrektsson 2001). Osteoinductivity plays an essential role in bone formation, thereby making this property a significant one during scaffold fabrication. In combination with growth factors, scaffolds must enhance the proliferation of bone marrow stem cells and their differentiation into osteoblasts (Wu et al. 2017). Through this property, the scaffold can command the surrounding to form ectopic bone. This can mainly be due to adsorption of the osteoinductive factors and presenting them to the surrounding tissues.

In addition, calcium phosphate release can influence the differentiation phenotype of the bone cells (Motamedian et al. 2015; Blokhuis et al. 2011 and Barradas et al. 2011). This property influences the migration of mesenchymal cells and pluripotent osteoprogenitor cells towards the bone injury site. For large defects, scaffolds alone cannot be sufficient for the regeneration, so to enhance the osteoinductivity they should be combined with agents such as recombinant human bone morphogenetic proteins (rhBMP-2 and rhBMP-7). The 3D structure and porosity of the scaffold are also significant to maintain its osteogenic property, allowing bone cells to grow on the surface and internal pores. Therefore, osteoinductivity can be improved by changing the physicochemical properties and supplementing with inductive agents (Subramaniam et al. 2015). The pore structure is also important for enhancing the osteoconductivity of the scaffold. The need is to make the scaffold vascularized for the proper supply of nutrients to the bone cells. For regenerating complex orthopedic structures such as osteochondral compartment, a nonhomogenous, multi-layered gradient scaffold is required (Grigolo et al. 2015; Manferdini et al. 2016 and Shimomura et al. 2014). In this regard, use of self-assembling material such as peptides in designing scaffolds is the main requirement for larger defects (Kyle et al. 2010; Kirkham et al. 2007 and Semino 2008).

Another feature is surface topography, which is influenced by the incorporation of extracellular matrix and bioactive agents such as growth factors. They enhance bone regeneration and also modulate the immune system. The biomaterial surface plays an important role in the host's acute immune system as the growth factors should be designed in such a way as to prevent macrophage adhesion and activation. This can be achieved by modifying the surface through protein adsorption. Also, making the scaffolds through computer-assisted design (CAD) can be useful for complex bone regenerations. Scaffolds with high porosity and interconnected pores are more beneficial as they can allow penetration of cells, nutrient diffusion, and cell colonization. The pore size of about 200 to 500 μm is ideal due to the large void volume and area-to-volume ratio, thereby maximizing the area for cell and blood vessel penetration (Liu et al. 2007; Thavornyutikarn et al. 2014 and Akeda et al. 2006). Ideally, the porosity of the designed scaffold is desired to be above 90% for promoting cell adhesion, bone in-growth, and transport of materials and discharge of metabolites formed during the reaction. The interconnected pores improve nutrient supply, waste removal, and cell viability at the scaffold's center

(Liu et al. 2007). Therefore, porosity, pore structure, pore size, pore connectivity, homogenous pore distribution, and twist in connected channels need to be considered properly, as they can affect the mechanical and biological properties of the scaffold (Wu et al. 2017).

The scaffold's mechanical properties, such as tensile strength, elastic modulus, fatigue, fracture toughness, and elongation percentage, are crucial for bone tissue engineering and should be modulated in a manner to match the implant site, thereby reducing the risk of stress shielding, osteopenia, and re-fracture (Bouët et al. 2014; Henkel et al. 2013; Atala et al. 2012 and Neel et al. 2014). During *in vitro* and *in vivo* studies, the scaffolds should be mechanically strong, withstanding hydrostatic pressure, as bone naturally faces large physiological stress such as compression, torsion, tension, and bending. Also, it should provide space for cell penetration and bone in-growth to match the properties of living bone (Leong et al. 2003). The stiffness should not be a way to avoid high-stress shielding, insufficient load transfer, filling of autogenous bone and cancellous bone, and insufficient mechanical stimulation (Wang et al. 2015). On the other hand, stiffness should not be too low, as it can reduce the carrying capacity, making the bone more prone to fracture (Qin et al. 2014). The biomaterial should also have good plasticity, that is, it should retain its shape in the body to overcome the different effects on the mechanical performance of the scaffold. The scaffolds must also be reproducible, sterilizable to reduce infections, and should not be deformed by sterilization techniques.

These requirements for the scaffold are the key to designing biomaterials for bone tissue engineering without failure, and with successful biomedical application.

8.4 BIONANOCOMPOSITE: A FACSIMILED BONE

For the development of better orthopedic materials, they should mimic the microstructure, composition, and properties of natural bone. As mentioned in the previous section, bone is a nanostructured tissue made up of polymer materials reinforced with nano-sized ground substances. Therefore, the thrust of recent research has shifted towards the fabrication of materials resembling bone in terms of nanoscale features for better osteoconductivity. Nowadays, the paradigm of tissue engineering involves cells, scaffolds, growth factors, and the extracellular microenvironment (Gaharwar et al. 2014). Through these, the design and development of materials in orthopedics is evolving. Novel micro-scale technologies have emerged in order to fabricate polymeric-based structures with additional nanofillers, thus providing researchers with a material inspired by nature (Du et al. 1999). These naturally occurring engineered materials encompass the nano- and micro-scale components which manifest macro-scale functions at tissue level. This new generation of material is an emerging field establishing the relation between broad areas of research such as life science, material science, and nanoscience (Ruiz-Hitzky et al. 2008). In the last decade, BnCs have been considered to be typical composite material wherein one or more naturally occurring polymers (biopolymers) and inorganic substances are involved that show at least one component at nanoscale level (Darder et al. 2007). That the materials are nanostructures is the critical factor in the advancement of novel properties and in controlling the structure at the nano level (Pande et al. 2017). BnCs, or biopolymer-based nanocomposite, exhibit several beneficial properties derived from synthetic polymers, such as mechanical, thermal properties, and biological moieties enhancing biocompatibility, biodegradability, and bioactivity (Darder et al. 2007 and Bergaya et al. 2006). Orthopedic implants with surface properties that promote cell adhesion and proliferation in the presence of the microenvironment that leads to implant osseointegration are needed. Surface properties such as area, topography, and charge depend on the size of a material. In this regard, nano-phase materials possessing higher surface area with increased portions of surface defects have an advantage in biomedical applications. Surface roughness and crystallinity determined by nanomaterials influence interactions of selected proteins and subsequent cell adhesion. Enhanced osteoblast adhesion and subsequent deposition of the calcium-containing mineral were observed by some researchers using nanoscale materials as compared to their microscale counterparts (Liu et al. 2005). Further, increased adhesion of osteoblasts, their activity, cytocompatibility, and simultaneously decreased adhesion of fibroblasts was observed with nano-phase alumina, titania, and HA (Wei et al. 2004).

The underlying mechanism was revealed to be adsorption of proteins such as vitronectin, the conformation of which plays a crucial role in osteoblast adhesion to the nanophase materials and which was found to be 10% greater compared to conventional alumina (Webster et al. 2001). The protein unfolding can expose cell-binding epitopes such as RGD sequences for subsequently enhanced osteoblast adhesion. BnCs also possess stiffness, compression, and tensile strength, mimicking the physiological properties of the bone but different from the conventional materials. Compared to traditional biomaterials, nano-phase biomaterials possess increased surface roughness resulting from both smaller grain and pore size. Moreover, nanophase materials possess enhanced surface wettability due to higher surface roughness and more significant numbers of grain boundaries on their surfaces. BnCs exhibit improved mechanical properties and biocompatibility providing a new gateway in the field of orthopedics where this composite material remains unexplored. For future improved orthopedic efficacy they can be designed to possess similar micro-architecture to that of healthy and physiological bone, mimicking the local natural environment.

8.5 COMPONENTS, PROPERTIES, AND CLASSIFICATION OF BNCS

Biomaterials and nanomaterials combined constitute BnCs, as presented in Figure 8.4. Biomaterial in itself is comprised of biopolymer and biomolecular substances. Biopolymers are considered to be polymers obtained through natural sources, such as animals, plants, and micro-organisms such as bacteria or fungi, as they can be degraded by the action of biocatalysts (Fomin et al. 2001). Biomolecular substances, on the other hand, comprise proteins, carbohydrates, nucleic acids, cellular components, or any other biologically active entity. The biomaterials mainly contain starch, chitin, lignin, cellulose, hemicellulose, and chitosan, depending on the source material. Starch, also known as amylum, is the storage form of energy in

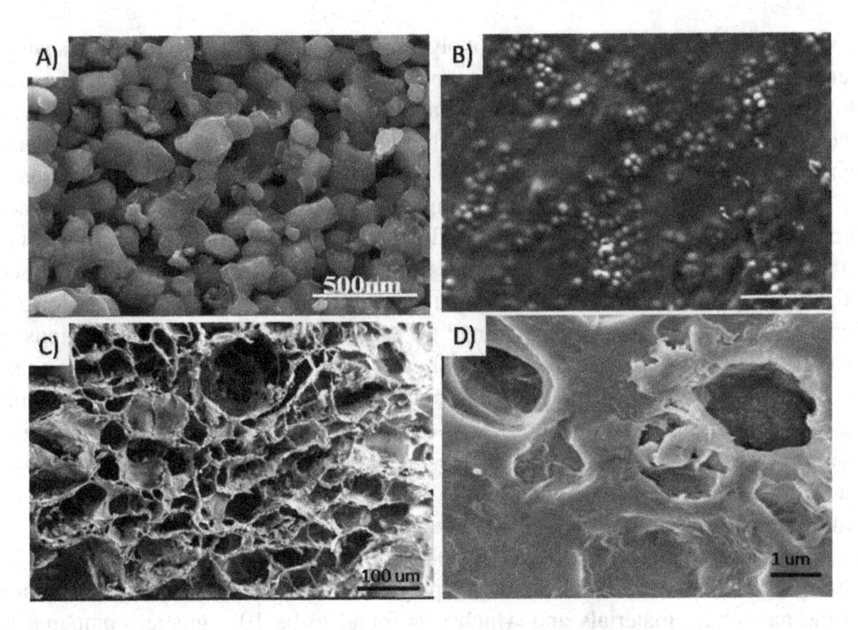

FIGURE 8.4 Different BnCs constituting nanomaterials in the form of A) hydroxyapatite B) bioglass C) silver D) zirconium dioxide. (Reprinted with permission from A) Govindaraj et al. "From waste to high-value product: Jackfruit peel derived pectin/apatite bionanocomposites for bone healing applications". *International Journal of Biological Macromolecules* 106 (2018): 293–301. B) Li et al. "Biodegradable multifunctional bioactive glass-based nanocomposite elastomers with controlled biomineralization activity, real-time bioimaging tracking, and decreased inflammatory response". *ACS Applied Materials & Interfaces* 10, no. 21 (2018): 17722–17731. C) Hasan et al. "Nano-biocomposite scaffolds of chitosan, carboxymethyl cellulose and silver nanoparticle modified cellulose nanowhiskers for bone tissue engineering applications". *International Journal of Biological Macromolecules* 111 (2018): 923–934. D) Bhowmick et al. "Development of bone-like zirconium oxide nanoceramic modified chitosan based porous nanocomposites for biomedical application". *International Journal of Biological Macromolecules* 95 (2017): 348–356.)

plants and micro-organisms comprising the large chain of glucose moieties with two components, amylose, and amylopectin. They store these polymeric carbohydrates in granules of size 0.5 to 175 μm, which vary depending on its source (Averous and Boquillon 2004). Starch is mainly extracted from maize, potato, wheat, and tapioca with a composition of 20% to 30% amylose and 70% to 80% amylopectin (Mousa et al. 2016). Pure starch has a very high melting point of 220°C and can be easily degraded during biopolymeric fabrication. Hence some plasticizers such as water, glycerol, and polyols are blended with starch during the gelatinization process to form a thermoplastic starch, thereby decreasing its melting point for successful fabrication. This blending procedure distorts the granular organization of starch by disruption of hydrogen bonding, making an amorphous polymer. The degree of disruption governs the mechanical properties by increasing the elongation and tensile strength of the polymer and subsequently decreasing the elastic modulus. Due to the limitations of using water in blends with starch, a composite polymer is mainly explored through the utilization of polylactic acid (PLA), polyhydroxybutyrate (PHB), and poly (ε-caprolactone) (PCL) (Averous 2004).

Cellulose is an abundant agro-based polymer and the main constituent of the cell wall of a plant. This polysaccharide polymer contains liner chains of sugar molecules, making it a robust natural polymer with varying properties depending on the chain length and the degree of polymerization (Reddy et al. 2012 and Averous 2004). Cellulose fibers are a commonly available, eco-friendly, inexpensive, recyclable material and require less energy-intensive manufacturing processes. The fabrication of nanofibers of cellulose through nano reinforcement leads to the development of two cellulose types, which are microfibrils and whiskers (Oksman et al. 2016 and De Azeredo 2009). Esterifying with respective acids and anhydride leads to the synthesis of thermoplastic cellulose, namely. propionate, cellulose acetate butyrate, and cellulose acetate (Reddy et al. 2012 and Averous 2004). For plasticization of cellulose, cotton and wood are used as raw materials and then esterified by different additives. The plasticizers used in the synthesis of plastic cellulose are commonly utilized but pose severe health issues, such as eye irritation and liver cancer; they have recently been replaced by citrate-based materials (De Azeredo 2009).

Other than from plants, cellulose can also be synthesized through bacterial species such as *Azotobacter*, *Acetobacter*, *Alcaligenes*, *Salmonella*, *Rhizobium*, *Pseudomonas*, and *Sarcina ventriculi*. Among them, *Acetobacter xylinum*, *Gluconacetobacter hansenii*, and *Acetobacter pasteurianus* are the most effective cellulose producers, of which *A. xylinum* is considered the model bacteria for cellulose production with good mechanical properties (De Azeredo 2009). Cellulose modification leads to the formation of biopolymer consisting of N-acetylglucosamine units (Averous 2004 and Mousa et al. 2016). This high molecular weight biopolymer is prominently known as chitin, which is abundantly found in invertebrates with the property of crystalline nanostructures similar to that of cellulose with improved biodegradability, biocompatibility, and mechanical properties (Pande et al. 2017). Deacetylation of chitin forms a linear polymer, chitosan, consisting of glucosamine and N-acetyl glucosamine units. It is a natural semi-crystalline polymer where its crystallinity depends on the degree of acetylation (De Azeredo 2009). It has numerous hydroxyl and amine groups generating a high molecular weight copolymer with medical applications for sutures and wound dressings (Mousa et al. 2016).

Polylactic acid (PLA) is a linear, nontoxic, biodegradable thermoplastic formed through the lactic acid most commonly present in corn or sugarbeet (Jem et al. 2010). It is mainly processed through polycondensation or ring-opening polymerization of lactic acid or lactide respectively. PLA is thermally unstable and inflexible and can be improved by using poly(ethylene oxide), polyethylene glycol, oligomer lactic acid, or citrate ester (Avérous 2004 and Martin et al. 2001).

Polyhydroxyalkanoate (PHA) is a polyester available in various forms and considered hydrophobic, biocompatible, and biodegradable. It is stored by microorganisms as a source of energy in the form of granules, for example, *Alcaligenes eutrophus*, which commonly store PHA as an energy source (Hakkarainen 2002). It can vary in chain length and, based on that, it can be divided into three types – short-chain, medium-chain, and long-chain PHA (Ahmad et al. 2017). PHA can be commonly divided into two broad types – polyhydroxybutyrate (PHB) and poly(hydroxybutyrate-co-hydroxyvalerate) (PHBV), where PHB is mainly stored by most of the micro-organisms. PHA degradation is dependent on microbial activity, crystallinity level, pH, and moisture (Reddy et al. 2013). PHA has a high melting point, stiffness, degree of crystallinity, resistance to ultraviolet radiation, glass transition temperature, lower toughness, and lower solvent resistance than polypropylene, which is known to have properties similar to that of PHA (Hakkarainen 2002). Its properties can be enhanced by making a composite with polybutyrate adipate terephthalate thereby improving toughness and crystallinity (Javadi et al. 2010).

Polycaprolactone (PCL) is a polymerization product of ε-caprolactone mainly polymerized through ring-opening phenomenon in the presence of metal alkoxides (Reddy et al. 2013). PCL is a flexible polymer with a high modulus of elasticity, but its melting point is very low, which can be improved through blending or cross-linking modification (Averous et al. 2000 and Mousa et al. 2016). A polyester known as polyhydroxy ester ethers is formed by combining diglycidil ether of bisphenol A and adipic acid, which is biodegradable but is toxic to the environment.

Poly (vinyl alcohol) (PVA) is a thermoplastic and water-soluble polymer formed by the hydrolysis of poly (vinyl acetate) or condensation of acetaldehyde (Baker et al. 2012). To form a low molecular weight PVA, polymerization of the metal compound or vinyl ester takes place first followed by a saponification or esterification process (Jose et al. 2014). The molecular weight of PVA is significant, as its properties are governed by this and by the degree of hydrolysis. On increasing the molecular weight and hydrolysis, there is an observed increase in tensile strength and solvent resistance while there is a decrease in flexibility, water sensitivity, and solubility (Oviedo et al. 2008). The wide range of properties of PVA, which are its biocompatibility, water solubility, high mechanical strength, resistance towards organic solvents, and inexpensive availability, have attracted its extensive application as an implant (Gross et al. 2002).

Another component of BnCs are nanofillers, which are nanosized materials formed from large particles into small particles of spherical, rod, triangle, star, and platelet shapes. The particles at the nanoscale are considered to have strength, elastic modulus, and stability. Layered silicate being natural or synthetic is the most popularly used nanofiller in BnCs. It constitutes clay minerals such as mica, laponites, fluorohectorite, bentonite, and magadiite. They comprise the stack of layers where van der Waal forces connect each layer via opposite ions, varying based on source, type, and fabrication technique. These nano-layers are 1 nm thick with a 30 nm length comprising tetrahedral, octahedral sheets where silicon, aluminum atoms are surrounded by four and eight oxygen atoms respectively (Chiu et al. 2014).

Kaolinite is a nano-silicate layer with a 1:1 ratio of silicon tetrahedral and an aluminum octahedral sheet with a common sharing plane for oxygen atoms, whereas phyllosilicate has the respective ratio of 2:1 (Miranda-Trevino et al. 2003 and Beyer 2002). Pyrophyllite is a layered silicate with a respective ratio of 2:1 and any substitution in these layers leads to the development of a new nanoclay called mica. Neither of these clays is easily miscible in water whereas pristine layered silicate with a similar ratio is miscible in water-soluble polymers such as PVA or PEO due to the presence of sodium and calcium ions in the layers (Mousa et al. 2016 and Aranda et al. 1999). To fabricate a hydrophobic polymer soluble nanoclay, ion substitution to some organophilic groups such as alkylammonium, phosphonium, and sulfonium ion is mostly required. Apart from this, functionalization is essential for increasing bond formation between clay and polymer (Manias et al. 2001 and Mousa et al. 2016).

Cellulose forming micro-fibrillate, nanocrystal, or nano-fibrillate lead to the development of nanocellulose having low density, coefficient of thermal expansion, high aspect ratio, and tensile strength (Postek et al. 2010). Micro-fibrillate has a high aspect ratio due to its dimensions of 10–100 nm diameter and 0.5–10 μm length. They are obtained in a wide variety of natures ranging from amorphous to crystalline. In contrast, nano-fibrillate dimensions vary from 4–20 nm in diameter to a length of 200–500 nm. They can be fabricated through different techniques such as cryo-milling, high-pressure homogenization, and micro fluidization (Johnson et al. 2009 and Saito et al. 2009). On the other hand, cellulose nanocrystal has a diameter of 8–20 nm and a length of 500 nm. They are also known as nanowhiskers and are synthesized through acid hydrolysis (Bai et al. 2009 and de Souza Lima et al. 2004). They are considered over carbon nanotubes due to superior mechanical properties but have the limitation of being hydrophilic, restricting their reinforcement capability with hydrophobic polymers (Gardner et al. 2008). If they are somehow mixed with the hydrophobic polymers, they separate out due to weak bonding between the molecules. Therefore, there is a requirement for surface modification for such nanocellulose (Freire et al. 2008).

The most commonly used nanotubes to date are the carbon nanotubes (CNT), which can be easily folded into allotrophic graphite sheets (Baughman et al. 2002). CNTs used as nanofillers render the material lightweight, with high tensile strength and Young's modulus (Coleman et al. 2006). CNTs can be divided into single-walled or multi-walled CNTs based on the number of sheets required for their fabrication. Their properties are defined by the direction of sheet rolling and the diameter of the sheets. CNTs can be synthesized through laser ablation, chemical vapor deposition, and carbon arc discharge (Baughman et al. 2002). Laser ablation and carbon arc discharge methods involve the evaporation of carbon and later condensation-generating single-walled tubes at a small scale. But these methods require special equipment and high energy, highlighting the challenges behind their use (Joselevich et al. 2007). While chemical vapor-deposition technique involves deposition of carbon-containing gas in the presence of metal ions for CNT synthesis (Ishigami et al. 2008), through this method, the morphology and structure of CNT can be controlled and modified according to requirements (Dai 2002). Apart from providing mechanical stability, CNT can also improve biodegradation of crystalline polymer (Wu et al. 2010).

Halloysite nanotubes (HNTs) are formed by a hydrothermal reaction in aluminosilicate minerals forming a tubular clay (Liu et al. 2014). Omalius d'Halloy discovered this kind of mineral in 1826; it can be present in the form of plates, tubes, or spheroids varying according to the geological and crystallization conditions (Du et al. 2010). The surface of HNT has siloxane groups rather than the hydroxyl group and the space between the layers is occupied with a large number of hydroxyl groups connected by aluminum ions. This property inhibits inter-tube bonding and increases the interaction with polymers and solvents, thereby enhancing their solubility (Ismail et al. 2013 and Wei et al. 2014). HNTs have recently been more researched, due to their very high mechanical strength, aspect ratio, heat resistance, cost-effectiveness, and their easy availability, along with being eco-friendly (Yuan et al. 2012; Liu et al. 2014 and Joo et al. 2012).

Further, BnCs can be classified based on composition, size, shape of fillers, and type of polymer used. Particulate, elongated, and layered BnC are the three types of BnCs according to configuration. Particulate BnCs consist of reinforcements of some particle size and shape which enhance activities, such as making them less permeable, enhancing inflammability, and being inexpensive. Elongated BnCs represent reinforcements present in elongated forms such as cellulose nanofibril and carbon nanotubes. They possess a high aspect ratio and perform biomechanically. Layered BnC incorporates reinforcement of layered particles such as nanoclays which can be further subdivided into intercalated, micro-composites, exfoliated, and flocculated/phase-separated BnCs (Arora et al. 2018). When the sheets are formed within the interstitial spaces between the nanoparticles, intercalated BnCs are formed. Microcomposites include microparticles uniformly dispersed in the polymeric matrix. The partitioning of individual layers results in exfoliated BnCs while no partition due to particle interaction forms flocculated BnC (Kim et al. 2003). BnCs can also be classified, based on the type of matrix used for fabrication, into micro- or nanocrystal matrix BnCs. Further, the composition of BnCs also leads to their classification into polymer- and non-polymer-based BnCs.

Polymeric BnCs are the most commonly used and researched biomaterials in biomedical applications. They contain fewer filler materials, empowering them to not behave as foreign material during applications and to maintain the matrix stable (Wagner et al. 2004). They can be classified into five different classes depending on the type of polymeric material used: polymer–polymer, polymer–layered silicate, inorganic–organic half and half, polymer–ceramic and inorganic–organic polymer-based BnCs (Hasnain et al. 2019). Nonpolymeric BnCs comprise of mainly metallic material rather than any polymer. They can be classified into metal–metal, ceramic–ceramic, and metal–ceramic BnCs. In metal BnCs, the matrix of metal ions or alloys such as ZnO–CuO and Ag–TiO_2 is reinforced with a small amount of nano-sized materials (Li et al. 2010). These BnCs are suitable for applications where shear stresses and temperature stability are required. Under metal–ceramic BnCs, metal and ceramic components are present in equal ratio, presenting nanoscale properties. These BnCs are generally fabricated through chemical processes such as sol–gel and co-precipitation methods (Lim et al. 2006). Within ceramic–ceramic BnCs, two or more ceramic materials are blended at the nanoscale level to exhibit improved strength, creep resistance, toughness, and hardness, as compared to the ceramic-only matrix (She et al. 2000).

For a material to be successfully utilized in the biological sciences, it should be eco-friendly and safe (Arora et al. 2018). BnCs are becoming popular in research areas due to the advantages of their properties for different applications. One of the significant properties of any material for it to be applied to the field of medicine is that it be biocompatible (Ratner et al. 2004). BnCs exhibit excellent properties such as biocompatibility, biodegradability, being antimicrobial, anti-inflammatory, and having superior mechanical strength, simultaneously with being safe, environmentally nontoxic, inexpensive, and recyclable which are favorable with regard to it being a green composite material for biomedical application (Ruiz-Hitzky et al. 2010, Rhim et al. 2013 and Ruiz-Hitzky et al. 2013). The nanofillers used in BnCs along with the biopolymers usually enhance the thermal, permeable characteristics and mechanical properties of the biopolymer matrix (Zafar et al. 2016). Green materials are nowadays replacing petroleum-based materials and more discoveries in the area of bio- and nano-technology are increasing the synthesis of greener materials, such as BnCs, for their utilization in various fields, based on the following properties.

8.5.1 Morphological Structure

Merely mixing biopolymers and nanofillers does not lead to the formation of BnCs; proper bonding and reaction between the two are necessary. A loose bond between the components leads to the separation of the phases and results in microcomposites instead of nanocomposites (Pavlidou and Papaspyrides 2008). For miscible systems, mainly intercalated and exfoliated BnCs are formed which are specifically dependent on the synthesis method and chemical composition of the precursors. Intercalated structures are formed when biopolymer layers are formed within the inter-nano spacing of nanomaterial whereas the exfoliated BnCs achieve the proper dispersity of nanoclay enhancing the mechanical properties.

8.5.2 Mechanical Properties

Reinforcements in the form of nanofillers to the polymers make the matrix stiffer, provided that the bonding between the two is strong enough. As the size of nanofillers decreases, the surface area increases and leads to more significant mechanical properties. In layered BnCs, adding nanoclays to the polymer matrix can increase Young's modulus and tensile strength significantly (Gorrasi et al. 2003 and Rhim et al. 2013). In some other cases, the level of exfoliation can be influenced by the nanofillers. Apart from nanofillers, the molecular weight of the polymer also plays an important part in determining the tensile strength of the BnCs (Pavlidou and Papaspyrides 2008). As the molecular weight increases, high shear stress can convert the clay material into platelets. Morphological heterogeneity in BnCs leads to phase separation, formation of voids, and cracks, affecting their toughness and making them fragile (Wang et al. 2014). This can be overcome by the incorporation of nanofillers into the matrix (Carli et al. 2011). There is evidence that incorporation of carbon nanotubes into the polymer increases ductility. However, some reports also represent the requirement of using uniform dispersion of nanofillers for the enhancement of properties (Meng et al. 2010; Paiva et al. 2004 and Yang et al. 2013). Dynamic mechanical testing is done to determine the mechanical behavior and deformation of BnCs in the presence of high temperature. Dynamic mechanical analysis incorporates three parameters that are mostly studied, which are loss modulus, storage modulus, and loss tangent. Loss modulus is the amount of energy released due to deformation, whereas storage modulus defines the energy stored during deformation. Loss tangent is the measure of the angle between stress and strain that is used to determine glass transition temperature (Pavlidou and Papaspyrides 2008). Within the BnC system, on increasing the concentration of nanofillers, there is a monotonic increase in the storage modulus since nanofillers limit the movement of polymeric chains leading to more storage of energy (Krikorian and Pochan 2003 and Ray and Bousmina 2005).

8.5.3 Thermal Stability

Heating the material in the presence of an inert gas or air leads to nonoxidative and oxidative degradation, respectively. This can be measured through thermogravimetric analysis wherein change in mass is recorded as a function of temperature. In the case of layered silicate BnCs, the layers of silicate act as the insulator and prevent any mass transfer reaction, thereby providing high thermal stability (Becker et al. 2004). In metal nanomaterial-based chitosan BnCs, the incorporation of nanomaterial leads to the rise in degradation temperature by more than 80°C as compared to chitosan matrix alone (Wang et al. 2005).

8.5.4 Biodegradability

Biodegradation involves the decomposition of materials in the presence of biocatalysts from microbial or any other source (Mousa et al. 2016). The same can be enhanced by the utilization of nanomaterials in composites. For instance, the use of metallic nanomaterials with biopolymeric composites such as PLA can improve the biodegradation rate (Tetto et al. 1999). The degradation mechanism involves first the hydrolysis of PLA chains to its low molecular weight components, then conversion to carbon dioxide, water, and humus by microbial action. The nanomaterials enhance the hydrolysis rate and subsequent biodegradation (Ray et al. 2003). Similarly, in the case of PHA, degradation of PHA BnCs is more than the PHA matrix alone (Fomin et al. 2001).

8.6 METHODS OF BnC SYNTHESIS

Nanomaterials are defined as materials of structural units ranging between 1 and 100 nm, exhibiting enhanced catalytic, magnetic, electrical, optical, and mechanical properties as compared to other formulations (Li et al. 2003). Until now, BnCs has not been explored much for orthopedic applications due to limited knowledge about its synthesis processes. Hence, the prime requirement remains to review different manufacturing methods used for the successful synthesis of BnCs. The synthetic techniques have been frequently used for developing uniform-sized BnC, including synthesis based on solution, vapor, and gas phases (Stankic et al. 2016). These synthetic approaches can be broadly divided into two categories, that is, top-down and bottom-up approaches (Oliveira and Machado 2013). The top-down approach is the traditional and the most common one where pre-designed scaffolds of suitable size and shape are used temporarily for regeneration. These scaffolds should replicate the structure and function of the extracellular matrix with features of requisite pore size, biodegradability, and mechanical properties. The scaffold should support the attachment of the cells, thereby populating it with the cells, and produce an appropriate extracellular matrix for restoring the tissue. The top-down approach includes methods such as chemical etching, nanolithography, electrospinning, template-assisted, and phase separation (Darder et al. 2007). Apart from several advantages, this approach has its own limitations, such as nonuniform cell distribution, inadequate diffusion, abnormal vascularization, uncontrollable cell density, and lack of osteoinductivity (Pavlidou and Papaspyrides 2008). The bottom-up approach includes self-assembly of the molecules from the nanoscale to macroscale and further rearrangement into structures functionally resembling the tissue. This approach includes manual manipulation, microfluidic-directed assembly, multilayer photo-patterning, layer pyrolysis, plasma spray synthesis, photolithography, cell-sheeting, 3D bioprinting, and random assembly (Kim et al. 2003 and Adeosun et al. 2012). This technology involves fabrication of the scaffold through biomolecules and chemicals of an organized structure resembling the architectural milieu of the bone. For example, different peptides can be considered as the building molecules for peptide-based scaffolds, assisting cell attachment, proliferation, and differentiation (Teeri et al. 2007). However, this method has numerous drawbacks, such as difficulty in manipulation, lack of scalability, technical sensitivity, poor mechanical stability, and a random biodegradation pattern (Satyanarayana 2015). Therefore, nowadays research needs to be focused on BnC fabrication techniques based on a knowledge of molecular structure to overcome the existing limitations.

The most appropriate manufacturing technique must be cost-effective, scalable, and mimic the 3D structure of the desired tissues. These different methods adopted for the synthesis of BnC are presented in Figure 8.5 and listed in the following sections.

8.6.1 Solvent-Casting with Particulate-Leaching

Solvent-casting with particulate-leaching is the one most widely used for the fabrication of three-dimensional porous scaffolds. In this technique, a particulate/porogen is used primarily for generating pores in the scaffolds. Salt is the most commonly used porogen due to its easy availability and cost-effectiveness. The salt is sieved or milled into fine particles of micron sizes and cast into the mold. Then the required polymer is suspended in the suitable solvent and transferred into the salt-containing mold. Later, the solvent is allowed to evaporate under air and/or in a vacuum and the salt is leached in the presence of water-generating pores of the required size depending upon the size of the salt crystals. Porosity can also be controlled by varying the ratio of salt to polymer. Researchers have reported the use of gelatin and waxy hydrocarbons as porogen during scaffold synthesis. Using them instead of salt has proved to provide better porosity, pore size, cell attachment, and proliferation (Suh et al. 2002 and Shastri et al. 2000). Through this method, thick samples with interconnected pores can be synthesized with increased strength and electrical conductivity by adding the second phase to the casting mold. This technique is easy, inexpensive, simple, and can be efficiently combined with other methods to form BnCs but the use of organic solvents becomes a great disadvantage due to the cytotoxicity they pose.

8.6.2 Gas-Foaming with Particulate-Leaching

To overcome the limitations of the solvent-casting method, a new approach of gas foaming was developed that circumvented the use of an organic solvent (Liu et al.

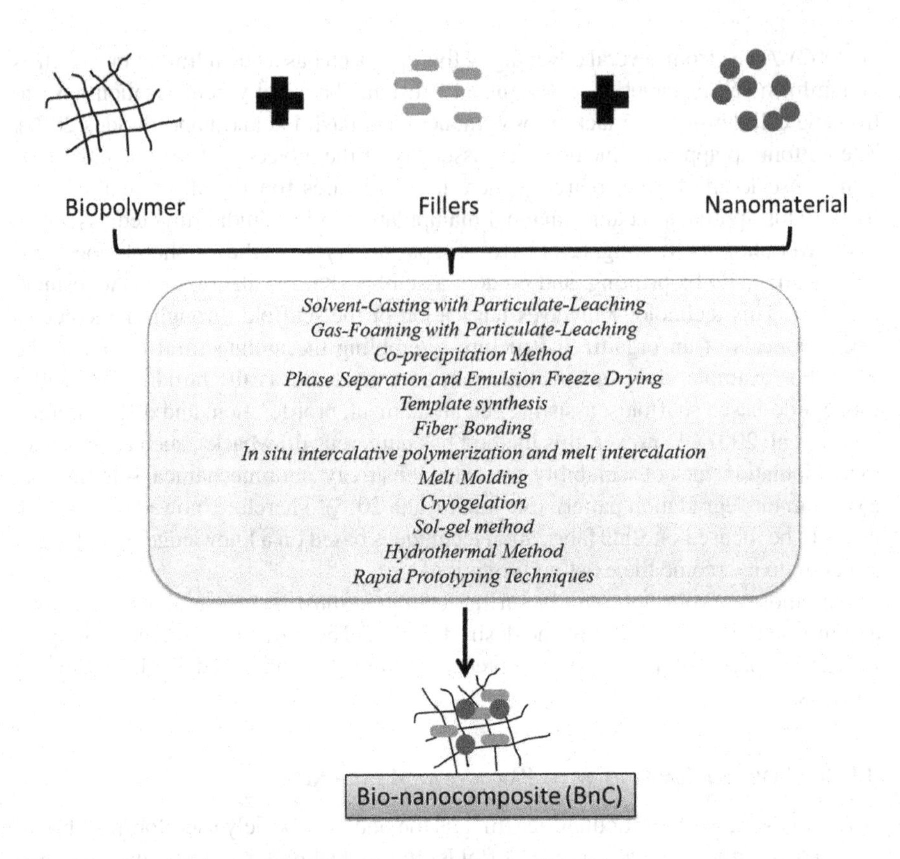

FIGURE 8.5 Fabrication technique for the synthesis of BnC.

2007). Under this technique, the scaffolds made of polymer and porogen were compressed at room temperature in the presence of carbon dioxide gas under high pressure. After the gas is brought back to atmospheric pressure, its solubility in the scaffold decreases, leading to the development of gas pores within the polymer, followed by porogen leaching. This method presents a better compressive and tensile modulus as compared to the solvent-casting process.

8.6.3 Co-Precipitation Method

In the co-precipitation technique, a homogenous mixture of inorganic salt and a solvent is prepared where the salt precipitates out as hydroxides after the critical concentration and leads to nucleation during scaffold synthesis (Stankic et al. 2016). After this precipitation process in the presence of sodium hydroxide, ammonium hydroxide, and sodium carbonate, the hydroxide ions were calcinated into oxides to form crystalline structures (Kalantari et al. 2013 and Sadegh et al. 2014). The concentration of salt, temperature, pH of the solution, and surfactants determines the size and shape of the particles involved in calcination. The method is advantageous in terms of the way the process parameters can be easily controlled and the process is simple and inexpensive.

8.6.4 Phase Separation and Emulsion Freeze-Drying

Phase separation and emulsion freeze-drying is based on the principle that to make a system thermodynamically stable, different components separate into phases when used under suitable conditions. Following this, during fabrication of the scaffold, mainly liquid–liquid and solid–liquid phase separations take place. The solid–liquid phase separation is done under subzero temperatures where the solvent crystallizes and is removed after sublimation, generating pores in its place. The scaffold thus generated is finally lyophilized for complete water and solvent removal. This process is also known as emulsion freeze-drying. The process generates a scaffold with high porosity, small pore size, and greater pore distribution (Whang et al. 1995). The pore structure can be controlled by controlling the molecular weight of the polymer, water-to-polymer concentration ratio, and emulsion viscosity.

8.6.5 Template Synthesis

In the template synthesis method, inorganic molecules, through their precursors, generate a scaffold, considering different biomolecules and cells as a template. This method is versatile, simple, easy, and can be applied to large-scale production. However, the chances of cross-contamination and toxicity exist due to side products generated during the synthesis (Siqueira et al. 2010).

8.6.6 Fiber Bonding

Nonwoven scaffolds are synthesized through fiber meshes wherein the polymers are crosslinked to form a mesh-like structure. Since these nonwoven meshes can easily be deformed when cells attach to them for proliferation, this leads to the development of a new technology called a fiber bonding technique for increasing the mechanical properties of the scaffold. In this method, a polymer dissolved in a suitable solvent is cast onto the mesh of another polymer followed by solvent evaporation and the composite being heated above the melting temperature of the latter polymer. Finally, the polymer cast was removed by dissolving it in the solvent resulting in the fibers with better mechanical properties (Liu et al. 2007).

8.6.7 *In Situ* Intercalative Polymerization and Melt Intercalation

Scaffolds can be synthesized through intercalative polymerization by the formation of the polymeric sheet through heating and/or radiation in-between the nanoparticle dispersion in the solution. The organic catalyst is used to initiate the polymerization process (Shchipunov 2012). This method provides irregular pores due to the invariable assembly of the nanoparticles. The method has now been overtaken by the melt intercalation technique, which has become the standard for the synthesis of BnC. In this process, the polymeric mass is heated beforehand at some specific temperature and finally mixed with nanoparticle suspension.

8.6.8 Melt Molding

Melt molding incorporates the utilization of a mold in which the polymeric powder is placed with porogen micro- or nanospheres and allowed to be heated at the glass transition temperature of the polymer under pressure (Liu et al. 2007). Later the porogen is allowed to leach out to form a linked polymeric scaffold. Specifically, the incorporation of nHA fibers into the polymer is targeted through this technique, generating a more specific targeting scaffold for regeneration.

8.6.9 Cryogelation

In cryogelation, the polymer and nanomaterial, for example, nHA, are mixed and allowed to crosslink in the presence of a crosslinker such as glutaraldehyde under subzero temperatures (Kim et al. 2003). The cryogels formed through subsequent crosslinking under freezing conditions develop pores of about 400 to 500 μm and with porosity of around 90%. The addition of a greater concentration of nanofillers decreases the pore size, and makes the surface of the scaffold rough. The cryogels have very high water absorption capability, even after the addition of nHA, and can improve strength and elastic modulus (Kumar et al. 2010).

8.6.10 Sol–Gel Method

The sol–gel method involves mild reaction conditions, fabricating scaffolds from precursor molecules with micro- or nanostructures (Guglielmi et al. 2014). This process incorporates hydrolysis and condensation, leading to the formation of sol which later, upon aging, form a gel-like network. The inorganic precursors undergoing chemical reactions result in this three-dimensional, interconnecting network. The reaction parameters influence the shape, size, and structure of the final product since the slow reactions taking place at room temperature define excellent structural characteristics at the nanoscale.

8.6.11 Hydrothermal Method

The chemical reactions in the solvent during the hydrothermal method take place at high temperature and pressure in the presence of some stabilizers (Byrappa and Yoshimura 2012). Recently, this technique has been used in combination with the sol–gel method, which incorporates variation in structural and physicochemical properties, stability, and single-phase fabrication (Li et al. 2005 and Wang et al. 2007).

8.6.12 Rapid Prototyping Techniques

The fabrication techniques discussed so far do not control the pore structure precisely, leading to the development of a new method known as rapid prototyping, which is mainly based on computer-aided designs. This involves different techniques such as 3D printing, fused deposition modeling, and stereolithography. Out of these, 3D printing, first developed at Massachusetts Institute of Technology, is the most commonly used rapid prototyping process for tissue engineering. The scaffold is generated through inkjet printing using different binders with polymeric materials by controlling the operation parameters through the computer-aided program. The growth factors or cells can also be incorporated into the scaffolds through this rocess (Liu et al. 2007). Even though this is advanced technology, there remain limitations to polymer selection, of insufficient resolution, low porosity, and inadequate mechanical properties. This technique can further be divided into two types based on the inkjet mechanism, namely droplet-based and continuous ink. Inks are generally prepared by using polymeric or polyelectrolyte molecules and heating them to form a stable homogenous ink with desirable viscosity (Raina et al. 2019). These include either wax-based inks, which form droplets on being heated and solidify upon being cooled, or less viscous solutions which can be removed during evaporation. Clogging of the inkjet is also an issue, establishing the requirement of appropriate ink with specific parameters of viscosity, speed, and flow rate.

8.7 BnCs IN ORTHOPEDICS: PRESENT AND FUTURE

The field of tissue engineering in regenerative medicine is an ever more demanding area of biomedical research for repairing or replacing damaged tissues of the body (Nayak and Pal 2012). In 1988, the National Science Foundation defined tissue engineering as an area utilizing the principles of engineering and life science for understanding the structural and functional aspects of the tissues and further development of substitutes for maintaining or enhancing the respective functions (Hasnain et al. 2010). This approach further led to the development of the scaffolds made of a variety of biomaterials, providing the structural environment where the extracellular matrix, cell-to-growth factor interaction, and cell-to-cell interaction facilitate the regeneration of damaged tissues when these scaffolds are implanted into the body (Nayak and Pal 2012 and Kim et al. 2000). Therefore, tissue engineering reciprocates the properties of cellular matrix and assists the regeneration capability through signaling for cell attachment, proliferation, and differentiation (Hasnain et al. 2010).

In this regard, around two decades ago, a research group studied the composite made of nano-hydroxyapatite (nHA) and collagen for its mimicking capacities during regeneration. They investigated the mechanical behavior of the composite through the Knoop microhardness test and discovered that it resembled bone, due to its low crystallinity and nanometer-sized structures, and also that it was bioactive and biodegradable. The material can be resorbed by the bone's metabolism thereby restricting its use as a permanent implant. But the lack of hierarchical organization related to the bone and invariability to localized pressure necessitates further advancement in these materials (Du et al. 1998). At the same time, the same group researched the nHA/collagen (nHAC) composite for mimicking bone structurally and compositionally. They prepared the sheets of nHAC and developed a three-dimensional construct *in vitro* by incorporating osteogenic cells into the scaffold. The porous structure of the scaffold resembles that of an *in vivo* bone model, and the cells can migrate and proliferate into the network of scaffold coils generating new bone matrix at the junction of bone and the scaffold (Du et al. 1999).

Another group synthesized nHAC composite through a coprecipitation method wherein composite presented self-organized nanostructure similar to bone. It resisted high pressure, thereby exhibiting prominent mechanical properties with the same biological activities as those experienced by the bone, concluding with the replacement of autologous bone grafts by the composite (Kikuchi et al. 2001). In the same year, collagen-gel-mixed nHA was used as a composite between the graft and the bone. Histological studies revealed that the test group showed that direct contact between the new bone and collagen fibers could penetrate new bone within four weeks as compared to the control, which did not show these effects even after 16 weeks (Ishikawa et al. 2001). When the composite was prepared by adding different concentrations of bioglass to highly porous poly(D,L-lactide) and poly(lactide-co-glycolide) there was a decrease in the pore volume with subsequent increase in the mechanical properties. By adding bioglass, the water absorption and HA formation increased while the reduction in molecular weight in the presence of a different microenvironment was slower than that of the noncomposite material. The study provided evidence that adding bioactive fillers can reduce the degradation of the composite with improved bioactive properties (Maquet et al. 2004).

The HA/gelatin nanocomposite has also shown better osteoblast activity, improving its potential for bone regeneration (Kim et al. 2005a). When collagen was added to this BnC, MG63 osteoblast cells showed better attachment, proliferation, and differentiation as compared to conventional HA/gelatin nanocomposite. It was observed to have a higher ionic release, adsorption of serum protein, and smaller apatite crystal, and the well-developed pore size configuration provided an improved cellular response with higher osteocalcin and alkaline phosphatase (ALP) activity (Kim et al. 2005b). HA/gelatin foam composite was fabricated using 30% (w/w) gelatin with HA powder through freeze-drying followed by cross-linking. Noncomposite gelatin developed pore configuration with pore size and porosity of 400 to 500 μm and 90% respectively but by adding HA, porosity decreased and pore shape become irregular with a pore size of 2 to 5 nm distributed within the gelatin network. This foam represented a hydrogel kind of structure where the addition of HA made the foam stiffer and stronger. The composite shows higher coating rate on titanium discs and cell proliferation, suggesting application in hard-tissue engineering (Kim et al. 2005c).

Another group developed BnCs based on starch and multiwall carbon nanotubes (MWCNTs), which were considered as good reinforcement (Famá et al. 2011). They showed high tensile strength due to the wrapping of MWCNT with the complex of starch-iodine. The dispersion of nanofiller led to the increment in stiffness (70%), strength (35%), deformation without a break (80%), and biaxial impact (100%). Later bovine-HA-diopside BnCs was developed using HA from bovine bones and by mixing it with diopside powder formed by a mechanical milling method. This BnC is coated onto titanium implants using an electrophoretic deposition technique. The BnC-coated implants represented better bioactivity, hardness, and wettability (Khandan et al. 2016). Vegetable-oil-based polyurethane and MWCNTs functionalized with rapeseed protein BnCs showed great biocompatibility, MG63 cell differentiation, adhesion, proliferation, alkaline phosphatase activity, neo bone formation, non-immunogenicity, biodegradability, and load-bearing capability (Das et al. 2015).

Yet another group fabricated BnCs with polyvinyl alcohol (PVA) reinforced with titanium oxide (TiO_2) nanoparticles, with an enhancement of thermal, mechanical, and biological activities. The BnC was synthesized through the solvent casting and crosslinking technique in which TiO_2 are imbibed in the PVA matrix by hydroxyl groups. The flexibility of PVA, as well as the hardness of the metal nanoparticles, enhanced the surface properties and osteoblastic adhesion. The BnC exhibited good antibacterial activity, as being positively charged they electrostatically interact with the negatively charged membrane of the bacteria, leading to cell disruption (Mohanapriya et al. 2016). Recently, β-tricalcium phosphate (TCP) with iron- and silver-based BnC was fabricated through an attrition milling technique followed by cold sintering at 2.5 GPa. TCP is used, since it is comparable to the chemical structure of the bone and regarded as significant in the bioresorbable scaffold. This BnC represented high mechanical strength and ductility mainly due to the incorporation of silver. The degradation capability of the BnC scaffold was investigated by immersing it in Ringer's solutions and saline solutions. The composite was able to retain its bending and compression strength up to 50% even after a month of the study. This BnC showed no toxicity and a high differentiation rate on osteoblast cell lines as compared to a non-substituted TCP scaffold, suggesting its successful utilization in implants (Swain et al. 2017).

A group developed a scaffold with chitosan and nHA through the sol-gel method with formic acid as a new solvent during the method. The self-assembly of nHA onto the chitosan matrix exhibited a homogenously arranged pore-size BnC to be used in bone tissue engineering (Nazeer et al. 2017). At the same time, another group developed a composite scaffold of chitosan–nHA with or without amine group modification. Its biodegradability in the presence of lysozyme was determined and observed to be higher in the non-amine modified chitosan–nHA BnC, while cell proliferation and biocompatibility were better in the presence of the amine-modified chitosan–nHA BnC studied against human bone mesenchymal stem cells. The osteoinductivity of amine-modified chitosan-nHA BnC was twice as high as other scaffolds. The best result was with the amine-modified chitosan–nHA BnC, which needs to be further studied *in vivo* (Atak et al. 2017).

BnC made of poly (butylene succinate) and cellulose nanocrystals through electrospinning under the solvent range of methanol and chloroform showed increased thermostability, crystallinity, tensile strength, hydrophilicity, Young's modulus, biodegradability, biocompatibility, and 3T3 fibroblast cell proliferation as compared to noncomposite poly (butylene succinate) scaffold (Huang et al. 2018). When bioactive glass nanoparticles and/or nanospheres (Bg) were blended with chitosan–gelatin polymer (CG), better bioactivity, cell proliferation, and osteoinductivity were observed *in vitro* and *in vivo*. This BgCG-based BnC showed accelerated apatite crystal deposition *in vitro*. The Bg makes BgCG-based BnCs more active in promoting osteogenic differentiation due to Bg's dissolution into osteogenic products (Covarrubias et al. 2018). Also, the same group investigated that the incorporation of nanocrystals of Bg during in situ polymerizations of the polyurethane (PU) enabled the development of Bg/PU BnC with enhanced bioactivity to stimulate bone regeneration (Covarrubias et al. 2019). BnC consisting of N, O-carboxymethyl chitosan (CC), and fucoidan (FU) were synthesized through covalent bonding between the amine groups. These hydrogels presented interconnected pores and when they were biomineralized for HA crystal growth, the mechanical and osteogenic properties increased. nHA-mineralized BnCs exhibited better alkaline phosphatase activity and nHA/CC/FU-based BnC was considered a versatile scaffold for tissue engineering (Lu et al. 2018).

Another group focused on the antibacterial activity of nHA and the mechanical properties of sodium alginate (SA) polymer in nHA/SA BnC. Incorporation of nHA improved the bactericidal property of the BnCs. Also, nHA filler enhanced the tensile strength, elongation, and moisture content of BnC films (Gholizadeh et al. 2018). Some researchers developed a BnC hydrogel blend of hydroxypropyl guar (HPG), polyvinyl alcohol (PVA), and nHA. Incorporation of PVA and nHA comparatively improved thermal stability, porosity, flexibility, biomineralization, biocompatibility, and cell viability (Parameswaran-Thankam et al. 2018). Recently, fluorescent erbium-doped hydroxyapatite-chitosan BnC has been developed. Doping made the scaffold biocompatible, bactericidal, and suitable for bioimaging purposes. Better biodegradability and mineralization in simulated body fluid is acceptable for up to two weeks (Banerjee et al. 2018). BnC fabricated through gelatin and cellulose nanofibrils for regeneration showed efficient pore size, interconnection, cell colonization, and proliferation. The study was mainly focused on determining the change in the mechanical behavior of gelatin matrix on the incorporation of cellulose nanofibrils. The designed, three-dimensional BnC obtained through freeze-drying provided better mechanical, chemical, and biological properties (Campodoni et al. 2019). Apart from these BnCs, some others, investigated in bone tissue engineering applications during the last 15 years, are presented in Table 8.1.

8.8 FINAL REMARKS

Recently, several BnCs have been fabricated and evaluated for industrial application. BnCs have been developed during recent decades for their successful utilization in the field of drug delivery, tissue engineering, cancer therapy, stem cell therapy, enzyme immobilization, wound dressings, biosensors, ion-exchanger, and cardiac prosthesis. Recently, BnCs have been accepted in orthopedic applications for the delivery of orthopedic drugs, hormones, proteins, and cells as well as in bone tissue engineering for the treatment of diseases such as osteoarthritis, osteoporosis, osteosarcoma, osteomyelitis, and bone metastasis. BnCs are developed as nanobiomaterials with improved properties of mechanical strength, biocompatibility, cell adherence, bioactivity, and osteoconductivity, which are beneficial for tissue engineering applications. Traditionally, synthetic or petroleum-derived polymers have been utilized for tissue engineering applications but they have now been overtaken by environmentally friendly biodegradable polymers. Later advancements have led to the reinforcement of these biomaterials with nanomaterials for improving physicochemical and mechanical properties leading to the development of BnCs for applications of these biohybrids in regenerative medicine. Currently, research is mainly focused on designing HAP-based BnC implants for bone tissue engineering. Apart from regeneration applications, BnCs are also used for drug delivery purposes, mainly targeting DNA non-viral vectors for DNA-based gene therapy. Therefore, BnCs are chosen

TABLE 8.1
BnC Investigated in Bone Tissue Engineering

BnC Composition	Nano-Component	Method of Synthesis	Reference
Poly(propylene fumarate)/single-walled carbon nanotube	Single-walled carbon nanotube	Crosslinking and functionalization	Shi et al. (2005)
Hydroxyapatite/gelatin matrix	Hydroxyapatite	Electrospinning	Kim et al. (2005a)
Hydroxyapatite/poly(lactic acid)	Hydroxyapatite	Electrospinning	Kim et al. (2006)
Poly(propylene fumarate)/diacrylate/ single-walled carbon nanotube	Single-walled carbon nanotube	Crosslinking and functionalization	Shi et al. (2008)
Poly(D,L-lactic-co-glycolic acid)/hydroxyapatite	Hydroxyapatite	Gas foaming and particulate leaching	Kim et al. (2008)
Poly(lactide-co-glycolide)/ tricalcium phosphate nanoparticles	Tricalcium phosphate nanoparticles	Electrospinning	Schneider et al. (2008)
Nano-hydroxyapatite rods/ poly(L-lactide acid)	Hydroxyapatite rod	Wet chemical	Nejati et al. (2008)
Calcium-deficient carbonate-containing hydroxyapatite/bacterial cellulose	Hydroxyapatite	Bio-mineralization	Shi et al. (2009)
Poly(lactide-co-glycolide)/hyd roxyapatite/L-lactic acid oligomer	Hydroxyapatite	Melt-molding particulate leaching	Cui et al. (2009)
Gelatin/hydroxyapatite/cartilage powder	Hydroxyapatite	Biomimetic precipitation and copolymerization	Haroun and Migonney (2010)
Otoliths/Collagen/Bacterial Cellulose	Otoliths	Wet immersion and soft drying	Olyveira et al. (2011)
PLA (poly lactic acid)/CDHA (carbonated calcium deficient hydroxyapatite)	Hydroxyapatite	Electrospinning	Zhou et al. (2011)
Hydroxyapatite/titania	Hydroxyapatite	High gravity precipitation	Nathanael et al. (2012)
Chitosan/halloysite nanotubes	Halloysite nanotubes	Solution casting	Liu et al. (2012)
Cellulose/Maleic Anhydride Grafted PLA	Cellulose	Electrospinning	Zhou et al. (2013)
Poly(xylitol sebacate)/ hydroxyapatite	Hydroxyapatite	Melt polycondensation	Ma et al. (2014)

(*Continued*)

TABLE 8.1 (CONTINUED)
BnC Investigated in Bone Tissue Engineering

BnC Composition	Nano-Component	Method of Synthesis	Reference
Magnesium/ hydroxy-apatite/ anatase/ periclase	Magnesium-anatase	Ball milling	(Khalajabadi et al. 2015)
Poly(ethylene glycol)/cellulose / poly(lactic acid)	Cellulose	Electrospinning	Zhang et al. (2015)
Chitin/chitosan/poly vinyl alcohol/hydroxyapatite	Chitin-chitosan-poly vinyl alcohol	Copolymerization	Pangon et al. (2016)
Hydroxyapatite-fucoidan	Hydroxyapatite	Wet chemical	Young et al. (2016)
Silane / poly (lactic acid)/ cellulose	Cellulose	Electrospinning	Rahmat et al. (2016)
Carbon dot/peptide/ polyurethane	Carbon dot	Hydrothermal coupling	Gogoi et al. (2017)
Chitosan/poly(ethylene glycol)/ hydroxyapatite/zirconium oxide	Hydroxyapatite-zirconium oxide	Wet chemical functionalization	Bhowmick et al. (2017)
Hydroxyapatite/ poly (vinyl alcohol)/cellulose	Hydroxyapatite-cellulose	Electrospinning	Enayati et al. (2018)
Polycaprolactone/Gelatin/ hydroxyapatite/ vitamin D3	Hydroxyapatite	Electrospinning	Sattary et al. (2018)
Epigallocatechin gallate/duck's feet collagen/hydroxyapatite	Hydroxyapatite	Freeze-drying and wet immersion	Kook et al. (2018)
Bioactive glass/ poly(citrate-siloxane)	Bioactive glass	Chemical crosslinking	Li et al. (2018)
Jackfruit pectin/hydroxyapatite	Hydroxyapatite	Chemical crosslinking	Govindaraj et al. (2018)
Polyhydroxybutyrate/poly (ε-caprolactone)/ sol-gel silica/ levofloxacin	Silica	Electrospinning	Ding et al. (2018)
Chitosan/carboxymethyl cellulose/ silver nanoparticle	Cellulose-silver nanoparticles	Micro-crystallization	Hasan et al. (2018)
Magnesium calcium silicate / gliadin/polycaprolactone	Magnesium calcium silicate	3D printing	Zhang et al. (2018)
Hydroxyapatite/gelatin/copper oxide	Copper oxide nanoparticle-Hydroxyapatite	Space holder technique	Sahmani et al. (2019)
Genipin/ fucoidan/ hydroxyapatite/hydroxypropyl chitosan	Hydroxyapatite	Wet co-precipitation	Lu et al. (2019)
Nanoclay/ titanium oxide	$Al_2Si_2O_5(OH)_4.2H_2O$	Space holder technique	Sahmani et al. (2019)

over biocomposites and nanocomposites due to the excellent material characteristics obtained. Although the reinforcements have tremendously improved the functionality of the developed BnCs, some future developments are still required in order to enhance properties, bioactivity, and multifunctionality; envisaging an emerging field of research around a variety of biopolymers and synergistic interaction of nanomaterials with them. Even though several promising research studies have been conducted during the last few decades, still the commercialization of BnCs has faced a challenge to develop material to meet user applications. Hence, research in the area of fabricating functionally active BnCs is always demanding and there is an open platform for emerging researchers to determine their applications.

REFERENCES

Adeosun, Samson O., G. I. Lawal, Sambo A. Balogun, and Emmanuel I. Akpan. "Review of green polymer nanocomposites." *Journal of Minerals and Materials Characterization and Engineering* 11, no. 04 (2012): 385.

Ahmad, Mudasir, Kaiser Manzoor, Sandeep Singh, and Saiqa Ikram. "Chitosan centered bionanocomposites for medical specialty and curative applications: a review." *International Journal of Pharmaceutics* 529, no. 1–2 (2017): 200–217.

Akeda, K., H. S. An, M. Okuma, M. Attawia, K. Miyamoto, EJ-MA. Thonar, M. E. Lenz, R. L. Sah, and K. Masuda. "Platelet-rich plasma stimulates porcine articular chondrocyte proliferation and matrix biosynthesis." *Osteoarthritis and Cartilage* 14, no. 12 (2006): 1272–1280.

Albrektsson, T., and C. Johansson. "Osteoinduction, osteoconduction and osseointegration." *European Spine Journal* 10, no. 2 (2001): S96–S101.

Amini, Ami R., Cato T. Laurencin, and Syam P. Nukavarapu. "Bone tissue engineering: recent advances and challenges." *Critical Reviews in Biomedical Engineering* 40, no. 5 (2012): 363.

Aranda, Pilar, and Eduardo Ruiz-Hitzky. "Poly (ethylene oxide)/NH4+-smectite nanocomposites." *Applied Clay Science* 15, no. 1–2 (1999): 119–135.

Arora, Bharti, Rohit Bhatia, and Pankaj Attri. "Bionanocomposites: green materials for a sustainable future." In *New Polymer Nanocomposites for Environmental Remediation*, pp. 699–712. Elsevier, 2018.

Atak, Besir Hakan, Berna Buyuk, Merve Huysal, Sevim Isik, Mehmet Senel, Wolfgang Metzger, and Guven Cetin. "Preparation and characterization of amine functional nano-hydroxyapatite/chitosan bionanocomposite for bone tissue engineering applications." *Carbohydrate Polymers* 164, (2017): 200–213.

Atala, Anthony, F. Kurtis Kasper, and Antonios G. Mikos. "Engineering complex tissues." *Science Translational Medicine* 4, no. 160 (2012): 160rv12–160rv12.

Averous, L., and N. Boquillon. "Biocomposites based on plasticized starch: thermal and mechanical behaviours." *Carbohydrate Polymers* 56, no. 2 (2004): 111–122.

Averous, L., L. Moro, P. Dole, and C. Fringant. "Properties of thermoplastic blends: starch–polycaprolactone." *Polymer* 41, no. 11 (2000): 4157–4167.

Avérous, Luc. "Biodegradable multiphase systems based on plasticized starch: a review." *Journal of Macromolecular Science, Part C: Polymer Reviews* 44, no. 3 (2004): 231–274.

Bai, Wen, James Holbery, and Kaichang Li. "A technique for production of nanocrystalline cellulose with a narrow size distribution." *Cellulose* 16, no. 3 (2009): 455–465.

Baker, Maribel I., Steven P. Walsh, Zvi Schwartz, and Barbara D. Boyan. "A review of polyvinyl alcohol and its uses in cartilage and orthopedic applications." *Journal of Biomedical Materials Research Part B: Applied Biomaterials* 100, no. 5 (2012): 1451–1457.

Banerjee, Somtirtha, Biswajoy Bagchi, Suman Bhandary, Arpan Kool, Nur Amin Hoque, Prosenjit Biswas, Kunal Pal et al. "Antimicrobial and biocompatible fluorescent hydroxyapatite-chitosan nanocomposite films for biomedical applications." *Colloids and Surfaces B: Biointerfaces* 171 (2018): 300–307.

Barradas, A. M., Huipin Yuan, Clemens A. van Blitterswijk, and Pamela Habibovic. "Osteoinductive biomaterials: current knowledge of properties, experimental models and biological mechanisms." *European Cells & Materials* 21, no. 407 (2011): 29.

Baughman, Ray H., Anvar A. Zakhidov, and Walt A. De Heer. "Carbon nanotubes: the route toward applications." *Science* 297, no. 5582 (2002): 787–792.

Becker, Ole, Russell J. Varley, and George P. Simon. "Thermal stability and water uptake of high performance epoxy layered silicate nanocomposites." *European Polymer Journal* 40, no. 1 (2004): 187–195.

Bergaya, F., B. K. G. Theng, and G. Lagaly. "Chapter 5." In *Handbook of Clay Science*, Vol. 1, pp. 141–245. Elsevier, 2006.

Beyer, Günter. "Nanocomposites: a new class of flame retardants for polymers." *Plastics, Additives and Compounding* 4, no. 10 (2002): 22–28.

Bhowmick, Arundhati, Nilkamal Pramanik, Piyali Jana, Tapas Mitra, Arumugam Gnanamani, Manas Das, and Patit Paban Kundu. "Development of bone-like zirconium oxide nanoceramic modified chitosan based porous nanocomposites for biomedical application." *International Journal of Biological Macromolecules* 95 (2017): 348–356.

Blokhuis, T. J., and J. J. Chris Arts. "Bioactive and osteoinductive bone graft substitutes: definitions, facts and myths." *Injury* 42 (2011): S26–S29.

Bouët, Guénaëlle, David Marchat, Magali Cruel, Luc Malaval, and Laurence Vico. "In vitro three-dimensional bone tissue models: from cells to controlled and dynamic environment." *Tissue Engineering Part B: Reviews* 21, no. 1 (2014): 133–156.

Byrappa, Kullaiah, and Masahiro Yoshimura. *Handbook of Hydrothermal Technology*. William Andrew, 2012.

Campodoni, Elisabetta, Ellinor B. Heggset, Ahmad Rashad, Gloria B. Ramírez-Rodríguez, Kamal Mustafa, Kristin Syverud, Anna Tampieri, and Monica Sandri. "Polymeric 3D scaffolds for tissue regeneration: evaluation of biopolymer nanocomposite reinforced with cellulose nanofibrils." *Materials Science and Engineering C* 94 (2019): 867–878.

Carli, Larissa N., Janaina S. Crespo, and Raquel S. Mauler. "PHBV nanocomposites based on organomodified montmorillonite and halloysite: the effect of clay type on the morphology and thermal and mechanical properties." *Composites Part A: Applied Science and Manufacturing* 42, no. 11 (2011): 1601–1608.

Charles, P., E. F. Eriksen, L. Mosekilde, F. Melsen, and F. T. Jensen. "Bone turnover and balance evaluated by a combined calcium balance and 47calcium kinetic study and dynamic histomorphometry." *Metabolism* 36, no. 12 (1987): 1118–1124.

Chen, Q., J. A. Roether, and A. R. Boccaccini. "Tissue engineering scaffolds from bioactive glass and composite materials." *Topics in Tissue Engineering* 4, no. 6 (2008): 1–27.

Chiu, Chih-Wei, Ting-Kai Huang, Ya-Chi Wang, Bryan G. Alamani, and Jiang-Jen Lin. "Intercalation strategies in clay/polymer hybrids." *Progress in Polymer Science* 39, no. 3 (2014): 443–485.

Coleman, Jonathan N., Umar Khan, Werner J. Blau, and Yurii K. Gun'ko. "Small but strong: a review of the mechanical properties of carbon nanotube–polymer composites." *Carbon* 44, no. 9 (2006): 1624–1652.

Covarrubias, Cristian, Amaru Agüero, Miguel Maureira, Emmanuel Morelli, Gisselle Escobar, Felipe Cuadra, Cristian Peñafiel, and Alfredo Von Marttens. "In situ preparation and osteogenic properties of bionanocomposite scaffolds based on aliphatic polyurethane and bioactive glass nanoparticles." *Materials Science and Engineering: C* 96 (2019): 642–653.

Covarrubias, Cristian, Monserrat Cádiz, Miguel Maureira, Isabel Celhay, Felipe Cuadra, and Alfredo von Marttens. "Bionanocomposite scaffolds based on chitosan–gelatin and nanodimensional bioactive glass particles: in vitro properties and in vivo bone regeneration." *Journal of Biomaterials Applications* 32, no. 9 (2018): 1155–1163.

Cui, Yang, Yi Liu, Yi Cui, Xiabin Jing, Peibiao Zhang, and Xuesi Chen. "The nanocomposite scaffold of poly (lactide-co-glycolide) and hydroxyapatite surface-grafted with L-lactic acid oligomer for bone repair." *Acta Biomaterialia* 5, no. 7 (2009): 2680–2692.

Dai, Hongjie. "Carbon nanotubes: opportunities and challenges." *Surface Science* 500, no. 1–3 (2002): 218–241.

Darder, Margarita, Pilar Aranda, and Eduardo Ruiz-Hitzky. "Bionanocomposites: a new concept of ecological, bioinspired, and functional hybrid materials." *Advanced Materials* 19, no. 10 (2007): 1309–1319.

Das, Beauty, Pronobesh Chattopadhyay, Somnath Maji, Aadesh Upadhyay, Manashi Das Purkayastha, Charu lata Mohanta, Tapas Kumar Maity, and Niranjan Karak. "Bio-functionalized MWCNT/hyperbranched polyurethane bionanocomposite for bone regeneration." *Biomedical Materials* 10, no. 2 (2015): 025011.

Davies, J. E. "In vitro modeling of the bone/implant interface." The Anatomical Record: *An Official Publication of the American Association of Anatomists* 245, no. 2 (1996): 426–445.

De Azeredo, Henriette M. C. "Nanocomposites for food packaging applications." *Food Research International* 42, no. 9 (2009): 1240–1253.

de Souza Lima, M. Miriam, and Redouane Borsali. "Rodlike cellulose microcrystals: structure, properties, and applications." *Macromolecular Rapid Communications* 25, no. 7 (2004): 771–787.

Ding, Yaping, Wei Li, Alexandra Correia, Yuyun Yang, Kai Zheng, Dongfei Liu, Dirk W. Schubert, Aldo R. Boccaccini, Hélder A. Santos, and Judith A. Roether. "Electrospun Polyhydroxybutyrate/Poly (ε-caprolactone)/Sol–Gel-Derived Silica Hybrid Scaffolds with Drug Releasing Function for Bone Tissue Engineering Applications." *ACS Applied Materials & Interfaces* 10, no. 17 (2018): 14540–14548.

Du, C., F. Z. Cui, Q. L. Feng, X. D. Zhu, and K. De Groot. "Tissue response to nano-hydroxyapatite/collagen composite implants in marrow cavity." *Journal of Biomedical Materials Research: An Official Journal of The Society for Biomaterials, The Japanese Society for Biomaterials, and the Australian Society for Biomaterials* 42, no. 4 (1998): 540–548.

Du, C., F. Z. Cui, X. D. Zhu, and K. De Groot. "Three-dimensional nano-HAp/collagen matrix loading with osteogenic cells in organ culture." *Journal of Biomedical Materials Research: An Official Journal of The Society for Biomaterials, The Japanese Society for Biomaterials, and The Australian Society for Biomaterials* 44, no. 4 (1999): 407–415.

Du, Mingliang, Baochun Guo, and Demin Jia. "Newly emerging applications of halloysite nanotubes: a review." *Polymer International* 59, no. 5 (2010): 574–582.

Enayati, Mohammad Saied, Tayebeh Behzad, Pawel Sajkiewicz, Mohammad Rafienia, Rouhollah Bagheri, Laleh Ghasemi-Mobarakeh, Dorota Kolbuk, Zari Pahlevanneshan, and Shahin H. Bonakdar. "Development of electrospun poly (vinyl alcohol)-based bionanocomposite scaffolds for bone tissue engineering." *Journal of Biomedical Materials Research Part A* 106, no. 4 (2018): 1111–1120.

Eriksen, Erik Fink. "Normal and pathological remodeling of human trabecular bone: three dimensional reconstruction of the remodeling sequence in normals and in metabolic bone disease." *Endocrine Reviews* 7, no. 4 (1986): 379–408.

Eriksen, E. F., A. Vesterby, M. Kassem, F. Melsen, and L. Mosekilde. "Bone remodeling and bone structure." In *Physiology and Pharmacology of Bone*, pp. 67–109. Springer, Berlin, Heidelberg, 1993.

Famá, Lucía M., Valeria Pettarin, Silvia N. Goyanes, and Celina R. Bernal. "Starch/multi-walled carbon nanotubes composites with improved mechanical properties." *Carbohydrate Polymers* 83, no. 3 (2011): 1226–1231.

Fomin, V. A., and V. V. Guzeev. "Biodegradable polymers, their present state and future prospects." *International Polymer Science and Technology* 28, no. 11 (2001): 76–84.

Fornes, T. D., P. J. Yoon, D. L. Hunter, H. Keskkula, and D. R. Paul. "Effect of organoclay structure on nylon 6 nanocomposite morphology and properties." *Polymer* 43, no. 22 (2002): 5915–5933.

Freire, Carmen S. R., Armando J. D. Silvestre, Carlos Pascoal Neto, Alessandro Gandini, Loli Martin, and Iñaki Mondragon. "Composites based on acylated cellulose fibers and low-density polyethylene: effect of the fiber content, degree of substitution and fatty acid chain length on final properties." *Composites Science and Technology* 68, no. 15–16 (2008): 3358–3364.

Frost, Harold M. *Bone remodelling dynamics*. Thomas, 1963.

Gaharwar, Akhilesh K., Nicholas A. Peppas, and Ali Khademhosseini. "Nanocomposite hydrogels for biomedical applications." *Biotechnology and Bioengineering* 111, no. 3 (2014): 441–453.

Gardner, Douglas J., Gloria S. Oporto, Ryan Mills, and My Ahmed Said Azizi Samir. "Adhesion and surface issues in cellulose and nanocellulose." *Journal of Adhesion Science and Technology* 22, no. 5–6 (2008): 545–567.

Gholizadeh, Bahador Safikhani, Foad Buazar, Seyed Mehdi Hosseini, and Seyed Mohammad Mousavi. "Enhanced antibacterial activity, mechanical and physical properties of alginate/hydroxyapatite bionanocomposite film." *International Journal of Biological Macromolecules* 116 (2018): 786–792.

Gogoi, Satyabrat, Somnath Maji, Debasish Mishra, K. Sanjana P. Devi, Tapas Kumar Maiti, and Niranjan Karak. "Nano-Bio engineered carbon dot-peptide functionalized water dispersible hyperbranched polyurethane for bone tissue regeneration." *Macromolecular Bioscience* 17, no. 3 (2017): 1600271.

Goldberg, Victor M., and Arnold I. Caplan, eds. *Orthopedic Tissue Engineering: Basic Science and Practice*. Informa Health Care, 2004.

Gorrasi, Giuliana, Mariarosaria Tortora, Vittoria Vittoria, Eric Pollet, Bénédicte Lepoittevin, Michael Alexandre, and Philippe Dubois. "Vapor barrier properties of polycaprolactone montmorillonite nanocomposites: effect of clay dispersion." *Polymer* 44, no. 8 (2003): 2271–2279.

Govindaraj, Dharman, Mariappan Rajan, Ashraf A. Hatamleh, and Murugan A. Munusamy. "From waste to high-value product: jackfruit peel derived pectin/apatite bionanocomposites for bone healing applications." *International Journal of Biological Macromolecules* 106 (2018): 293–301.

Grigolo, Brunella, Carola Cavallo, Giovanna Desando, Cristina Manferdini, Gina Lisignoli, Andrea Ferrari, Nicoletta Zini, and Andrea Facchini. "Novel nano-composite biomimetic biomaterial allows chondrogenic and osteogenic differentiation of bone marrow concentrate derived cells." *Journal of Materials Science: Materials in Medicine* 26, no. 4 (2015): 173.

Gross, Richard A., and Bhanu Kalra. "Biodegradable polymers for the environment." *Science* 297, no. 5582 (2002): 803–807.

Guglielmi, Massimo, Guido Kickelbick, and Alessandro Martucci. *Sol-gel Nanocomposites*. Springer, New York/Heidelberg/Dordrecht/London 2014.

Hakkarainen, Minna. "Aliphatic polyesters: abiotic and biotic degradation and degradation products." In *Degradable Aliphatic Polyesters*, pp. 113–138. Springer, Berlin, Heidelberg, 2002.

Haroun, Ahmed A., and V. Migonney. "Synthesis and in vitro evaluation of gelatin/hydroxyapatite graft copolymers to form bionanocomposites." *International Journal of Biological Macromolecules* 46, no. 3 (2010): 310–316.

Hasan, Abshar, Gyan Waibhaw, Varun Saxena, and Lalit M. Pandey. "Nano-biocomposite scaffolds of chitosan, carboxymethyl cellulose and silver nanoparticle modified cellulose nanowhiskers for bone tissue engineering applications." *International Journal of Biological Macromolecules* 111 (2018): 923–934.

Hasnain, M. S., A. K. Nayak, R. Singh, and F. Ahmad. "Emerging trends of natural-based polymeric systems for drug delivery in tissue engineering applications." *Science Journal of UBU* 1 (2010): 1–13.

Hasnain, M. Saquib, Syed Anees Ahmad, Mohammad Akram Minhaj, Tahseen Jahan Ara, and Amit Kumar Nayak. "Nanocomposite materials for prosthetic devices." In *Applications of Nanocomposite Materials in Orthopedics*, pp. 127–144. Woodhead Publishing, 2019.

Henkel, Jan, Maria A. Woodruff, Devakara R. Epari, Roland Steck, Vaida Glatt, Ian C. Dickinson, Peter FM Choong, Michael A. Schuetz, and Dietmar W. Hutmacher. "Bone regeneration based on tissue engineering conceptions—a 21st century perspective." *Bone Research* 1 (2013): 216.

Hofmann, Sandra, and Marcos Garcia-Fuentes. "Bioactive scaffolds for the controlled formation of complex skeletal tissues." In *Regenerative Medicine and Tissue Engineering-Cells and Biomaterials*. IntechOpen, 2011.

Huang, An, Xiangfang Peng, Lihong Geng, Lingli Zhang, Keqing Huang, Binyi Chen, Zhipeng Gu, and Tairong Kuang. "Electrospun poly (butylene succinate)/cellulose nanocrystals bio-nanocomposite scaffolds for tissue engineering: preparation, characterization and in vitro evaluation." *Polymer Testing* 71 (2018): 101–109.

Ishigami, Naoki, Hiroki Ago, Kenta Imamoto, Masaharu Tsuji, Konstantin Iakoubovskii, and Nobutsugu Minami. "Crystal plane dependent growth of aligned single-walled carbon nanotubes on sapphire." *Journal of the American Chemical Society* 130, no. 30 (2008): 9918–9924.

Ishikawa, Hiroyuki, Tomihisa Koshino, Ryohei Takeuchi, and Tomoyuki Saito. "Effects of collagen gel mixed with hydroxyapatite powder on interface between newly formed bone and grafted Achilles tendon in rabbit femoral bone tunnel." *Biomaterials* 22, no. 12 (2001): 1689–1694.

Ismail, H., S. Z. Salleh, and Z. Ahmad. "Properties of halloysite nanotubes-filled natural rubber prepared using different mixing methods." *Materials & Design* 50 (2013): 790–797.

Javadi, Alireza, Adam J. Kramschuster, Srikanth Pilla, Jungjoo Lee, Shaoqin Gong, and Lih-Sheng Turng. "Processing and characterization of microcellular PHBV/PBAT blends." *Polymer Engineering & Science* 50, no. 7 (2010): 1440–1448.

Jem, K. Jim, Johan F. van der Pol, and Sicco de Vos. "Microbial lactic acid, its polymer poly (lactic acid), and their industrial applications." In *Plastics from Bacteria*, pp. 323–346. Springer, Berlin, Heidelberg, 2010.

Johnson, Richard K., Audrey Zink-Sharp, Scott H. Renneckar, and Wolfgang G. Glasser. "A new bio-based nanocomposite: fibrillated TEMPO-oxidized celluloses in hydroxypropylcellulose matrix." *Cellulose* 16, no. 2 (2009): 227–238.

Joo, Yongho, Yangjun Jeon, Sang Uck Lee, Jae Hyun Sim, Jungju Ryu, Sungyoung Lee, Hoik Lee, and Daewon Sohn. "Aggregation and stabilization of carboxylic acid functionalized halloysite nanotubes (HNT-COOH)." *The Journal of Physical Chemistry C* 116, no. 34 (2012): 18230–18235.

Jose, Thomasukutty, Soney C. George, Hanna J. Maria, Runcy Wilson, and Sabu Thomas. "Effect of bentonite clay on the mechanical, thermal, and pervaporation performance of the poly (vinyl alcohol) nanocomposite membranes." *Industrial & Engineering Chemistry Research* 53, no. 43 (2014): 16820–16831.

Joselevich, Ernesto, Hongjie Dai, Jie Liu, Kenji Hata, and Alan H. Windle. "Carbon nanotube synthesis and organization." In *Carbon Nanotubes*, pp. 101–165. Springer, Berlin, Heidelberg, 2007.

Kalantari, Katayoon, Mansor Bin Ahmad, Kamyar Shameli, and Roshanak Khandanlou. "Synthesis of talc/Fe3O4 magnetic nanocomposites using chemical co-precipitation method." *International Journal of Nanomedicine* 8 (2013): 1817.

Kargozar, Saeid, Peiman Brouki Milan, Francesco Baino, and Masoud Mozafari. "Nanoengineered biomaterials for bone/dental regeneration." In *Nanoengineered Biomaterials for Regenerative Medicine*, pp. 13–38. Elsevier, 2019.

Khalajabadi, Shahrouz Zamani, Mohammed Rafiq Abdul Kadir, Sudin Izman, and Mohd Zamri Mohd Yusop. "Facile fabrication of hydrophobic surfaces on mechanically alloyed-Mg/HA/TiO2/MgO bionanocomposites." *Applied Surface Science* 324 (2015): 380–392.

Khandan, Amirsalar, Majid Abdellahi, Neriman Ozada, and Hamid Ghayour. "Study of the bioactivity, wettability and hardness behaviour of the bovine hydroxyapatite-diopside bio-nanocomposite coating." *Journal of the Taiwan Institute of Chemical Engineers* 60 (2016): 538–546.

Kikuchi, Masanori, Soichiro Itoh, Shizuko Ichinose, Kenichi Shinomiya, and Junzo Tanaka. "Self-organization mechanism in a bone-like hydroxyapatite/collagen nanocomposite synthesized in vitro and its biological reaction in vivo." *Biomaterials* 22, no. 13 (2001): 1705–1711.

Kim, Byung-Soo, Carlos E. Baez, and Anthony Atala. "Biomaterials for tissue engineering." *World Journal of Urology* 18, no. 1 (2000): 2–9.

Kim, Hae-Won, Hae-Hyoung Lee, and J. C. Knowles. "Electrospinning biomedical nanocomposite fibers of hydroxyapatite/poly (lactic acid) for bone regeneration." *Journal of Biomedical Materials Research Part A: An Official Journal of The Society for Biomaterials, The Japanese Society for Biomaterials, and The Australian Society for Biomaterials and the Korean Society for Biomaterials* 79, no. 3 (2006): 643–649.

Kim, Hae-Won, Hyoun-Ee Kim, and Vehid Salih. "Stimulation of osteoblast responses to biomimetic nanocomposites of gelatin–hydroxyapatite for tissue engineering scaffolds." *Biomaterials* 26, no. 25 (2005b): 5221–5230.

Kim, Hae-Won, Jonathan C. Knowles, and Hyoun-Ee Kim. "Hydroxyapatite and gelatin composite foams processed via novel freeze-drying and crosslinking for use as temporary hard tissue scaffolds." *Journal of Biomedical Materials Research Part A: An Official Journal of The Society for Biomaterials, The Japanese Society for Biomaterials, and The Australian Society for Biomaterials and the Korean Society for Biomaterials* 72, no. 2 (2005c): 136–145.

Kim, H-W., J-H. Song, and H-E. Kim. "Nanofiber generation of gelatin–hydroxyapatite biomimetics for guided tissue regeneration." *Advanced Functional Materials* 15, no. 12 (2005a): 1988–1994.

Kim, Jae-Yoo, Moonhee Kim, HeonMo Kim, Jinsoo Joo, and Jong-Ho Choi. "Electrical and optical studies of organic light emitting devices using SWCNTs-polymer nanocomposites." *Optical Materials* 21, no. 1–3 (2003): 147–151.

Kim, Sinae, Sang-Soo Kim, Soo-Hong Lee, Seong Eun Ahn, So-Jung Gwak, Joon-Ho Song, Byung-Soo Kim, and Hyung-Min Chung. "In vivo bone formation from human embryonic stem cell-derived osteogenic cells in poly (d, l-lactic-co-glycolic acid)/hydroxyapatite composite scaffolds." *Biomaterials* 29, no. 8 (2008): 1043–1053.

Kirkham, J., A. Firth, D. Vernals, N. Boden, C. Robinson, R. C. Shore, S. J. Brookes, and A. Aggeli. "Self-assembling peptide scaffolds promote enamel remineralization." *Journal of Dental Research* 86, no. 5 (2007): 426–430.

Kook, Yeon Ji, Jingwen Tian, Yoo Shin Jeon, Min Jung Choi, Jeong Eun Song, Chan Hum Park, Rui L. Reis, and Gilson Khang. "Nature-derived epigallocatechin gallate/duck's feet collagen/hydroxyapatite composite sponges for enhanced bone tissue regeneration." *Journal of Biomaterials Science, Polymer Edition* 29, no. 7–9 (2018): 984–996.

Krikorian, Vahik, and Darrin J. Pochan. "Poly (L-lactic acid)/layered silicate nanocomposite: fabrication, characterization, and properties." *Chemistry of Materials* 15, no. 22 (2003): 4317–4324.

Kumar, Ashok, Ruchi Mishra, Yvonne Reinwald, and Sumrita Bhat. "Cryogels: freezing unveiled by thawing." *Materials Today* 13, no. 11 (2010): 42–44.

Kumbar, Sangamesh, Cato Laurencin, and Meng Deng, eds. *Natural and Synthetic Biomedical Polymers*. Newnes, 2014.

Kyle, Stuart, Amalia Aggeli, Eileen Ingham, and Michael J. McPherson. "Recombinant self-assembling peptides as biomaterials for tissue engineering." *Biomaterials* 31, no. 36 (2010): 9395–9405.

Leong, K. F., C. M. Cheah, and C. K. Chua. "Solid freeform fabrication of three-dimensional scaffolds for engineering replacement tissues and organs." *Biomaterials* 24, no. 13 (2003): 2363–2378.

Li, Benxia, and Yanfen Wang. "Facile synthesis and photocatalytic activity of ZnO–CuO nanocomposite." *Superlattices and Microstructures* 47, no. 5 (2010): 615–623.

Li, Ping, Donald E. Miser, Shahryar Rabiei, Ramkuber T. Yadav, and Mohammad R. Hajaligol. "The removal of carbon monoxide by iron oxide nanoparticles." *Applied Catalysis B: Environmental* 43, no. 2 (2003): 151–162.

Li, Yannan, Yi Guo, Wen Niu, Mi Chen, Yumeng Xue, Juan Ge, Peter X. Ma, and Bo Lei. "Biodegradable multifunctional bioactive glass-based nanocomposite elastomers with controlled biomineralization activity, real-time bioimaging tracking, and decreased inflammatory response." *ACS Applied Materials & Interfaces* 10, no. 21 (2018): 17722–17731.

Li, Zhijie, Bo Hou, Yao Xu, Dong Wu, Yuhan Sun, Wei Hu, and Feng Deng. "Comparative study of sol–gel-hydrothermal and sol–gel synthesis of titania–silica composite nanoparticles." *Journal of Solid State Chemistry* 178, no. 5 (2005): 1395–1405.

Lim, Wen Pei, Chiong Teck Wong, Si Ling Ang, Hong Yee Low, and Wee Shong Chin. "Phase-selective synthesis of copper sulfide nanocrystals." *Chemistry of Materials* 18, no. 26 (2006): 6170–6177.

Listl, Stephan, J. Galloway, P. A. Mossey, and W. Marcenes. "Global economic impact of dental diseases." *Journal of Dental Research* 94, no. 10 (2015): 1355–1361.

Liu, Huinan, and Thomas J. Webster. "Bioinspired nanocomposites for orthopedic applications." In *Nanotechnology for the Regeneration of Hard and Soft Tissues*, pp. 1–51. 2007.

Liu, Huinan, Elliott B. Slamovich, and Thomas J. Webster. "Increased osteoblast functions on nanophase titania dispersed in poly-lactic-co-glycolic acid composites." *Nanotechnology* 16, no. 7 (2005): S601.

Liu, Mingxian, Yun Zhang, Chongchao Wu, Sheng Xiong, and Changren Zhou. "Chitosan/halloysite nanotubes bionanocomposites: structure, mechanical properties and biocompatibility." *International Journal of Biological Macromolecules* 51, no. 4 (2012): 566–575.

Liu, Mingxian, Zhixin Jia, Demin Jia, and Changren Zhou. "Recent advance in research on halloysite nanotubes-polymer nanocomposite." *Progress in Polymer Science* 39, no. 8 (2014): 1498–1525.

Lu, Hsien-Tsung, Tzu-Wei Lu, Chien-Ho Chen, and Fwu-Long Mi. "Development of genipin-crosslinked and fucoidan-adsorbed nano-hydroxyapatite/hydroxypropyl chitosan composite scaffolds for bone tissue engineering." *International Journal of Biological Macromolecules* 128 (2019): 973–984.

Lu, Hsien-Tsung, Tzu-Wei Lu, Chien-Ho Chen, Kun-Ying Lu, and Fwu-Long Mi. "Development of nanocomposite scaffolds based on biomineralization of N, O-carboxymethyl chitosan/fucoidan conjugates for bone tissue engineering." *International Journal of Biological Macromolecules* 120 (2018): 2335–2345.

Ma, Piming, Ting Li, Wei Wu, Dongjian Shi, Fang Duan, Huiyu Bai, Weifu Dong, and Mingqing Chen. "Novel poly (xylitol sebacate)/hydroxyapatite bio-nanocomposites via one-step synthesis." *Polymer Degradation and Stability* 110 (2014): 50–55.

Manferdini, Cristina, Carola Cavallo, Brunella Grigolo, Mauro Fiorini, A. Nicoletti, Elena Gabusi, Nicoletta Zini, Daniele Pressato, Andrea Facchini, and Gina Lisignoli. "Specific inductive potential of a novel nanocomposite biomimetic biomaterial for osteochondral tissue regeneration." *Journal of Tissue Engineering and Regenerative Medicine* 10, no. 5 (2016): 374–391.

Manias, E., A. Touny, L. Wu, K. Strawhecker, B. Lu, and T. C. Chung. "Polypropylene/montmorillonite nanocomposites. Review of the synthetic routes and materials properties." *Chemistry of Materials* 13, no. 10 (2001): 3516–3523.

Maquet, Véronique, Aldo R. Boccaccini, Laurent Pravata, I. Notingher, and Robert Jérôme. "Porous poly (α-hydroxyacid)/Bioglass® composite scaffolds for bone tissue engineering. I: preparation and in vitro characterisation." *Biomaterials* 25, no. 18 (2004): 4185–4194.

Martin, O., and L. Averous. "Poly (lactic acid): plasticization and properties of biodegradable multiphase systems." *Polymer* 42, no. 14 (2001): 6209–6219.

Meng, Z. X., W. Zheng, L. Li, and Y. F. Zheng. "Fabrication and characterization of three-dimensional nanofiber membrance of PCL–MWCNTs by electrospinning." *Materials Science and Engineering: C* 30, no. 7 (2010): 1014–1021.

Miranda-Trevino, Jorge C., and Cynthia A. Coles. "Kaolinite properties, structure and influence of metal retention on pH." *Applied Clay Science* 23, no. 1–4 (2003): 133–139.

Mohanapriya, S., M. Mumjitha, K. Purnasai, and V. Raj. "Fabrication and characterization of poly (vinyl alcohol)-TiO2 nanocomposite films for orthopedic applications." *Journal of the Mechanical Behavior of Biomedical Materials* 63 (2016): 141–156.

Motamedian, Saeed Reza, Sepanta Hosseinpour, Mitra Ghazizadeh Ahsaie, and Arash Khojasteh. "Smart scaffolds in bone tissue engineering: a systematic review of literature." *World Journal of Stem Cells* 7, no. 3 (2015): 657.

Mousa, Mohanad Hashim, Yu Dong, and Ian Jeffery Davies. "Recent advances in bionanocomposites: preparation, properties, and applications." *International Journal of Polymeric Materials and Polymeric Biomaterials* 65, no. 5 (2016): 225–254.

Nathanael, A. Joseph, Jun Hee Lee, D. Mangalaraj, S. I. Hong, and Y. H. Rhee. "Multifunctional properties of hydroxyapatite/titania bio-nano-composites: bioactivity and antimicrobial studies." *Powder Technology* 228 (2012): 410–415.

Nather, A. "24 Biology of healing of large deep-frozen cortical bone allografts." *The Scientific Basis of Tissue Transplantation* 5 (2001): 434.

Navarro, M., A. Michiardi, O. Castano, and J. A. Planell. "Biomaterials in orthopaedics." *Journal of the Royal Society Interface* 5, no. 27 (2008): 1137–1158.

Nayak, Amit Kumar, and D. Pal. "Natural polysaccharides for drug delivery in tissue engineering applications." *Everyman's Science* 46 (2012): 347–352.

Nazeer, Muhammad Anwaar, Emel Yilgör, and Iskender Yilgör. "Intercalated chitosan/hydroxyapatite nanocomposites: promising materials for bone tissue engineering applications." *Carbohydrate Polymers* 175 (2017): 38–46.

Neel, Ensanya Ali Abou, Wojciech Chrzanowski, Vehid M. Salih, Hae-Won Kim, and Jonathan C. Knowles. "Tissue engineering in dentistry." *Journal of Dentistry* 42, no. 8 (2014): 915–928.

Neighbour, T. "The global orthobiologics market: players, products and technologies driving change." *Espicom Business Intelligence* 2 (2008).

Nejati, E., H. Mirzadeh, and M. Zandi. "Synthesis and characterization of nano-hydroxyapatite rods/poly (l-lactide acid) composite scaffolds for bone tissue engineering." *Composites Part A: Applied Science and Manufacturing* 39, no. 10 (2008): 1589–1596.

Ning, Chengyun, Lei Zhou, and Guoxin Tan. "Fourth-generation biomedical materials." *Materials Today* 19, no. 1 (2016): 2–3.

Office of the Surgeon General (US). "Bone health and osteoporosis: a report of the Surgeon General." (2004).

Oksman, Kristiina, Yvonne Aitomäki, Aji P. Mathew, Gilberto Siqueira, Qi Zhou, Svetlana Butylina, Supachok Tanpichai, Xiaojian Zhou, and Saleh Hooshmand. "Review of the recent developments in cellulose nanocomposite processing." *Composites Part A: Applied Science and Manufacturing* 83 (2016): 2–18.

Oliveira, M., and A. V. Machado. "Preparation of polymer-based nanocomposites by different routes." *Nanocomposites: Synthesis, Characterization and Applications* (2013): 1–22.

Olyveira, G. M., Daisy Pereira Valido, Ligia Maria Manzine Costa, Plácia Barreto Prata Gois, Lauro Xavier Filho, and Pierre Basmaji. "First otoliths/collagen/bacterial cellulose nanocomposites as a potential scaffold for bone tissue regeneration." *Journal of Biomaterials and Nanobiotechnology* 2, no. 03 (2011): 239.

Oviedo, I. Rojas, N. A. Noguéz Méndez, M. P. Gómez Gómez, H. Cárdenas Rodríguez, and A. Rubio Martínez. "Design of a physical and nontoxic crosslinked poly (vinyl alcohol) hydrogel." *International Journal of Polymeric Materials* 57, no. 12 (2008): 1095–1103.

Paiva, M. C., B. Zhou, K. A. S. Fernando, Y. Lin, J. M. Kennedy, and Y-P. Sun. "Mechanical and morphological characterization of polymer–carbon nanocomposites from functionalized carbon nanotubes." *Carbon* 42, no. 14 (2004): 2849–2854.

Pande, V. V., and V. M. Sanklecha. "Bionanocomposite: a review." *Austin Journal of Nanomedicine & Nanotechnology* 5, no. 1 (2017): 1045.

Pangon, Autchara, Somsak Saesoo, Nattika Saengkrit, Uracha Ruktanonchai, and Varol Intasanta. "Hydroxyapatite-hybridized chitosan/chitin whisker bionanocomposite fibers for bone tissue engineering applications." *Carbohydrate Polymers* 144 (2016): 419–427.

Parameswaran-Thankam, Anil, Qudes Al-Anbaky, Zeiyad Al-karakooly, Ambar B. RanguMagar, Bijay P. Chhetri, Nawab Ali, and Anindya Ghosh. "Fabrication and characterization of hydroxypropyl guar-poly (vinyl alcohol)-nano hydroxyapatite composite hydrogels for bone tissue engineering." *Journal of Biomaterials Science, Polymer Edition* 29, no. 17 (2018): 2083–2105.

Parfitt, A. Michael. "Quantum concept of bone remodeling and turnover: implications for the pathogenesis of osteoporosis." *Calcified Tissue International* 28, no. 1 (1979): 1–5.

Parfitt, A. Michael. "The coupling of bone formation to bone resorption: a critical analysis of the concept and of its relevance to the pathogenesis of osteoporosis." (1982): 1–6.

Parfitt, A. Michael. "The physiologic and clinical significance of bone histomorphometric data." *Bone Histomorphometry: Techniques and Interpretation* (1983): 143–223.

Pavlidou, S., and C. D. Papaspyrides. "A review on polymer–layered silicate nanocomposites." *Progress in Polymer Science* 33, no. 12 (2008): 1119–1198.

Postek, Michael T., András Vladár, John Dagata, Natalia Farkas, Bin Ming, Ryan Wagner, Arvind Raman et al. "Development of the metrology and imaging of cellulose nanocrystals." *Measurement Science and Technology* 22, no. 2 (2010): 024005.

Qin, Mian, Yaxiong Liu, Jiankang He, Ling Wang, Qin Lian, Dichen Li, Zhongmin Jin, Sanhu He, Gang Li, and Z. Wang. "Application of digital design and three-dimensional printing technique on individualized medical treatment." *Chinese Journal of Reparative and Reconstructive Surgery* 28, no. 3 (2014): 286–291.

Rahmat, M., M. Karrabi, I. Ghasemi, M. Zandi, and H. Azizi. "Silane crosslinking of electrospun poly (lactic acid)/nanocrystalline cellulose bionanocomposite." *Materials Science and Engineering: C* 68 (2016): 397–405.

Raina, Deepak Bushan, Irfan Qayoom, David Larsson, Ming Hao Zheng, Ashok Kumar, Hanna Isaksson, Lars Lidgren, and Magnus Tägil. "Guided tissue engineering for healing of cancellous and cortical bone using a combination of biomaterial based scaffolding and local bone active molecule delivery." *Biomaterials* 188 (2019): 38–49.

Ratner, Buddy D., Allan S. Hoffman, Frederick J. Schoen, and Jack E. Lemons. *Biomaterials Science: An Introduction to Materials in Medicine*. Elsevier, 2004.

Ray, Suprakas Sinha, and Mosto Bousmina. "Biodegradable polymers and their layered silicate nanocomposites: in greening the 21st century materials world." *Progress in Materials Science* 50, no. 8 (2005): 962–1079.

Ray, Suprakas Sinha, Kazunobu Yamada, Masami Okamoto, and Kazue Ueda. "New polylactide-layered silicate nanocomposites. 2. Concurrent improvements of material properties, biodegradability and melt rheology." *Polymer* 44, no. 3 (2003): 857–866.

Ray, Suprakas Sinha. "Rheology of polymer/layered silicate nanocomposites." *Journal of Industrial and Engineering Chemistry* 12, no. 6 (2006): 811–842.

Reddy, Murali M., Manjusri Misra, and Amar K. Mohanty. "Bio-based materials in the new bio-economy." *Chemical Engineering Progress* 108, no. 5 (2012): 37–42.

Reddy, Murali M., Singaravelu Vivekanandhan, Manjusri Misra, Sujata K. Bhatia, and Amar K. Mohanty. "Biobased plastics and bionanocomposites: Current status and future opportunities." *Progress in Polymer Science* 38, no. 10–11 (2013): 1653–1689.

Rhim, Jong-Whan, Hwan-Man Park, and Chang-Sik Ha. "Bio-nanocomposites for food packaging applications." *Progress in Polymer Science* 38, no. 10–11 (2013): 1629–1652.

Rho, Jae-Young, Liisa Kuhn-Spearing, and Peter Zioupos. "Mechanical properties and the hierarchical structure of bone." *Medical Engineering & Physics* 20, no. 2 (1998): 92–102.

Ruiz-Hitzky, Eduardo, Katsuhiko Ariga, and Yuri M. Lvov, eds. *Bio-inorganic Hybrid Nanomaterials: Strategies, Synthesis, Characterization and Applications*. John Wiley & Sons, 2008.

Ruiz-Hitzky, Eduardo, Margarita Darder, Francisco M. Fernandes, Bernd Wicklein, Ana CS Alcântara, and Pilar Aranda. "Fibrous clays based bionanocomposites." *Progress in Polymer Science* 38, no. 10–11 (2013): 1392–1414.

Ruiz-Hitzky, Eduardo, Pilar Aranda, Margarita Darder, and Giora Rytwo. "Hybrid materials based on clays for environmental and biomedical applications." *Journal of Materials Chemistry* 20, no. 42 (2010): 9306–9321.

Sachlos, E., and J. T. Czernuszka. "Making tissue engineering scaffolds work. Review: the application of solid freeform fabrication technology to the production of tissue engineering scaffolds." *European Cells & Materials* 5, no. 29 (2003): 39–40.

Sadegh, Hamidreza, Ramin Shahryari-ghoshekandi, and Maryam Kazemi. "Study in synthesis and characterization of carbon nanotubes decorated by magnetic iron oxide nanoparticles." *International Nano Letters* 4, no. 4 (2014): 129–135.

Sahmani, S., M. Shahali, M. Ghadiri Nejad, A. Khandan, M. M. Aghdam, and S. Saber-Samandari. "Effect of copper oxide nanoparticles on electrical conductivity and cell viability of calcium phosphate scaffolds with improved mechanical strength for bone tissue engineering." *The European Physical Journal Plus* 134, no. 1 (2019): 7.

Sahmani, S., S. Saber-Samandari, A. Khandan, and M. M. Aghdam. "Nonlinear resonance investigation of nanoclay based bio-nanocomposite scaffolds with enhanced properties for bone substitute applications." *Journal of Alloys and Compounds* 773 (2019): 636–653.

Saito, Tsuguyuki, Masayuki Hirota, Naoyuki Tamura, Satoshi Kimura, Hayaka Fukuzumi, Laurent Heux, and Akira Isogai. "Individualization of nano-sized plant cellulose fibrils by direct surface carboxylation using TEMPO catalyst under neutral conditions." *Biomacromolecules* 10, no. 7 (2009): 1992–1996.

Salgado, António J., Olga P. Coutinho, and Rui L. Reis. "Bone tissue engineering: state of the art and future trends." *Macromolecular Bioscience* 4, no. 8 (2004): 743–765.

Saltzman, W. M. *Tissue Engineering: Engineering Principles for the Design of Replacement Organs and Tissues*. Oxford University Press, 2004.

Sattary, Mansoureh, Mohammad Taghi Khorasani, Mohammad Rafienia, and Hossein Salehi Rozve. "Incorporation of nanohydroxyapatite and vitamin D3 into electrospun PCL/Gelatin scaffolds: the influence on the physical and chemical properties and cell behavior for bone tissue engineering." *Polymers for Advanced Technologies* 29, no. 1 (2018): 451–462.

Satyanarayana, K. G. "Recent developments in'green'composites based on plant fibers-preparation, structure property studies." *Journal of Bioprocessing & Biotechniques* 5, no. 2 (2015): 1.

Schneider, Oliver D., Stefan Loher, Tobias J. Brunner, Patrick Schmidlin, and Wendelin J. Stark. "Flexible, silver containing nanocomposites for the repair of bone defects: antimicrobial effect against E. coli infection and comparison to tetracycline containing scaffolds." *Journal of Materials Chemistry* 18, no. 23 (2008): 2679–2684.

Semino, C. E. "Self-assembling peptides: from bio-inspired materials to bone regeneration." *Journal of Dental Research* 87, no. 7 (2008): 606–616.

Shastri, Venkatram Prasad, Ivan Martin, and Robert Langer. "Macroporous polymer foams by hydrocarbon templating." *Proceedings of the National Academy of Sciences* 97, no. 5 (2000): 1970–1975.

Shchipunov, Yury. "Bionanocomposites: green sustainable materials for the near future." *Pure and Applied Chemistry* 84, no. 12 (2012): 2579–2607.

She, Jihong, Takahiro Inoue, Masato Suzuki, Satoshi Sodeoka, and Kazuo Ueno. "Mechanical properties and fracture behavior of fibrous Al2O3/SiC ceramics." *Journal of the European Ceramic Society* 20, no. 12 (2000): 1877–1881.

Shi, Shuaike, Shiyan Chen, Xiang Zhang, Wei Shen, Xin Li, Weili Hu, and Huaping Wang. "Biomimetic mineralization synthesis of calcium-deficient carbonate-containing hydroxyapatite in a three-dimensional network of bacterial cellulose." *Journal of Chemical Technology & Biotechnology: International Research in Process, Environmental & Clean Technology* 84, no. 2 (2009): 285–290.

Shi, Xinfeng, Balaji Sitharaman, Quynh P. Pham, Patrick P. Spicer, Jared L. Hudson, Lon J. Wilson, James M. Tour, Robert M. Raphael, and Antonios G. Mikos. "In vitro cytotoxicity of single-walled carbon nanotube/biodegradable polymer nanocomposites." *Journal of Biomedical Materials Research Part A* 86, no. 3 (2008): 813–823.

Shi, Xinfeng, Jared L. Hudson, Patrick P. Spicer, James M. Tour, Ramanan Krishnamoorti, and Antonios G. Mikos. "Rheological behaviour and mechanical characterization of injectable poly (propylene fumarate)/single-walled carbon nanotube composites for bone tissue engineering." *Nanotechnology* 16, no. 7 (2005): S531.

Shimomura, Kazunori, Yu Moriguchi, Christopher D. Murawski, Hideki Yoshikawa, and Norimasa Nakamura. "Osteochondral tissue engineering with biphasic scaffold: current strategies and techniques." *Tissue Engineering Part B: Reviews* 20, no. 5 (2014): 468–476.

Siqueira, Gilberto, Julien Bras, and Alain Dufresne. "Cellulosic bionanocomposites: a review of preparation, properties and applications." *Polymers* 2, no. 4 (2010): 728–765.

Skalak, Richard, C. Fred Fox, C. Fred Fox, and C. Fred Fox. 1988. *Tissue engineering: proceedings of a workshop held at Granlibakken, Lake Tahoe, California*, February 26–29, 1988; *eds.: Richard Skalak, C. Fred Fox*. New York: Liss.

Stankic, Slavica, Sneha Suman, Francia Haque, and Jasmina Vidic. "Pure and multi metal oxide nanoparticles: synthesis, antibacterial and cytotoxic properties." *Journal of Nanobiotechnology* 14, no. 1 (2016): 73.

Subramanian, Gayathri, Callan Bialorucki, and Eda Yildirim-Ayan. "Nanofibrous yet injectable polycaprolactone-collagen bone tissue scaffold with osteoprogenitor cells and controlled release of bone morphogenetic protein-2." *Materials Science and Engineering: C* 51 (2015): 16–27.

Suh, Soo Won, Ji Youn Shin, Jinhoon Kim, Jhingook Kim, Chung Hwan Beak, Dong-Ik Kim, Hojoong Kim, Seong Soo Jeon, and In-Wook Choo. "Effect of different particles on cell proliferation in polymer scaffolds using a solvent-casting and particulate leaching technique." *ASAIO Journal* 48, no. 5 (2002): 460–464.

Swain, S. K., I. Gotman, R. Unger, and E. Y. Gutmanas. "Bioresorbable β-TCP-FeAg nanocomposites for load bearing bone implants: high pressure processing, properties and cell compatibility." *Materials Science and Engineering: C* 78 (2017): 88–95.

Teeri, Tuula T., Harry Brumer III, Geoff Daniel, and Paul Gatenholm. "Biomimetic engineering of cellulose-based materials." *TRENDS in Biotechnology* 25, no. 7 (2007): 299–306.

Tetto, J. A., D. M. Steeves, E. A. Welsh, and B. E. Powell. "Biodegradable poly (1-caprolactone)/clay nanocomposites." In *ANTEC*, vol. 99, pp. 1628–1632, 1999.

Thavornyutikarn, Boonlom, Nattapon Chantarapanich, Kriskrai Sitthiseripratip, George A. Thouas, and Qizhi Chen. "Bone tissue engineering scaffolding: computer-aided scaffolding techniques." *Progress in Biomaterials* 3, no. 2–4 (2014): 61–102.

Trippel, Stephen B. "Potential role of insulinlike growth factors in fracture healing." *Clinical Orthopaedics and Related Research* 355 (1998): S301–S313.

Wagner, H. Daniel, and Richard A. Vaia. "Nanocomposites: issues at the interface." *Materials Today* 7, no. 11 (2004): 38–42.

Walmsley, Graham G., Adrian McArdle, Ruth Tevlin, Arash Momeni, David Atashroo, Michael S. Hu, Abdullah H. Feroze et al. "Nanotechnology in bone tissue engineering." *Nanomedicine: Nanotechnology, Biology and Medicine* 11, no. 5 (2015): 1253–1263.

Wang, Rongpeng, Thomas Schuman, Ramabhadraraju R. Vuppalapati, and K. Chandrashekhara. "Fabrication of bio-based epoxy–clay nanocomposites." *Green Chemistry* 16, no. 4 (2014): 1871–1882.

Wang, S. F., L. Shen, Y. J. Tong, L. Chen, I. Y. Phang, P. Q. Lim, and T. X. Liu. "Biopolymer chitosan/montmorillonite nanocomposites: preparation and characterization." *Polymer Degradation and Stability* 90, no. 1 (2005): 123–131.

Wang, W-W., Y-J. Zhu, and L-X. Yang. "ZnO–SnO2 hollow spheres and hierarchical nanosheets: hydrothermal preparation, formation mechanism, and photocatalytic properties." *Advanced Functional Materials* 17, no. 1 (2007): 59–64.

Wang, YanEn, XinPei Li, Ming Ming Yang, QingHua Wei, ChuanChuan Li, WeiFan Zhang, Yi Wei, and ShengMin Wei. "Three dimensional fabrication custom-made bionic bone preoperative diagnosis models for orthopaedics surgeries." *Scientia Sinica Informationis* 45, no. 2 (2015): 235.

Weatherholt, Alyssa M., Robyn K. Fuchs, and Stuart J. Warden. "Specialized connective tissue: bone, the structural framework of the upper extremity." *Journal of Hand Therapy* 25, no. 2 (2012): 123–132.

Webster, Thomas J. "Nanophase ceramics as improved bone tissue engineering materials." *American Ceramic Society Bulletin* 82, no. 6 (2003): 23–28.

Webster, Thomas J., Linda S. Schadler, Richard W. Siegel, and Rena Bizios. "Mechanisms of enhanced osteoblast adhesion on nanophase alumina involve vitronectin." *Tissue Engineering* 7, no. 3 (2001): 291–301.

Wei, Guobao, and Peter X. Ma. "Structure and properties of nano-hydroxyapatite/polymer composite scaffolds for bone tissue engineering." *Biomaterials* 25, no. 19 (2004): 4749–4757.

Wei, Wenbo, Renata Minullina, Elshad Abdullayev, Rawil Fakhrullin, David Mills, and Yuri Lvov. "Enhanced efficiency of antiseptics with sustained release from clay nanotubes." *RSC Advances* 4, no. 1 (2014): 488–494.

Whang, K., C. H. Thomas, K. E. Healy, and G. Nuber. "A novel method to fabricate bioabsorbable scaffolds." *Polymer* 36, no. 4 (1995): 837–842.

Wu, Defeng, Liang Wu, Weidong Zhou, Ming Zhang, and Tao Yang. "Crystallization and biodegradation of polylactide/carbon nanotube composites." *Polymer Engineering & Science* 50, no. 9 (2010): 1721–1733.

Wu, Tong, Suihuai Yu, Dengkai Chen, and Yanen Wang. "Bionic design, materials and performance of bone tissue scaffolds." *Materials* 10, no. 10 (2017): 1187.

Yang, Guanghui, Chengzhen Geng, Juanjuan Su, Weiwei Yao, Qin Zhang, and Qiang Fu. "Property reinforcement of poly (propylene carbonate) by simultaneous incorporation of poly (lactic acid) and multiwalled carbon nanotubes." *Composites Science and Technology* 87 (2013): 196–203.

Young, Ahn Tae, Jeong Han Kang, Dong Jun Kang, Jayachandran Venkatesan, Hee Kyung Chang, Ira Bhatnagar, Kwan-Young Chang et al. "Interaction of stem cells with nano hydroxyapatite-fucoidan bionanocomposites for bone tissue regeneration." *International Journal of Biological Macromolecules* 93 (2016): 1488–1491.

Yuan, Peng, Daoyong Tan, Faïza Annabi-Bergaya, Wenchang Yan, Mingde Fan, Dong Liu, and Hongping He. "Changes in structure, morphology, porosity, and surface activity of mesoporous halloysite nanotubes under heating." *Clays and Clay Minerals* 60, no. 6 (2012): 561–573.

Zafar, Rabia, Khalid Mahmood Zia, Shazia Tabasum, Farukh Jabeen, Aqdas Noreen, and Mohammad Zuber. "Polysaccharide based bionanocomposites, properties and applications: a review." *International Journal of Biological Macromolecules* 92 (2016): 1012–1024.

Zhang, Chunmei, Max R. Salick, Travis M. Cordie, Tom Ellingham, Yi Dan, and Lih-Sheng Turng. "Incorporation of poly (ethylene glycol) grafted cellulose nanocrystals in poly (lactic acid) electrospun nanocomposite fibers as potential scaffolds for bone tissue engineering." *Materials Science and Engineering: C* 49 (2015): 463–471.

Zhang, Yiqun, Wei Yu, Zhaoyu Ba, Shusen Cui, Jie Wei, and Hong Li. "3D-printed scaffolds of mesoporous bioglass/gliadin/polycaprolactone ternary composite for enhancement of compressive strength, degradability, cell responses and new bone tissue ingrowth." *International Journal of Nanomedicine* 13 (2018): 5433.

Zhou, Chengjun, Qingfeng Shi, Weihong Guo, Lekeith Terrell, Ammar T. Qureshi, Daniel J. Hayes, and Qinglin Wu. "Electrospun bio-nanocomposite scaffolds for bone tissue engineering by cellulose nanocrystals reinforcing maleic anhydride grafted PLA." *ACS Applied Materials & Interfaces* 5, no. 9 (2013): 3847–3854.

Zhou, Huan, Ahmed H. Touny, and Sarit B. Bhaduri. "Fabrication of novel PLA/CDHA bionanocomposite fibers for tissue engineering applications via electrospinning." *Journal of Materials Science: Materials in Medicine* 22, no. 5 (2011): 1183.

9 Life Cycle Assessment and Future Perspectives of Green Polymeric Nanocomposites

Sharanya Sarkar, Khushboo Gulati, and Krishna Mohan Poluri

CONTENTS

9.1 INTRODUCTION

Nanotechnology is a highly emerging field with widespread applications in every sphere of life. It is estimated that the global market value for nanotechnology will reach USD 11.3 billion by 2020 (Inshakova and Inshakov 2017). This demand can be met by improvement of the properties of the nanocomposites such as enhanced thermal stability (Njuguna et al. 2009), improved electrical conductivity (Hussain et al. 2006), resistance to scratch, mar and corrosion (Devaprakasam et al. 2008, Dasari et al. 2009, Kalaivasan and Shafi 2017), resistance to water, dirt and microbes (Upreti et al. 2015), increased modulus (Dorairaju 2008, Njuguna et al. 2011), flame retardancy (Berta et al. 2006), etc. For instance, it has been observed that the introduction of 5% of nanoparticles by weight into polymers considerably improves the behaviour of the material as a whole, as compared to using micro fillers (Marzuki and Jaafar 2016, Rafiee et al. 2011, Goffin 2010). This is owing to the fact that nanoparticles in fillers have a high volume-to-surface ratio, due to which the contact area between the fibre and the matrix is increased to a greater extent and the stress surrounding the fillers is reduced (Luo and Daniel 2003). Nanotechnology is now being widely utilized in engineered batteries, medical equipment, devices for diagnosis and therapy, nanocoatings, solar cells and processes such as wastewater treatment, air and water purification, etc. (Mullaney 2007, Koo et al. 2005, Watlington 2005, Sinha et al. 2007, Theron et al. 2008).

However, along with advances in the era of nanotechnology, the risks associated with human health and the environment need to be addressed. For example, the safety impacts of a commercial product that involves nanotechnology must be taken into consideration to resolve ethical issues. The synthesis of nanomaterials suffers from issues of sustainability, as it is often a multistep process involving several ecotoxic solvents. Moreover, the production protocols are often energy-intensive and rely on scarce resources that potentially impact the environment (Kushnir and Sandén 2011). Even after the production phase, the nanomaterials in the finished product can still possess hazardous threats to human health and environment (Oberdörster 2010). Therefore, a comprehensive assessment needs to be carried out in order to analyze how a product – starting from its manufacture until its end of life – impacts human health and ecosystems. Life cycle assessment (LCA) is thus taken up as a tool by product designers, commercial industries, academics and engineers to understand the benefits and risks associated with a nanoproduct, as well as to suggest alternate improvement strategies. LCA usually covers all stages of the life cycle of a product in order to have a holistic analysis. LCA involves an intensive methodology that has been categorized into four phases/stages according to the ISO. The first stage, "goal and scope definition", provides the functional unit and the system boundaries. The second stage, "life cycle inventory", compiles and calculates flow of data within the concerned system. The third stage, "life cycle impact assessment", organizes the collected data and analyzes the potential impacts associated with the product. The last stage, "interpretation of results", derives critical conclusions from the analyses and provides suitable recommendations. There is immense potential to influence the trajectories of development of the nano-based product via LCA, although the available data, in its early stages, is scarce and not completely accurate (Som et al. 2010).

Since manufactured nanomaterials have different physical and chemical properties as compared to standard organic chemicals, it is difficult to evaluate their fate or ultimate outcome in different environmental media (Caballero-Guzman and Nowack 2016, Meesters et al. 2014). Given the discrepancies, researchers have suggested adapting the fate models designated to conventional chemicals to suit the behavioural patterns of nanomaterials (Arvidsson et al. 2011, Praetorius et al. 2014). LCA is quite established or accepted as a methodology, yet there are challenges associated with its applications to nanomaterials. Therefore, it is necessary to bridge the knowledge gaps and redesign the protocol, as needed, to have a more cogent analysis (Gilbertson et al. 2015).

This chapter defines all the concepts related to the life cycle of a product and discusses the significance of evaluating life cycles. It also delineates all the life cycle stages of a nanomaterial, with special emphasis on nanocomposites and describes the methodology or protocol followed by LCA. Moreover, it throws light on the challenges that are faced while performing LCA and provides suggestions and recommendations that can be adopted to obtain better results. Lastly, it discusses several interesting LCA studies on green polymer nanocomposites.

9.2 LIFE CYCLE ASSESSMENT

The life span of any product, which comprises various stages starting from its manufacture to disposal, is usually referred to as its "life cycle". However, the term may be interpreted differently in varying scenarios (Christensen and Olsen 2004). For example, a particular nanomaterial can form a part of several nanocomposites, or it can be used in any commercial product alone or in combination. Consequently, the life cycle stages of the nanomaterial, such as the usage phase and the disposal phase, would vary in these two cases. There are many terms, namely "life cycle thinking", "life cycle perspective", "life cycle concepts" and "life cycle approach", which are widely used in literature to delineate the life cycle of a certain product. Several methods based on these concepts, such as life cycle costing, eco-design, product road-mapping, value chain analysis, material flow analysis, life cycle management, etc., are being adopted to further describe a product's life cycle (Davis 2007). These methods have been utilized in the field of nanotechnology to understand the effects of exposure to nanomaterials. Most of these methods focus on all the life cycle stages of the nanoproduct, namely manufacture, transport, use and disposal, while some methods, such as life cycle assessment, focus only on the environmental and human health impacts.

The term "life cycle assessment" is often thought to be equivalent to "life cycle concepts". However, unlike life cycle concepts that provide a holistic view of the impacts of the nanomaterials, LCA is merely a procedure that furnishes comprehensive information on their environmental sustainability. According to the definition of the ISO 14040:2006 standard, LCA is a methodological framework that compiles and evaluates inputs (energy, land use, materials, chemicals, etc.) as well as outputs (wastes, emissions, by-products, etc.) through all the life cycle stages of the nanoproduct (Standard 2006). In other words, it analyzes the total impact of a product system on the environment (both natural and man-made), human health as well as other natural resources. The information provided by LCA can be categorized in terms of global warming, ecotoxicity, resource use, ozone layer depletion, acidification, climate change, water use, toxicological stress, etc.

However, it should be borne in mind that LCA differs from methods such as risk assessment (RA) and environmental performance evaluation (EPE), as it does not provide absolute or specific quantification of the actual impacts of the product or process, either temporally or spatially. RA yields absolute information on the impacts of certain chemicals on target organs at particular exposure concentrations. However, LCA is a comparative approach, which essentially offers relative comparisons of the possibility of causing damage to human health and the environment. It is based on a functional unit and comprises all the inputs and outputs associated with that unit. Usually, companies manufacturing nanoproducts and/or nanomaterials conduct the LCA studies, although they might be done by academicians or hired consultants too. Sometimes institutions that are government-funded aid in these LCA studies via secondary work such as funding and creation of databases.

9.3 IMPORTANCE OF LCA

According to the Nanotechnology Consumer Products Inventory, the majority of products (60%) originating from the nanotechnology-based industries are associated with the health and fitness sector (Som et al. 2010). They also form part of frequently used objects such as electronic gadgets, food packaging and coatings. However, it is important to note that this database suffers from the problem of insufficiency. Since existing legislation has not yet made it obligatory for the nano-based companies to label their products to indicate the inclusion of nanocomponents, there may be some products containing nanomaterials that are not included in the database (Som et al. 2012). Moreover, there is no regulation in the European Union (EU) or anywhere else in the world that is strictly nano-specific (Hodge et al. 2009). All policies to date revolve around conventional chemicals and do not cover any aspects of nanomaterials, including their physicochemical characteristics (Breggin and Pendergrass 2007, Chatterjee 2008, Franco et al. 2007). This leads to the creation of knowledge gaps regarding the adverse effects these products have on human health and the environment (Nowack et al. 2012, Maynard and Aitken 2007, Borm et al. 2006, Chaudhry et al. 2008, Handy et al. 2008, Klaine et al. 2008, Maynard et al. 2006, Wiesner et al. 2009). The development of regulatory frameworks that govern the use and disposal of nanomaterials lags far behind the expanding commercialization of these materials.

LCA for nanoproducts and/or nanomaterials is vital, primarily due to two reasons. First, in order to make nanotechnology widely acceptable to consumers around the globe, it is important that concerns related to safety issues are addressed (Ostertag and Hüsing 2008). Nanomaterials can have a multitude of harmful effects, including toxicology, accumulation, pollution, etc. Second, products from nanotechnology-based industries may also have the potential to release nano-sized particles during their life cycle, owing to mechanical, physical or chemical stress and/or aging (Nowack et al. 2012, Köhler and Som 2008). The nanoparticles released may be very different from the pristine nanocomponents that were originally included in the product and they may pose potential risks that were unknown about or were not assigned to the original product. This unanticipated release of nanoparticles might, in turn, expose industry workers, academics, product consumers and even the environment to potential risk. Since the quantity of released nanoparticles depends on factors such as the manufacture, lifetime, and usage of the nanoproduct (Köhler et al. 2008), LCA becomes pivotal to alleviating the adverse effects associated with the release. LCA, in combination with other life cycle concept methods, can thus provide the basis for rational decision-making as well as precise risk assessment during the early stages of the product manufacture in order to foster nanoproducts that are both environmentally sustainable and safe for human health (Klöpffer et al. 2007, Shatkin 2008, Wardak et al. 2008). LCA performed even for the initial life cycle stages of a nanoproduct can help in developing the trajectory for the subsequent stages. LCA assists researchers in identifying which of the product's life cycles has the greatest contribution to the impacts of the product (Klöpffer et al. 2007). LCA can also be utilized to compare two similar products, one containing nanocomponents and the other excluding them.

Differences in parameters such as resource use, energy consumption and environmental performance can be determined for the product that includes nanocomponents and the conventional counterpart that does not (Khanna et al. 2008, Kushnir and Sandén 2011). LCA can determine the extent to which enhancement in the energy efficiency of a nanoproduct, compared to a conventional equivalent, is offset by energy consumption during its manufacture. LCA also takes into account the environmental sustainability of a nanoproduct by considering factors such as material and energy consumption, waste generation, etc. and may help prevent the unplanned shifting of environmental impacts from one impact category to another (Klöpffer et al. 2007, Meyer et al. 2009, Şengül et al. 2008). LCA aids in managing issues related to the end-of-life stage of the nanoproduct and offers solutions concerning its reuse or recycling. Moreover, with the help of LCA data, companies and/or researchers can not only provide nanoproducts for critical applications such as medical use with a highly functional and safe design but can also use it to feed into adaptive regulation.

9.4 EXPOSURE SCENARIOS OF NANOMATERIALS DURING DIFFERENT LIFE CYCLE STAGES OF NANOCOMPOSITES

It is crucial to assess the possibility of human and environmental exposure to nanomaterials in order that preventive, as well as corrective, measures for protection can be adopted. For this, it is important to take into account the nature of released nanoparticles, the conditions of release and the technosphere of release (air filter, incinerator, etc.). Exposure of humans and the environment to nanocomposites can occur during different stages of the life cycle of the product. For example, workers in an industry setting are usually exposed during the manufacture/recycling phase of the nanoproduct, whereas consumers are generally exposed during its usage phase. Analyzing consumer and occupational exposures paves the way for understanding environmental exposure (Som et al. 2010). The life cycle of a nanoproduct can be divided into design phase, manufacture phase, usage phase and disposal phase. In the following sections, the potential of exposure of nanomaterials from each of the phases have been outlined with specific emphasis on nanocomposites (see Figure 9.1).

9.4.1 Product Design Phase

During this phase, the materials required for the manufacture of the product as well as its protocols are strategized. LCA practitioners search for better

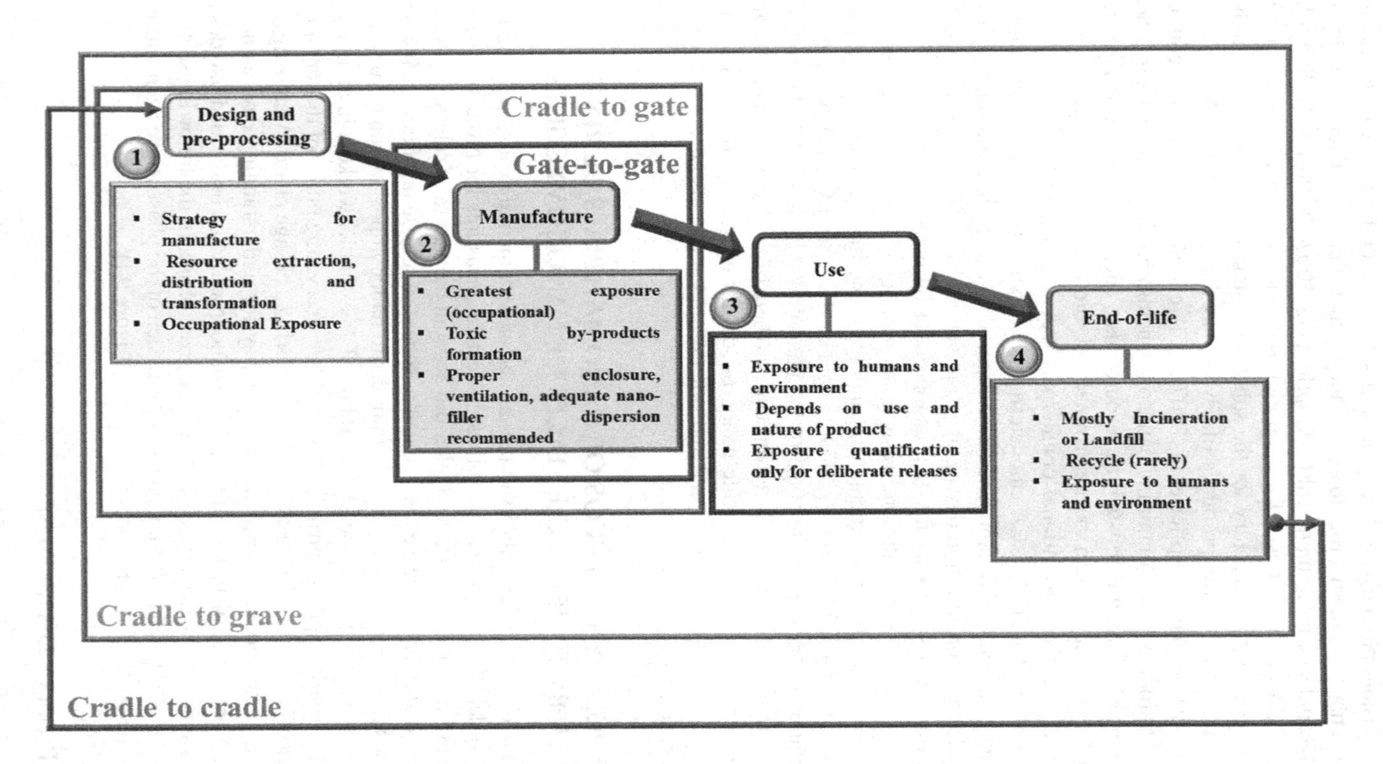

FIGURE 9.1 Schematic illustrating the exposure potential of nanomaterials from different life cycle stages.

alternatives to engineered nanomaterials in a product, although available knowledge on this is quite scarce. Researchers attempt to incorporate principles of green chemistry during the design of the products, such that the production has increased energy efficiency and the finished products have decreased toxicity and waste generation (Anastas 1998). Previous LCA studies assist significantly in the product design stage.

9.4.2 Product Manufacture Phase

There is awareness about safety protocols during the product manufacturing phase since the possibility of exposure to free nanomaterials is at its greatest during this stage (Helland et al. 2007). Previous studies have determined that occupational exposure usually occurs during the synthesis and handling of dry powders (Bello et al. 2008, Fujitani et al. 2008, Han et al. 2008, Mazzuckelli et al. 2007, Yeganeh et al. 2008). Despite this, a survey has found that most companies do not conduct any kind of risk assessment (Helland et al. 2007). A major problem in assessing exposure scenarios in this phase is that the analysis methods do not distinguish between released nanoparticles and the other ultrafine particles present in the atmosphere. This problem can be tackled by employing techniques such as electron microscopy.

Manufacturing of nanoproducts takes place either by a top-down approach (e.g. grinding of bulk material) or by a bottom-up approach (e.g. nucleation followed by growth) (Kuhlbusch et al. 2011). In the second approach, the nanoparticles are usually produced by the gas phase or the liquid phase (Wang et al. 2011, Park et al. 2009). Production of nanoparticles via a liquid phase is quite safe due to the reduced possibilities of inhalation, although there are some exceptions (Park et al. 2009). Previous studies have also determined that good enclosure and proper ventilation can limit occupational exposure to released nanomaterials during manufacture (Han et al. 2008, Maynard et al. 2004, Tsai et al. 2009).

Studies on occupational exposure to nanomaterials during the manufacturing stage of nanocomposites have thrown light on nano-additivated material preparation (Vogel et al. 2013, Nowack et al. 2012, Wohlleben et al. 2014, Vance and Marr 2015). The processes of waste management become easy once nano-additivation is done and the nanomaterials are embedded in a matrix (Nowack et al. 2013). Nanoparticles are released from nanocomposites during machining and the released material is analyzed by performing simulations of real operations. Thereby, it was concluded that released matrix from the nanocomposites usually contains the embedded nanomaterials. It was also observed that if nanofillers were well-dispersed in the matrix, the rate of release of the isolated nanoparticles was notably reduced (Golanski et al. 2012).

During the synthesis of the nanoproduct, other toxic by-products are also produced, although they have not yet been completely characterized (Plata et al. 2008). Moreover, there are no regulations regarding the disposal of these by-products (Breggin and Pendergrass 2007, Chaudhry et al. 2008). However, some countries have taken the initiative and progressed in this regard (Gendre et al. 2015). For example, The British Standard Institution published guidelines in 2012 regarding the disposal of by-products containing nano-wastes. In the United States, the Resource Conservation and Recovery Act, subtitle C, regulates the management of nano-waste.

9.4.3 Product Use Phase

The exposure to nanomaterials during the usage phase of the product can either be deliberate (as in sunscreens) or inadvertent (as in release from textiles). The quantity of release, as well as the source, is known only for intentional releases. It is critical that products containing nanocomponents be labelled, as this would make the consumer aware of the potential for exposure. There are very few studies on the release of nanomaterials during usage and most of them are based on laboratory simulation (Gendre et al. 2015). The release of nanoparticles during this phase depends on how they are incorporated into the product as well as the extent of use of the product (Köhler and Som 2008). For example, nanomaterials that are strongly fixated have fewer chances of release until disposal (Türk et al. 2005). On the other hand, rough handling of the product, such as intense cleaning and/or scrubbing, would make nanoparticle release more likely. Design changes that can lead to the release of relatively large particles can alleviate the toxicity associated with nanoparticles, to some extent (Reijnders 2009). Studies on polymeric nanocomposites have elucidated that the polymer matrix has the potential to get degraded during this phase, owing to photo and chemical degradation. In case the matrix is degraded, accumulated nanoparticles at the composite surface have greater possibilities of being released into the environment (Göhler et al. 2013, Hirth et al. 2013).

9.4.4 Product Disposal Phase

The risks associated with nanomaterial release during disposal of products have been calculated by the Royal Society and Royal Academy of Engineering (Royal 2004). Around 98% of the composites are either incinerated or buried in landfills (Halliwell 2006). However, it is not yet known how the released nanomaterials behave during this end-of-life stage: that is, what fraction of them is degraded at the incineration temperature, become airborne, or remain behind in the slag (Som et al. 2010). Research has been done in the field of material science to understand the temperatures that could potentially degrade nanomaterials by incineration. Studies done in this field have suggested that even if nanomaterials become airborne during the incineration process, there is a very high probability that they will be effectively removed by filters (Mueller and Nowack 2008). In the case of landfill, it has been observed that nanomaterials are capable of leaching into the surrounding environments. The mobility depends on the nature of the nanomaterial and the leachate composition (pH, ionic strength, etc.) (Lozano and Berge 2012, Khan et al. 2013). The release of nanomaterials into landfills also raises the concern as to whether or not they can influence biological activity. Recycling can be another end-of-life option for nanomaterials, although it is one that is rarely used. Moreover, during recycling, nanoparticles can undergo "shredding", which may result in occupational health hazards. For recycled composites, research shows that the presence of nanomaterials negatively impacts the quality of the recycled product (Sánchez et al. 2014, Touati et al. 2011).

9.5 METHODOLOGY OF LCA

LCA is essentially an iterative approach and consists of four fundamental steps that are interrelated. These steps are performed in accordance with the standardized framework (ISO) (Hischier and Walser 2012). The four basic steps in any LCA study are (i) goal and scope definition, (ii) life cycle inventory, (iii) life cycle impact assessment and (iv) interpretation of results (see Figure 9.2).

9.5.1 Goal and Scope Definition

Being the first step of any LCA study, goal definition becomes crucial since the subsequent steps in the study are dependent on it (Miseljic and Olsen 2014). Here, certain parameters of the LCA analysis are recognized and delineated. The parameters usually include the reasons behind the study, its drawbacks, preconceived applications, the target audience as well as the assumptions considered during the study. On the other hand, the scope of the study describes the system in question in the utmost detail and lays out the analytical provisions. Critical attention must be paid to ensure that the scope of the study is in line with the defined goal. The scope of the system elucidates the life cycle stages that should be a part of the system. Ideally, scope "from cradle to grave" or "from cradle to cradle", which comprises all life cycle stages should be considered. "Cradle to grave" of a product describes the phase starting from the selection of raw materials (also called "cradle" or "birth") for the product until its end-of-life ("grave"); whereas "cradle to cradle" considers the cycle from the birth of a product in one generation to the next, after the end-of-life stage in the first generation. In essence, it is iterative and operates in a closed-loop mode. "Gate to gate" essentially means "from factory gate to factory gate" and defines only the manufacturing stage of the product. For most practical purposes, however, a "cradle to gate" scope is used (see Figure 9.1). In this step, the functional unit (the qualitative and quantitative services of the product), reference flow (quantity of product required for the specified function)

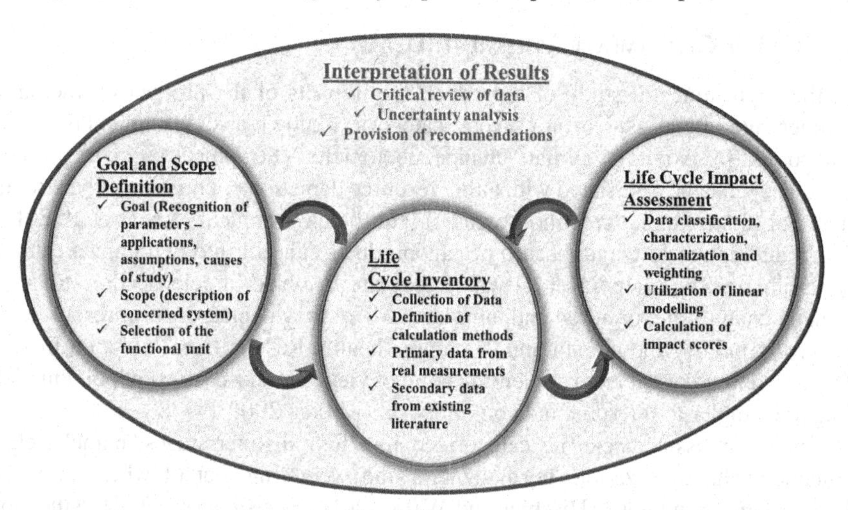

FIGURE 9.2 An overview of life cycle assessment (LCA) framework.

and other parameters such as threshold and allotment rules, environmental impact categories and system boundaries are outlined. Choosing the functional unit becomes a very crucial process. To define the functional unit, the functionalities that are improved in the nanoproduct as compared to the conventional counterpart, should be considered. In order to perform a comparative analysis of, say, nanomaterial-coated textiles and conventional ones, one has to specify the conditions of cleaning and wearing the textiles. However, for certain cases, such as nanoproducts in pharmaceuticals, it is difficult to find a functional equivalent (Klöpffer et al. 2007).

9.5.2 Life Cycle Inventory (LCI)

The aim of this step is to collect data and define the calculation methods in order to evaluate inputs (raw material, energy, water intake) and outputs (emissions to soil, water, air, etc.) via the system boundaries. A flow balance is performed by taking into account the inputs and outputs of each life cycle stage (Gottschalk et al. 2010). This can be done by fractionating the system into smaller, interconnected systems. Primary data, which arises from real-time monitoring and/or measurements, is usually preferred. For additional processes, secondary data from literature and databases (such as the ELCD [European Platform on Life Cycle Database] or commercial databases such as Ecoinvent or GABI) can be used. The data collection procedure should ensure that the data is reliable and consists of all assumptions made. Moreover, it should also be in alignment with the goal and scope defined prior to data collection. For example, a manufacturer of nanomaterials performing LCA should use data that is company-specific. However, LCA performed to determine government strategy could take into account the data averaged from various sectors (Klöpffer et al. 2007). In order to relate to impact assessment, the inventory analysis for nanomaterials should also describe additional characteristics that have an effect on the toxicity of the nanomaterials (for example, surface area, solubility, crystal structure, adhesive properties, surface charge, etc.)

9.5.3 Life Cycle Impact Assessment (LCIA)

In the environmental study of a product, the impacts of the product on the environment may be classified into several categories, known as environmental impact categories (for example, climate change, ecotoxicity, eutrophication of freshwater, etc.). They may be associated with either resource depletion or emission of hazardous material (for example, greenhouse gases) into the environment (Keller et al. 2013). The manufacture, use and discard of nanoproducts can fall into two impact categories: either at the midpoint level where they are involved in acidification, toxicity, climate change, etc. or at the endpoint level where they deplete natural resources or adversely modify the human and ecosystem health. Results from midpoint modelling are more relevant from a scientific point of view, but those from endpoint modelling are simpler to interpret and communicate (Kuiken 2009).

Impact assessment can be categorized into four distinct stages, namely, classification, characterization, normalization and weighting, out of which only the first two are mandatory (Hischier and Walser 2012) (see Figure 9.3). Classification groups and assigns the inventory data into their corresponding impact categories.

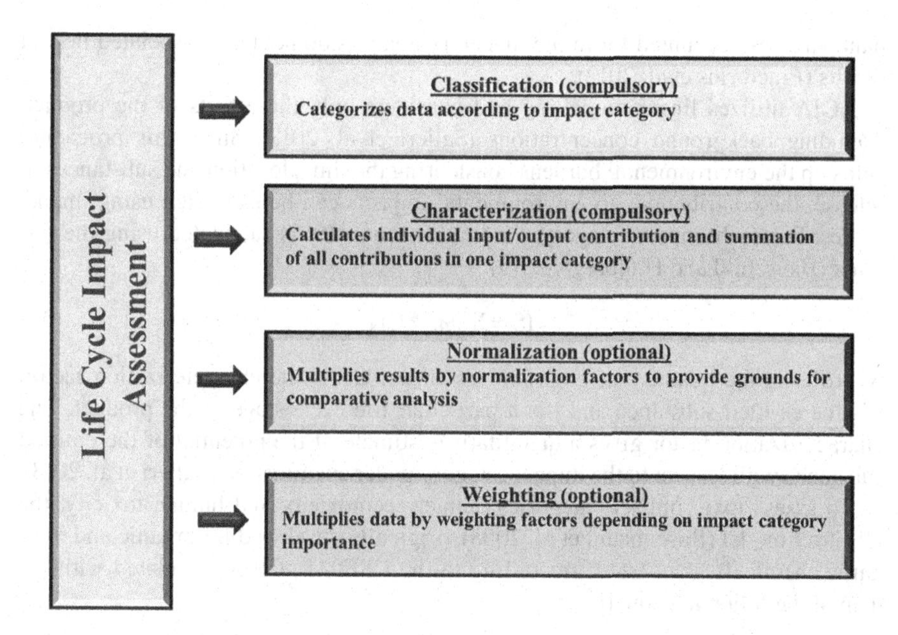

FIGURE 9.3 Schematic showing the stages involved in life cycle impact assessment (LCIA).

Characterization calculates the contribution of each input/output in its respective impact category and the summation of all the contributions in every category. Normalization comprises multiplying the results of the impact assessment by normalization factors in order to have a comparative analysis of the magnitude of their contribution to the concerned impact category. Weighting comprises multiplying the normalized data by weighting factors that are assigned depending on the relative importance of the impact category in question.

LCIA is a significant step because it also assesses resource depletion and signifies the necessity of reaching consensus on a methodology to characterize the resource use. For example, wide use of a very scarce metal such as Indium in nanomaterials can actually surpass the quantity of Indium usually used in conventional semiconductors. Impact assessment also elucidates the characteristics of nanomaterials that are generally relevant for particular impact categories. For example, large surface-to-volume ratios of nanoparticles have been related to categories such as ozone layer depletion and photochemical smog (Von der Kammer et al. 2012, Sun et al. 2014).

During impact assessment, factors such as structural change of nanoparticles after release, potency of dermal uptake, fate, and exposure models, etc. should be taken into consideration. To determine the toxic impacts of nanomaterials on humans, it is necessary that the nanomaterials are classified on the basis of composition, structure, mobility, dispersive capacity, degradability, etc. In the case of limited data availability, screening should be performed considering the worst possible implications of nanomaterials. Here, it should be assumed that (i) nanomaterials have an impact that is equivalent to the most toxic chemicals in use and (ii) the nanomaterials have the highest intake potency.

Currently, the impact assessment methods lack characterization factors (CFs) for release of nanoparticles, often resulting in the fact that the released nanoparticles accounted for in the inventory analysis do not have associated impact results (Praetorius et al. 2012).

LCIA utilizes linear modelling and considers only the effects of the product, excluding background concentrations (Salieri et al. 2018). Since this procedure sums up the environmental burdens considering the time, location and substances of release, the contributions to environmental impacts can be estimated using impact scores. For each impact category, the impact score (IS) is calculated using the formula (Hauschild and Huijbregts 2015):

$$IS = \Sigma_{i,j} m_{i,j} * CF_i \tag{9.1}$$

where m is the quantity of emission resources, CF is the characterization factor, i is the emitted substance and j is a particular life cycle stage of the product. The characterization factor gives a quantitative estimate of the potential of the emitted substance with respect to the impact category under examination (Salieri et al. 2018).

To assess toxic impact categories, namely ecotoxicity and human toxicity, the USEtox® model (Rosenbaum et al. 2008), originally developed for organic and inorganic chemicals, is utilized. According to the USEtox®, CF is calculated with the help of the following equation:

$$CF = FF \cdot EF \cdot XF \tag{9.2}$$

where FF denotes fate factor, or the ability of the chemical to persist in a particular media, XF denotes exposure factor, which relates the substance quantity to its intake (it is dimensionless for ecotoxicity studies, but has dimension of $days^{-1}$ for human toxicity studies) and EF denotes effect factor (see Figure 9.4). In human

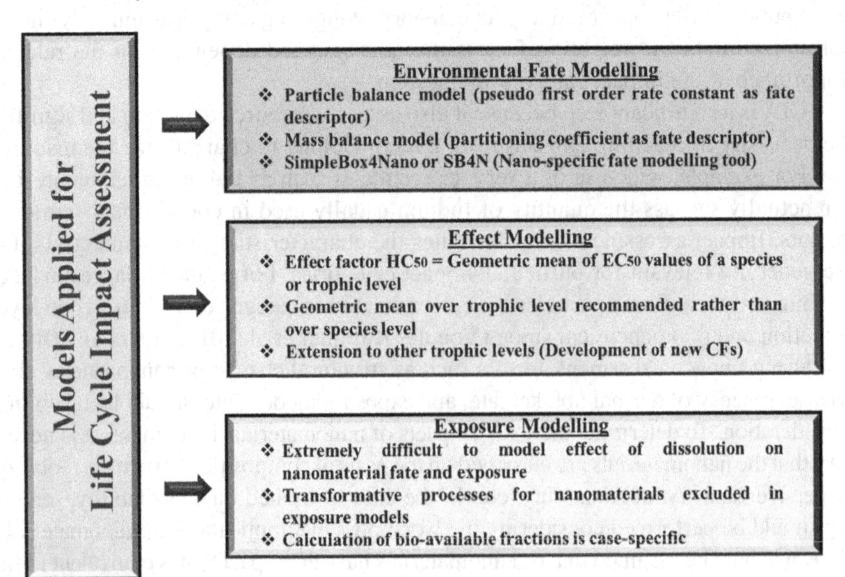

FIGURE 9.4 Schematic representation of types of modelling employed in life cycle impact Assessment (LCIA).

toxicity assessment studies, EF (in cases/kg intake) relates the substance concentration inhaled or ingested by the population to the probability of unfavourable effects, either carcinogenic or non-carcinogenic. On the other hand, EF for ecotoxicity studies (PAF m3/kg exposed) denotes the species fraction affected per mass of the exposed substance. Based on the USEtox® model, some recent studies have assessed CFs for manufactured nanomaterials such as carbon nanotubes (Salieri et al. 2018).

9.5.3.1 Environmental Fate Modelling

The USEtox® model assesses environmental fate by utilizing multimedia fate-modelling-based principles. Multimedia fate models redress the problems associated with chemical pollution. These models consider factors such as transport, source, sink etc. of the chemicals to elucidate the behaviour of the chemicals in multiple media in the environment, such that biological accumulation and exposure of these chemicals can be estimated (MacLeod et al. 2010, Dale et al. 2015, Di Guardo et al. 2018, Zhang et al. 2015). There are several well-known multimedia environmental fate models, namely fugacity-based models, SimpleBox and ELPOS models, the MAMI model, the SESAMe model, etc. (Su et al. 2019). However, USEtox® models comprising of multimedia fate modelling have only recently been applied to manufactured nanomaterials (Meesters et al. 2014, Praetorius et al. 2012). There is an inherent disadvantage associated with the multimedia fate models as they employ partitioning coefficients as fate descriptors. These descriptors are deemed inappropriate for manufactured nanomaterials, as they tend to form colloidal dispersions instead of thermodynamically stable solutions (Arvidsson et al. 2011, Praetorius et al. 2012, Mackay et al. 2006). This paves the way for the need of suitable fate descriptors, such as rate of change (determined by rate constant). However, the debate regarding whether pseudo–first-order rate constants should be applied for environmental fate modelling of manufactured nanomaterials remains open. A study by Dale et al. claims that particle balance models that utilize pseudo–first-order rate constants are more complicated than mass balance models that apply partitioning coefficients. Moreover, the former models do not make a significant contribution to the description of manufactured nanomaterial fate processes on a large scale (Dale et al. 2015).

FFs that are used to derive characterization factors have been calculated using approaches that involve both pseudo–first-order rate constants and partitioning coefficients. The first study that applied the particle balance model calculated FFs utilizing nano-specific or non-conventional fate descriptors such as particle sedimentation, re-suspension and hetero-aggregation (Salieri et al. 2015). In recent times, more fate modelling tools have been exclusively designed for manufactured nanomaterials. For example, SimpleBox4Nano (SB4N) is a USEtox®-based tool that has been utilized to quantitate specific CFs of nano titanium dioxide for impact categories corresponding to human toxicity and freshwater (Ettrup et al. 2017). This tool can account for a wide range of environmental media where such emissions are possible and can help to distinguish the fate of free, attached or aggregated forms of manufactured nanomaterial.

9.5.3.2 Effect Modelling

According to the USEtox® model, chemical toxicity is defined by EFs corresponding to HD_{50} and HC_{50} values (respectively for human toxicity and freshwater). HC_{50}

is defined as the concentration of a substance in kg/m^3 at which 50% of the species are vulnerable to a concentration greater than their EC_{50}. EC_{50} gives the drug concentration corresponding to half-maximal response. To calculate HC_{50}, the geometric mean of EC_{50} values of the species or a particular trophic level is calculated. However, owing to a shortage of data related to EC_{50} values among the trophic levels for many kinds of manufactured nanomaterials, researchers have tried to develop new CFs. One possible remedy was found by extending the assessment to new trophic levels. For example, in one study involving graphene oxide, a marine algal species was included (Deng et al. 2017).

It should be noted that the USEtox® model requires three different phyla for a minimum of three EC_{50} values (Henderson et al. 2011). Moreover, consideration of different taxonomic groups at the species level often causes an unequal distribution, which in turn may create a bias in the calculation of EF (Hauschild and Huijbregts 2015). Therefore, for HC_{50} calculation, geometric mean at the trophic level is recommended over a species-level-based geometric mean. To validate the calculation of the EFs for manufactured nanomaterials, LCA practitioners should peruse the quality of the input data or the EC_{50} values. Moreover, a more standardized approach to select such data should be introduced to bring about consistency in the calculations.

9.5.3.3 Exposure Modelling

Different approaches have been previously used to calculate Exposure Factor (XF) for manufactured nanomaterials. For freshwater ecotoxicity assessment, XF has been defined as the fraction of dissolved species in the aquatic domain. Until now, only one study has utilized the conventional XF method or the method defined by USEtox®. However, this USEtox® calculation methodology, which is fitting for organic substances, requires modification when applied to manufactured nanomaterials. This is primarily because it is difficult to evaluate and model the effect of dissolution on the exposure and fate of these nanomaterials (Pu et al. 2017). During the process of dissolution, there is a conversion of the nanomaterials from their solid phase to their ionic phase. Moreover, their dissolution property is dependent on a variety of factors including environmental conditions and properties of the solvent (temperature, pH, etc.) and the nanomaterial (surface properties, concentration, aggregation, etc.). This makes dissolution a highly variable process among different types of nanomaterials.

One major setback in exposure modelling of nanomaterials is that they undergo a wide range of modifications on exposure to the environment, namely, attachment, aggregation, oxidation, functionalization, sulfidation, etc. For example, aggregation often results in a combination of several nanomaterials that has the potential to exert a different, combined effect on the environment as compared to the effects exerted by the individual constituents prior to aggregation (Skjolding et al. 2016). Barring attachment and aggregation, the other transformative processes are not incorporated in the fate models specified for nanomaterials. This is owing to the underlying complexity of these processes (Jacobs et al. 2016). Therefore, exposure modelling becomes inaccurate if dissolution and the other transformative processes are not quantified, both in environmental and biological systems.

In order to approximate the nanomaterial bioavailable fraction, the recommendation of REACH (Registration, Evaluation, Authorisation and Restriction of Chemicals) for metals can be extrapolated. The bioavailable fraction of metals and their oxides, as defined by REACH, is the extent that can pass through a filter with a pore size less than 450 nm (Salieri et al. 2018). In accordance with this definition, various forms of manufactured nanomaterials including dissolved and aggregated species with a size of less than 450 nm should be termed bioavailable (Jacobs et al. 2016).

The exposure assessment of the manufactured nanomaterials should be taken up on a case-by-case basis, depending on the nature of the nanomaterials. For example, for a soluble nanomaterial, dissolution should be considered as the most notable transformation process and the exposure modelling can be done conventionally as for other chemicals in accordance with the USEtox® model. For any non-transformed fraction, nano-specific models can be employed. In order to calculate XF for insoluble nanomaterials, the bioavailable fractions are calculated by adding up the free and aggregated fractions, that is,

$$XF = \left[\left(M_{free}/M_{total}\right)+\left(M_{hetero\text{-}agg}/M_{total}\right)\right] \tag{9.3}$$

where M_{total} represents the sum of the free, attached and the hetero-aggregated mass of the species. The XF calculated by this equation, however, does not consider aggregation that results in the formation of coarse particles with sizes greater than 450 nm.

Exposure modelling is a daunting task because the toxicological tests that are performed experimentally do not essentially represent the exposure due to the bioavailable fractions. This is because the tests are carried out without the foresight of ultimately integrating their results into computational models. This makes the investigation process to link exposure and effect factors for nanomaterials a very arduous and intensive task. Moreover, in the absence of data on the bioavailable fractions, Equation 9.3 is not applicable. In such cases, the value of XF should be assumed to be 1.

9.5.4 Interpretation of Results

This step comprises of critically reviewing and analyzing the inventory data and impact results in order to draw conclusions and elucidate scope for improvement. Issues such as feasibility of upscaling a nanoproduct to society-wide use should also be discussed here. The interpretations are highly valuable as they not only update and improve the existing LCA databases but also make future LCA studies on other nanomaterials more feasible and comprehensive. The data is checked for its reliability, consistency, completeness and sensitivity such that correct inferences are drawn from them. Sensitivity analysis is performed by varying the most vital parameters to understand their relative influence in the results. According to the ISO standards, it is recommended to perform a critical review as well as an uncertainty analysis of the LCA-generated data (Owen et al. 2009). The former checks whether all the methodology requirements are properly fulfilled, while the latter determines the extent to which the LCA outcomes are influenced by parameter variations and uncertainties related to any of the steps in LCA. The data variation is represented by either a distribution or a standard deviation range. Monte Carlo techniques are employed to deal with the uncertainties in LCA steps (Hendren et al. 2013). Such analyses can thus pave the way to attenuating uncertainty in future LCA studies.

9.6 LCA CASE STUDIES

Particularly in the last decade, researchers have turned their attention towards performing LCA of nanomaterials. Although there have been quite a few studies on the LCA of nanoparticles and nanocomposites using non-green polymers, LCA studies involving green polymer nanocomposites are scarce. Indeed, most of them have used cellulose as the green polymer. The LCA of nanocellulose, for example, has been carried out comprehensively. The life cycle of nanocellulose was categorized into five phases, namely (i) cultivation of raw material, (ii) isolation of nanocellulose, (iii) manufacture or synthesis, (iv) use and (v) end of life. The first phase, or raw material cultivation, considers factors such as planting, logging, processing, transportation, etc. The second phase evaluates all chemical and mechanical methods that are employed for isolation. The third phase models the energy and resource inputs and the corresponding outputs, in terms of, for example, emissions. The phases of usage and end of life have very inadequate data. For this purpose, the data from the corresponding phases of glass fibres have been extrapolated to have a preliminary assessment (see Figure 9.5) (Yang et al. 2019).

Shaktin et al. conducted a life cycle risk assessment study on cellulose nanomaterials (CN) covering occupational, environmental and consumer exposure stages, discerned the possible exposure scenarios and assessed the potential risks. It also laid out a roadmap to characterize the data gaps that lead to uncertainty in the safety assessment. It was found that the knowledge gaps of highest priority stemmed from the uncertainty of occupational hazards involving inhalation while handling CN in dried powder form, challenges to quantify exposures and paucity of data regarding CN toxicity in commercial end products and consequent consumer safety (Shatkin and Kim 2015). Figueirêdo et al. performed a study on cellulose nanowhiskers, which are crystalline nanoscale rods of cellulose, and elucidated that nanowhiskers fabricated from white cotton fibres instead of unripe coconut fibres required fewer resources (water, energy, etc.) and also made a lesser contribution to the impact categories of climate change, eutrophication, human toxicity, etc. (de Figueirêdo et al. 2012).

Piccinno et al. produced cellulose nanofibres from food waste (vegetable waste) by first liberating the fibres using enzymes and then coating them and orienting them by spinning techniques. Piccinno et al. also performed a comparative LCA study to

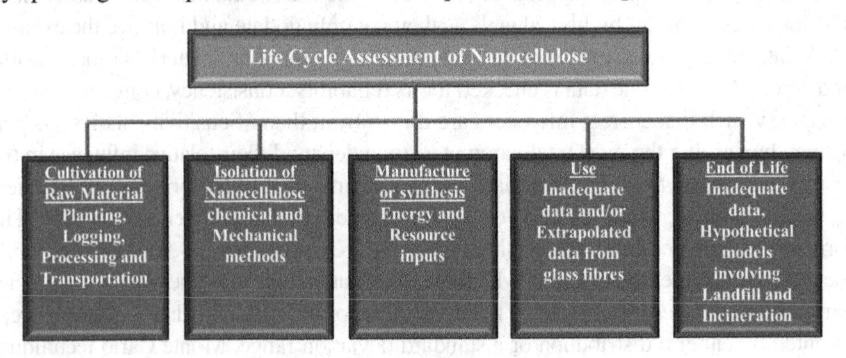

FIGURE 9.5 Schematic demonstrating the life cycle assessment (LCA) stages of nanocellulose.

determine the associated environmental impact of cellulose nanofibres. It was found that the related impact for the electrospinning process was higher than that for the wet spinning process (Piccinno et al. 2015). A similar comparative LCA study was performed by Arvidsson et al. to compare the cradle-to-gate environmental impacts of the production of cellulose nanofibrils from wood pulp via three different production schemes. The study found that as per expectations, the production route involving carboxymethylation in place of enzyme treatment or no pretreatment had the greatest associated impact owing to the great extent of crude-oil-based solvent use (Arvidsson et al. 2015). Another interesting study by Hevry et al. pointed out that nanocellulose-reinforced epoxy composites have greater possibility of abiotic depletion of fossil fuels (ADf) and global warming potential (GWP) than neat polylactide (PLA) and glass fibre-reinforced polypropylene (GF/PP) composites. However, the assessment further showed that both ADf and GWP of nanocellulose-reinforced epoxy composites can be lowered even more than their PLA and GF/PP counterparts on increasing the constituting quantity of nanocellulose in the composite to more than 60% (Hervy et al. 2015).

There are a few other studies based on green polymers other than cellulose. Cinelli et al. conducted LCA on materials composed of biodegradable polymer polylactic acid (PLA) and chitin nano fibrils (NC). It was a "cradle-to-gate" LCA that excluded the usage phase. It was mainly aimed at understanding the impacts of the PLA/NC-based materials on the environment and for the purpose of "eco-design", that is, to suggest improvement options for the production of these materials. The LCA included the master batch preparation of NC based on NC, plasticizer and water and the procedure by which these formulations were utilized to prepare flexible films and rigid packing. SimaPro7.3.3, software was employed to carry out the LCA. The primary data corresponded to the period of 2011 to 2014, while the secondary data taken from Ecoinvent v2. database represented data during the period of 2004 to 2010. The compilation and analysis of the data were done in accordance with the "International Reference Life Cycle Data System" Handbook (ILCD) as well as ISO standards (ISO 14040, 14044). The method employed involved 16 midpoint indicators that accounted for a wide range of environmental concerns, including emission possibilities of greenhouse gases (GHG). Moreover, the calculations were performed with another method called "Recipe" in order to gain a holistic view of the impact. The functional units in the study were 1 kg of nano chitin fibrils and 1 kg of pellets based on PLA/NC/PEG400. From the analysis, it was evident that there was a need to optimize chemical use and water use as well as to incorporate recycling whenever possible in the entire production process. The LCA further showed that it was beneficial to use PLA/NC/PEG products because the components, namely chitosan, glycerol, chitin, PLA arose from renewable sources. The sustainability of the PLA/NC materials was further evident due to the fact that the impacts associated with them were related to production, processing, transportation processes and land use and are the same for petroleum-derived polymers, but the compostability of these PLA/NC materials, as opposed to landfill or incineration, was a major environmental benefit (Cinelli et al. 2017).

Another interesting study involving PLA was carried out by Lorite et al., who evaluated PLA-nanocomposite packaging with respect to pure PLA and polyethylene terephthalate or PET (conventional plastic). The LCA pointed out that PLA-based packaging with nanoclays had the greatest performance with regard to environmental sustainability. Moreover, the impacts on human health were also considerably lower for PLA-based packaging (Lorite et al. 2017).

Pourzahedi et al. performed LCA on nanosilver-based consumer products (socks, T-shirts, baby blankets, towels, food containers, medical cloths, etc.). The increasing demand for products with antimicrobial properties and the potential of environmental burdens from silver nanoparticles made it necessary to assess the impacts of the nanoparticles in the context of the actual products that contain them. The assessment of environmental burdens was performed over several impact categories (acidification, global warming, ecotoxicity, ozone depletion, eutrophication, photochemical smog, human health carcinogens, fossil fuel depletion etc.) in accordance with the United States Environmental Protection Agency's Tool for the Reduction and Assessment of Chemical and Other Environmental Impacts (TRACI 2.1) method. The contribution of the silver nanoparticles to the impacts altogether varied from 1% to 99% depending on the nature of the product as well as the loading of silver. It was also seen that samples that were made of solid polymers had more silver loss during washing than those having materials of a fibrous nature. Interestingly, impacts for almost half of the products taken into consideration arose from the traditional material constituents such as fibres (cotton, cellulose, etc.) or plastic resins. For example, products such as towels and baby blankets that had small quantities of silver nanoparticles exhibited higher environmental impacts from polyester fabric than from silver nanoparticles. The impacts that resulted from direct release of silver nanoparticles were evaluated using characterization factors that had been defined in previous literature. The ecotoxicity impacts that resulted from direct silver nanoparticle releases were actually lower than those associated with the cradle-to-gate journey of the products. This indicated that inclusion of upstream impacts is essential for understanding the holistic impacts of the product. Conversely, it was also found that products having much fewer concentrations of silver had low upstream impacts but had relatively higher silver nanoparticle release rates. This proved that the LCA models and the release evaluations were complementary in nature and that inclusion of both is indispensable to an accurate assessment of the environmental impacts of the products (Pourzahedi et al. 2017).

9.7 CHALLENGES IN LCA

Until now, very few LCA studies have been performed for nanomaterials, and only a small fraction of them are complete and comprehensive (Bauer et al. 2008, Healy et al. 2008, Khanna et al. 2008, Kushnir and Sandén 2011, Lloyd et al. 2005, Hischier 2014). Conducting LCA of nanomaterials is a daunting task because it is faced with several challenges. First, the inventory data is not adequate owing to the fact that the manufacturing processes of the nanoproducts are either new, confidential or constantly evolving due to the current lack of protocol standardization (Meyer et al. 2009). Moreover, for lab-scale processes, the inventory data does not reflect the industrial-scale production. Even for industrial-scale inventories, the tremendous variability in materials and methods results in insufficiency of collected data. Additionally, the inventories exclude data on nanoparticle emission at different life cycle stages. Furthermore, the inventories do not take into account nanomaterial behaviour post disposal. For example, data on nanomaterial interaction with other substances during incineration or landfill is missing from the inventories (Klöpffer et al. 2007). Second, the lack of nano-specific CFs presents a great hurdle for conducting nanotechnology-based LCA studies (Bauer et al. 2008, Gavankar et al. 2012, Kim and Fthenakis 2013, Rosenbaum et al. 2008, Hetherington et al. 2014, Bare 2011). Third, inventory data cannot be utilized in the same manner as for conventional products because cut-offs solely based on mass are not logical for nanomaterials (Klöpffer et al. 2007). Fourth, the exact quantities of released nanomaterials into either the local or global environments are still not established (Brouwer et al. 2009, Meyer et al. 2009, Keller et al. 2013). Factors that affect the quantity of released nanoparticles (nanofiller used, matrix used, abrasion, handling, etc.) vary widely and are extremely case-specific. There is no standard methodology to determine the nature and quantity of the nano-components released during mechanical stress. The particles released during the use and/or end-of-life phases of the product are often very different from the pristine nanoparticles. Since the released nanoparticles are not classified, the actual nanomaterial release, in actuality, can be a mixture of free or embedded nanoparticles, aggregates, or even other contaminant nano-sized particles that are naturally occurring in the atmosphere or are produced during the manufacturing process. Background noises owing to ventilation facilities, presence of outdoor particles (Yeganeh et al. 2008), other production activities in the same plant area, release of carbon brushes from the motors inside the machines (Fujitani et al. 2008) are also impediments to conducting LCA, as they introduce variability in the results. Moreover, certain assumptions in LCA studies can lead to erroneous results. For example, the method that is utilized to measure the dimensions of the nanoparticles released into the air assumes that the particles are spherical, which is generally not the case. Furthermore, in most of the studies, only manufacture and usage stages of the life cycle are assessed, leaving out the transportation and end-of- life impacts of the product. Therefore, such studies do not fulfil the criteria of LCA as per the definition of the ISO standard (Salieri et al. 2018). Lastly, functional units that link or connect inputs and outputs of the employed model to compare LCA results by providing a common ground are often inadequate and material flow analysis (MFA) models that analyze the concentrations of manufactured nanomaterials in the environment face a number of problems such as inaccurate input data and high uncertain outputs (Gottschalk et al. 2013, Baalousha et al. 2016, Wang et al. 2016, Sun et al. 2017).

Apart from these major pitfalls, a few other drawbacks need to be addressed. First, the awareness to apply LCA to all nanotechnology-based products to answer the concerns regarding environmental impacts is still low. Second, due to the knowledge gap in bridging LCA to RA, the specific potential impacts resulting in (eco)toxicity are not yet fully elucidated. Third, researchers have not been able to standardize a common formula to compute all the assessments. Lastly, no legislation has yet been passed to compulsorily label inclusion of nanocomponents in a product.

9.8 CONCLUDING REMARKS

Presently, nanotechnology focusses on production protocols that replace chemicals of potential toxicity with bio-based materials. However, a mere "green synthesis" does not ensure that the process or the product is going to be environmentally sustainable (Seager et al. 2012). To provide a complete and effective solution to environmental sustainability, we need to consider it from a wider perspective, in terms of the life cycle of a product. Life cycle assessment (LCA) evaluates environmental sustainability by considering input factors such as resource and energy use, byproduct formation, etc. and therefore influences decisions towards more environmentally benign paths (Som et al. 2010).

Owing to the fact that there are several nanoproducts with different compositions, properties and usage patterns, LCA of nanomaterials cannot be generalized. LCA is extremely case-specific and there are no general rules. In spite of the complex nature of the nanoproducts and the paucity of available data on impacts, the ISO framework (ISO 14040:2006) for LCA is completely suited to nanomaterials. Moreover, since the life cycle stage which is responsible for the environmental impacts is not known, it is recommended that LCA be performed for all the life cycle stages of the product (Klöpffer et al. 2007).

There are quite a few challenges that are faced while performing LCA of nanomaterials. For example, the risks associated with nanomaterials are evaluated on the basis of two factors: exposure and toxicity. There are no specific regulations for the assessment of exposure to nanoparticles during the later stages of the life cycle, when released nanoparticles are likely to be different from the pristine nanoparticles. Similarly, the toxicity of nanomaterials, which depends on several parameters such as size, composition, shape, surface properties, etc. is markedly different from that of conventional chemicals, which depends solely on mass. Therefore, the current regulations, which estimate the toxicity of nanomaterials in proportion to the bulk of the material, need to be remodeled (Gendre et al. 2015).

The uncertainty associated with the effects of engineered nanomaterials on environmental and human health can be dealt with by both risk assessment (RA) and life cycle concepts. Therefore, the strategy for life cycle concepts should incorporate RA, and vice versa, to overcome the limitation of scarce data on the risks associated with nanomaterials (Som et al. 2010).

9.9 FUTURE PERSPECTIVE ON LCA

There is a lot of scope for improving the LCA studies of nanoproducts and/or nanomaterials. First, inventory analysis should be refined by gathering datasets about nanotechnology-based methodologies and subsequently publishing them in renowned LCA databases. Software should also be updated by incorporating nano-specific information. Industry, academia and governments should join hands to make confidential business information (CBI) of nano-based products easily available for conducting LCA studies (Pati et al. 2014). Simultaneously, an impact assessment methodology should be developed by choosing the most appropriate parameters as outputs (Seager et al. 2012). This can be achieved by collaboration between risk assessment personnel and consulting toxicologists. Second, toxicological models should be built to include toxicological characterization in LCA. Standard protocols and practical methods should also be developed to bring about uniform assessment. Third, risk assessment (RA) and LCA should be connected by the use of the same terminology, parameters and information flow in order to have a holistic view on the effects of engineered nanomaterials on environment and human health. RA takes into account the quantities of discharged pollutants/toxic substances and also their potential to affect human health via multiple exposure pathways (Upreti et al. 2015). Since the adverse effects of nanomaterials on human health depend on their quantity and on other characteristics such as surface area and particle size, combining LCA and RA is likely to provide better results (Savolainen et al. 2010). LCA can also be merged with other processes such as social and economic assessment in order to have a broader scope of sustainability management. Fourth, researchers should ideally try to address all the life cycle stages of a product, including qualitative and quantitative analyses. If they omit some life cycle stages, because they deem some background procedures insignificant or for lack of available data, they should report their results, explaining the reason for their omission (Klöpffer et al. 2007). This would make the data more transparent and credible. Moreover, data from similar technology/industry can be utilized and extrapolated to nanotechnology to fill in minor knowledge gaps. Researchers and/or companies should also not halt the LCA studies until entire data on all life cycle stages are made available. Instead, they should make rational assumptions with reasonable upper and lower boundaries.

Lastly, a full, time-consuming and expensive LCA study can be replaced by a procedure called "LCA streamlining". LCA streamlining attempts to find out data from various assessment techniques that are complementary to each other. This would help to minimize the time, cost and effort of the analysis. During streamlining, researchers usually focus on only one or two life cycle stages of interest or they restrict the environmental impact categories they deem relevant enough to include in their studies (Upreti et al. 2015). This streamlining approach is usually recommended for small and medium-sized enterprises (SMEs). In a nutshell, LCA studies are presently affected by several knowledge gaps. Based on several recommendations provided by existing literature, LCA practitioners have a lot of future scope to enhance impact assessment studies with respect to their uniformity, completeness and clarity (Salieri et al. 2018).

ACKNOWLEDGMENTS

KMP acknowledges the research support of DBT-IYBA fellowship, research grants from NMCG-MoWR, SERB-DST and MHRD-IITR from Govt. of India and BEST Pvt. Ltd.

REFERENCES

Anastas, P. 1998. *Jc Warner Green Chemistry: Theory and Practice*. Oxford University Press, Oxford.

Arvidsson, R., S. Molander, B. A. Sandén, and M. Hassellöv. 2011. Challenges in Exposure Modeling of Nanoparticles in Aquatic Environments. *Human and Ecological Risk Assessment* 17 (1): 245–62.

Arvidsson, R., D. Nguyen, and M. Svanström. 2015. Life Cycle Assessment of Cellulose Nanofibrils Production by Mechanical Treatment and Two Different Pretreatment Processes. *Environmental Science & Technology* 49 (11): 6881–90.

Baalousha, M., G. Cornelis, T. Kuhlbusch, I. Lynch, C. Nickel, W. Peijnenburg, and N. Van Den Brink. 2016. Modeling Nanomaterial Fate and Uptake in the Environment: Current Knowledge and Future Trends. *Environmental Science: Nano* 3 (2): 323–45.

Bare, J. 2011. Traci 2.0: The Tool for the Reduction and Assessment of Chemical and Other Environmental Impacts 2.0. *Clean Technologies and Environmental Policy* 13 (5): 687–96.

Bauer, C., J. Buchgeister, R. Hischier, W. Poganietz, L. Schebek, and J. Warsen. 2008. Towards a Framework for Life Cycle Thinking in the Assessment of Nanotechnology. *Journal of Cleaner Production* 16 (8–9): 910–26.

Bello, D., A. J. Hart, K. Ahn, M. Hallock, N. Yamamoto, E. J. Garcia, M. J. Ellenbecker, and B. L. Wardle. 2008. Particle Exposure Levels During Cvd Growth and Subsequent Handling of Vertically-Aligned Carbon Nanotube Films. *Carbon* 46 (6): 974–77.

Berta, M., C. Lindsay, G. Pans, and G. Camino. 2006. Effect of Chemical Structure on Combustion and Thermal Behaviour of Polyurethane Elastomer Layered Silicate Nanocomposites. *Polymer Degradation and Stability* 91 (5): 1179–91.

Borm, P. J., D. Robbins, S. Haubold, T. Kuhlbusch, H. Fissan, K. Donaldson, R. Schins, *et al.* 2006. The Potential Risks of Nanomaterials: A Review Carried out for Ecetoc. *Particle and Fibre Toxicology* 3 (1): 11.

Breggin, L., and J. Pendergrass. 2007. Where Does the Nano Go? End-of-Life Regulation of Nanotechnologies, Project on Emerging Nanotechnologies, Woodrow Wilson International Center for Scholars.

Brouwer, D., B. van Duuren-Stuurman, M. Berges, E. Jankowska, D. Bard, and D. Mark. 2009. From Workplace Air Measurement Results toward Estimates of Exposure? Development of a Strategy to Assess Exposure to Manufactured Nano-Objects. *Journal of Nanoparticle Research* 11 (8): 1867.

Caballero-Guzman, A., and B. Nowack. 2016. A Critical Review of Engineered Nanomaterial Release Data: Are Current Data Useful for Material Flow Modeling? *Environmental Pollution* 213: 502–17.

Chatterjee, R. 2008. *The Challenge of Regulating Nanomaterials.* ACS Publications.

Chaudhry, Q., M. Scotter, J. Blackburn, B. Ross, A. Boxall, L. Castle, R. Aitken, and R. Watkins. 2008. Applications and Implications of Nanotechnologies for the Food Sector. *Food Additives and Contaminants* 25 (3): 241–58.

Christensen, F. M., and S. I. Olsen. 2004. The Potential Role of Life Cycle Assessment in Regulation of Chemicals in the European Union. *The International Journal of Life Cycle Assessment* 9 (5): 327.

Cinelli, P., M. Coltelli, N. Mallegni, P. Morganti, and A. Lazzeri. 2017. Degradability and Sustainability of Nanocomposites Based on Polylactic Acid and Chitin Nano Fibrils. *Chemical Engineering Transactions* 60.

Dale, A. L., E. A. Casman, G. V. Lowry, J. R. Lead, E. Viparelli, and M. Baalousha. 2015. *Modeling Nanomaterial Environmental Fate in Aquatic Systems.* ACS Publications.

Dasari, A., Z.-Z. Yu, and Y.-W. Mai. 2009. Fundamental Aspects and Recent Progress on Wear/Scratch Damage in Polymer Nanocomposites. *Materials Science and Engineering: R: Reports* 63 (2): 31–80.

Davis, J. M. 2007. How to Assess the Risks of Nanotechnology: Learning from Past Experience. *Journal of Nanoscience and Nanotechnology* 7 (2): 402–09.

de Figueirêdo, M. C. B., M. de Freitas Rosa, C. M. L. Ugaya, M. d. S. M. de Souza, A. C. C. da Silva Braid, and L. F. L. de Melo. 2012. Life Cycle Assessment of Cellulose Nanowhiskers. *Journal of Cleaner Production* 35: 130–39.

Deng, Y., J. Li, M. Qiu, F. Yang, J. Zhang, and C. Yuan. 2017. Deriving Characterization Factors on Freshwater Ecotoxicity of Graphene Oxide Nanomaterial for Life Cycle Impact Assessment. *The International Journal of Life Cycle Assessment* 22 (2): 222–36.

Devaprakasam, D., P. Hatton, G. Möbus, and B. Inkson. 2008. Effect of Microstructure of Nano-and Micro-Particle Filled Polymer Composites on Their Tribo-Mechanical Performance. Paper presented at the Journal of Physics: Conference Series.

Di Guardo, A., T. Gouin, M. MacLeod, and M. Scheringer. 2018. Environmental Fate and Exposure Models: Advances and Challenges in 21 St Century Chemical Risk Assessment. *Environmental Science: Processes & Impacts* 20 (1): 58–71.

Dorairaju, G. 2008. *Additive Synergy in Flexible Pvc Nanocomposites for Wire and Cable Applications*. University of Massachusetts Lowell.

Ettrup, K., A. Kounina, S. F. Hansen, J. A. Meesters, E. B. Vea, and A. Laurent. 2017. Development of Comparative Toxicity Potentials of Tio2 Nanoparticles for Use in Life Cycle Assessment. Environmental Science & *Technology* 51 (7): 4027–37.

Franco, A., S. F. Hansen, S. I. Olsen, and L. Butti. 2007. Limits and Prospects of the "Incremental Approach" and the European Legislation on the Management of Risks Related to Nanomaterials. *Regulatory Toxicology and Pharmacology* 48 (2): 171–83.

Fujitani, Y., T. Kobayashi, K. Arashidani, N. Kunugita, and K. Suemura. 2008. Measurement of the Physical Properties of Aerosols in a Fullerene Factory for Inhalation Exposure Assessment. *Journal of Occupational and Environmental Hygiene* 5 (6): 380–89.

Gavankar, S., S. Suh, and A. F. Keller. 2012. Life Cycle Assessment at Nanoscale: Review and Recommendations. *The International Journal of Life Cycle Assessment* 17 (3): 295–303.

Gendre, L., K. Blackburn, J. Brighton, V. Marchante, and H. A. Rodriguez. 2015. Nanomaterials Life Cycle Analysis: Health 2 and Safety Practices, Standards and 3 Regulations–Past, Present and Future 4 Perspective 5. *NANOSCALE* 65: 66.

Gilbertson, L. M., B. A. Wender, J. B. Zimmerman, and M. J. Eckelman. 2015. Coordinating Modeling and Experimental Research of Engineered Nanomaterials to Improve Life Cycle Assessment Studies. *Environmental Science: Nano* 2 (6): 669–82.

Goffin, A.-L. 2010. Polymer Bionanocomposites Reinforced by Functionalized Nanoparticles: Impact of Nanofiller Size, Nature and Composition. PhD thesis, Université de Mons, Belgium.

Göhler, D., A. Nogowski, P. Fiala, and M. Stintz. 2013. Nanoparticle Release from Nanocomposites Due to Mechanical Treatment at Two Stages of the Life-Cycle. Paper presented at the Journal of Physics: Conference Series.

Golanski, L., A. Guiot, M. Pras, M. Malarde, and F. Tardif. 2012. Release-Ability of Nano Fillers from Different Nanomaterials (toward the Acceptability of Nanoproduct). *Journal of Nanoparticle Research* 14 (7): 962.

Gottschalk, F., R. W. Scholz, and B. Nowack. 2010. Probabilistic Material Flow Modeling for Assessing the Environmental Exposure to Compounds: Methodology and an Application to Engineered Nano-Tio2 Particles. *Environmental Modelling & Software* 25 (3): 320–32.

Gottschalk, F., T. Sun, and B. Nowack. 2013. Environmental Concentrations of Engineered Nanomaterials: Review of Modeling and Analytical Studies. *Environmental Pollution* 181: 287–300.

Halliwell, S. 2006. End of Life Options for Composite Waste: Recycle, Reuse or Dispose. National Composites Network Best Practice Guide.

Han, J. H., E. J. Lee, J. H. Lee, K. P. So, Y. H. Lee, G. N. Bae, S.-B. Lee, *et al.* 2008. Monitoring Multiwalled Carbon Nanotube Exposure in Carbon Nanotube Research Facility. *Inhalation Toxicology* 20 (8): 741–49.

Handy, R. D., F. Von der Kammer, J. R. Lead, M. Hassellöv, R. Owen, and M. Crane. 2008. The Ecotoxicology and Chemistry of Manufactured Nanoparticles. *Ecotoxicology* 17 (4): 287–314.

Hauschild, M. Z., and M. A. Huijbregts. 2015. Introducing Life Cycle Impact Assessment. *Life Cycle Impact Assessment*, 1–16: Springer.

Healy, M. L., L. J. Dahlben, and J. A. Isaacs. 2008. Environmental Assessment of Single-Walled Carbon Nanotube Processes. *Journal of Industrial Ecology* 12 (3): 376–93.

Helland, A., M. Scheringer, M. Siegrist, H. G. Kastenholz, A. Wiek, and R. W. Scholz. 2007. Risk Assessment of Engineered Nanomaterials: A Survey of Industrial Approaches. *Environmental Science & Technology* 42 (2): 640–46.

Helland, A., P. Wick, A. Koehler, K. Schmid, and C. Som. 2007. Reviewing the Environmental and Human Health Knowledge Base of Carbon Nanotubes. *Environmental Health Perspectives* 115 (8): 1125–31.

Henderson, A. D., M. Z. Hauschild, D. van de Meent, M. A. Huijbregts, H. F. Larsen, M. Margni, T. E. McKone, *et al.* 2011. Usetox Fate and Ecotoxicity Factors for Comparative Assessment of Toxic Emissions in Life Cycle Analysis: Sensitivity to Key Chemical Properties. *The International Journal of Life Cycle Assessment* 16 (8): 701.

Hendren, C. O., A. R. Badireddy, E. Casman, and M. R. Wiesner. 2013. Modeling Nanomaterial Fate in Wastewater Treatment: Monte Carlo Simulation of Silver Nanoparticles (Nano-Ag). *Science of the Total Environment* 449: 418–25.

Hervy, M., S. Evangelisti, P. Lettieri, and K.-Y. Lee. 2015. Life Cycle Assessment of Nanocellulose-Reinforced Advanced Fibre Composites. *Composites Science and Technology* 118: 154–62.

Hetherington, A. C., A. L. Borrion, O. G. Griffiths, and M. C. McManus. 2014. Use of Lca as a Development Tool within Early Research: Challenges and Issues across Different Sectors. *The International Journal of Life Cycle Assessment* 19 (1): 130–43.

Hirth, S., L. Cena, G. Cox, Ž. Tomović, T. Peters, and W. Wohlleben. 2013. Scenarios and Methods That Induce Protruding or Released Cnts after Degradation of Nanocomposite Materials. *Journal of Nanoparticle Research* 15 (4): 1504.

Hischier, R. 2014. Life Cycle Assessment of Manufactured Nanomaterials: Inventory Modelling Rules and Application Example. *The International Journal of Life Cycle Assessment* 19 (4): 941–43.

Hischier, R., and T. Walser. 2012. Life Cycle Assessment of Engineered Nanomaterials: State of the Art and Strategies to Overcome Existing Gaps. *Science of the Total Environment* 425: 271–82.

Hodge, G. A., D. Bowman, and K. Ludlow. 2009. *New Global Frontiers in Regulation: The Age of Nanotechnology*. Edward Elgar Publishing, Cheltenham.

Hussain, F., M. Hojjati, M. Okamoto, and R. E. Gorga. 2006. Polymer-Matrix Nanocomposites, Processing, Manufacturing, and Application: An Overview. *Journal of Composite Materials* 40 (17): 1511–75.

Inshakova, E., and O. Inshakov. 2017. World Market for Nanomaterials: Structure and Trends. Paper presented at the MATEC web of conferences.

Jacobs, R., J. A. Meesters, C. J. ter Braak, D. van de Meent, and H. van der Voet. 2016. Combining Exposure and Effect Modeling into an Integrated Probabilistic Environmental Risk Assessment for Nanoparticles. *Environmental Toxicology and Chemistry* 35 (12): 2958–67.

Kalaivasan, N., and S. S. Shafi. 2017. Enhancement of Corrosion Protection Effect in Mechanochemically Synthesized Polyaniline/Mmt Clay Nanocomposites. *Arabian Journal of Chemistry* 10: S127–S33.

Keller, A. A., S. McFerran, A. Lazareva, and S. Suh. 2013. Global Life Cycle Releases of Engineered Nanomaterials. *Journal of Nanoparticle Research* 15 (6): 1692.

Khan, I. A., N. D. Berge, T. Sabo-Attwood, P. L. Ferguson, and N. B. Saleh. 2013. Single-Walled Carbon Nanotube Transport in Representative Municipal Solid Waste Landfill Conditions. *Environmental Science & Technology* 47 (15): 8425–33.

Khanna, V., B. R. Bakshi, and L. J. Lee. 2008. Carbon Nanofiber Production: Life Cycle Energy Consumption and Environmental Impact. *Journal of Industrial Ecology* 12 (3): 394–410.

Kim, H. C., and V. Fthenakis. 2013. Life Cycle Energy and Climate Change Implications of Nanotechnologies. *Journal of Industrial Ecology* 17 (4): 528–41.

Klaine, S. J., P. J. Alvarez, G. E. Batley, T. F. Fernandes, R. D. Handy, D. Y. Lyon, S. Mahendra, M. J. McLaughlin, and J. R. Lead. 2008. Nanomaterials in the Environment: Behavior, Fate, Bioavailability, and Effects. *Environmental Toxicology and Chemistry* 27 (9): 1825–51.

Klöpffer, W., M. A. Curran, P. Frankl, R. Heijungs, A. Köhler, and S. I. Olsen. 2007. Nanotechnology and Life Cycle Assessment: A Systems Approach to Nanotechnology and the Environment. Woodrow Wilson International Center for Scholars.

Köhler, A. R., and C. Som. 2008. Environmental and Health Implications of Nanotechnology—Have Innovators Learned the Lessons from Past Experiences? *Human and Ecological Risk Assessment* 14 (3): 512–31.

Köhler, A. R., C. Som, A. Helland, and F. Gottschalk. 2008. Studying the Potential Release of Carbon Nanotubes Throughout the Application Life Cycle. *Journal of Cleaner Production* 16 (8–9): 927–37.

Koo, O. M., I. Rubinstein, and H. Onyuksel. 2005. Role of Nanotechnology in Targeted Drug Delivery and Imaging: A Concise Review. *Nanomedicine: Nanotechnology, Biology and Medicine* 1 (3): 193–212.

Kuhlbusch, T. A., C. Asbach, H. Fissan, D. Göhler, and M. Stintz. 2011. Nanoparticle Exposure at Nanotechnology Workplaces: A Review. *Particle and Fibre Toxicology* 8 (1): 22.

Kuiken, T. 2009. It's Time to Move Forward on Lca of Nanomaterials. Paper presented at the Chicago, IL (USA): Nanotechnology & Life Cycle Analysis Workshop.

Kushnir, D., and B. A. Sandén. 2011. Multi-Level Energy Analysis of Emerging Technologies: A Case Study in New Materials for Lithium Ion Batteries. *Journal of Cleaner Production* 19 (13): 1405–16.

Lloyd, S. M., L. B. Lave, and H. S. Matthews. 2005. Life Cycle Benefits of Using Nanotechnology to Stabilize Platinum-Group Metal Particles in Automotive Catalysts. *Environmental Science & Technology* 39 (5): 1384–92.

Lorite, G. S., J. M. Rocha, N. Miilumäki, P. Saavalainen, T. Selkälä, G. Morales-Cid, M. Gonçalves, *et al.* 2017. Evaluation of Physicochemical/Microbial Properties and Life Cycle Assessment (Lca) of Pla-Based Nanocomposite Active Packaging. *LWT* 75: 305–15.

Lozano, P., and N. D. Berge. 2012. Single-Walled Carbon Nanotube Behavior in Representative Mature Leachate. *Waste Management* 32 (9): 1699–711.

Luo, J.-J., and I. M. Daniel. 2003. Characterization and Modeling of Mechanical Behavior of Polymer/Clay Nanocomposites. *Composites Science and Technology* 63 (11): 1607–16.

Mackay, C. E., M. Johns, J. H. Salatas, B. Bessinger, and M. Perri. 2006. Stochastic Probability Modeling to Predict the Environmental Stability of Nanoparticles in Aqueous Suspension. *Integrated Environmental Assessment and Management: An International Journal* 2 (3): 293–98.

MacLeod, M., M. Scheringer, T. E. McKone, and K. Hungerbuhler. 2010. *The State of Multimedia Mass-Balance Modeling in Environmental Science and Decision-Making.* ACS Publications.

Marzuki, H. F. A., and M. Jaafar. 2016. Laminate Design of Lightweight Glass Fiber Reinforced Epoxy Composite for Electrical Transmission Structure. *Procedia Chemistry* 19: 871–78.

Maynard, A. D., and R. J. Aitken. 2007. Assessing Exposure to Airborne Nanomaterials: Current Abilities and Future Requirements. *Nanotoxicology* 1 (1): 26–41.

Maynard, A. D., R. J. Aitken, T. Butz, V. Colvin, K. Donaldson, G. Oberdörster, M. A. Philbert, *et al.* 2006. Safe Handling of Nanotechnology. *Nature* 444 (7117): 267.

Maynard, A. D., P. A. Baron, M. Foley, A. A. Shvedova, E. R. Kisin, and V. Castranova. 2004. Exposure to Carbon Nanotube Material: Aerosol Release During the Handling of Unrefined Single-Walled Carbon Nanotube Material. *Journal of Toxicology and Environmental Health, Part A* 67 (1): 87–107.

Mazzuckelli, L. F., M. M. Methner, M. E. Birch, D. E. Evans, B.-K. Ku, K. Crouch, and M. D. Hoover. 2007. Identification and Characterization of Potential Sources of Worker Exposure to Carbon Nanofibers During Polymer Composite Laboratory Operations. *Journal of Occupational and Environmental Hygiene* 4 (12): D125–D30.

Meesters, J. A., A. A. Koelmans, J. T. Quik, A. J. Hendriks, and D. van de Meent. 2014. Multimedia Modeling of Engineered Nanoparticles with Simplebox4nano: Model Definition and Evaluation. *Environmental Science & Technology* 48 (10): 5726–36.

Meyer, D. E., M. A. Curran, and M. A. Gonzalez. 2009. *An Examination of Existing Data for the Industrial Manufacture and Use of Nanocomponents and Their Role in the Life Cycle Impact of Nanoproducts.* ACS Publications.

Miseljic, M., and S. I. Olsen. 2014. Life-Cycle Assessment of Engineered Nanomaterials: A Literature Review of Assessment Status. *Journal of Nanoparticle Research* 16 (6): 2427.

Mueller, N. C., and B. Nowack. 2008. Exposure Modeling of Engineered Nanoparticles in the Environment. Environmental Science *& Technology* 42 (12): 4447–53.

Mullaney, M. 2007. Beyond Batteries: Storing Power in a Sheet of Paper. Rensselaer Polytechnic Institute: Troy, NY, USA: 2007–08.

Njuguna, J., I. Pena, H. Zhu, S. Rocks, M. Blázquez, and S. Desai. 2009. Opportunities and Environmental Health Challenges Facing Integration of Polymer Nanocomposites: Technologies for Automotive Applications. *International Journal of Polymers and Technologies* 2: 117–26.

Njuguna, J., F. Silva, and S. Sachse. 2011. Nanocomposites for Vehicle Structural Applications. *Nanofibers-Production, Properties and Functional Applications*: IntechOpen.

Nowack, B., C. Brouwer, R. E. Geertsma, E. H. Heugens, B. L. Ross, M.-C. Toufektsian, S. W. Wijnhoven, and R. J. Aitken. 2012. Analysis of the Occupational, Consumer and Environmental Exposure to Engineered Nanomaterials Used in 10 Technology Sectors. *Nanotoxicology* 7 (6): 1152–56.

Nowack, B., R. M. David, H. Fissan, H. Morris, J. A. Shatkin, M. Stintz, R. Zepp, and D. Brouwer. 2013. Potential Release Scenarios for Carbon Nanotubes Used in Composites. *Environment International* 59: 1–11.

Nowack, B., J. F. Ranville, S. Diamond, J. A. Gallego-Urrea, C. Metcalfe, J. Rose, N. Horne, A. A. Koelmans, and S. J. Klaine. 2012. Potential Scenarios for Nanomaterial Release and Subsequent Alteration in the Environment. *Environmental Toxicology and Chemistry* 31 (1): 50–59.

Oberdörster, G. 2010. Safety Assessment for Nanotechnology and Nanomedicine: Concepts of Nanotoxicology. *Journal of Internal Medicine* 267 (1): 89–105.

Ostertag, K., and B. Hüsing. 2008. Identification of Starting Points for Exposure Assessment in the Post-Use Phase of Nanomaterial-Containing Products. *Journal of Cleaner Production* 16 (8–9): 938–48.

Owen, R., M. Crane, K. Grieger, R. Handy, I. Linkov, and M. Depledge. 2009. Strategic Approaches for the Management of Environmental Risk Uncertainties Posed by Nanomaterials. *Nanomaterials: Risks and Benefits*, 369–84: Springer.

Park, J., B. K. Kwak, E. Bae, J. Lee, Y. Kim, K. Choi, and J. Yi. 2009. Characterization of Exposure to Silver Nanoparticles in a Manufacturing Facility. *Journal of Nanoparticle Research* 11 (7): 1705–12.

Pati, P., S. McGinnis, and P. J. Vikesland. 2014. Life Cycle Assessment of "Green" Nanoparticle Synthesis Methods. *Environmental Engineering Science* 31 (7): 410–20.

Piccinno, F., R. Hischier, S. Seeger, and C. Som. 2015. Life Cycle Assessment of a New Technology to Extract, Functionalize and Orient Cellulose Nanofibers from Food Waste. *ACS Sustainable Chemistry & Engineering* 3 (6): 1047–55.

Plata, D., P. Gschwend, and C. Reddy. 2008. Industrially Synthesized Single-Walled Carbon Nanotubes: Compositional Data for Users, Environmental Risk Assessments, and Source Apportionment. *Nanotechnology* 19 (18): 185706.

Pourzahedi, L., M. Vance, and M. J. Eckelman. 2017. Life Cycle Assessment and Release Studies for 15 Nanosilver-Enabled Consumer Products: Investigating Hotspots and Patterns of Contribution. *Environmental Science & Technology* 51 (12): 7148–58.

Praetorius, A., M. Scheringer, and K. Hungerbühler. 2012. Development of Environmental Fate Models for Engineered Nanoparticles a Case Study of Tio2 Nanoparticles in the Rhine River. *Environmental Science & Technology* 46 (12): 6705–13.

Praetorius, A., N. Tufenkji, K.-U. Goss, M. Scheringer, F. von der Kammer, and M. Elimelech. 2014. The Road to Nowhere: Equilibrium Partition Coefficients for Nanoparticles. *Environmental Science: Nano* 1 (4): 317–23.

Pu, Y., B. Laratte, and R. E. Ionescu. 2017. Influence of Dissolution on Fate of Nanoparticles in Freshwater. *International Journal of Environmental Science and Development* 8 (5): 347.

Rafiee, M. A., F. Yavari, J. Rafiee, and N. Koratkar. 2011. Fullerene–Epoxy Nanocomposites-Enhanced Mechanical Properties at Low Nanofiller Loading. *Journal of Nanoparticle Research* 13 (2): 733–37.

Reijnders, L. 2009. The Release of Tio2 and Sio_2 Nanoparticles from Nanocomposites. *Polymer Degradation and Stability* 94 (5): 873–76.

Rosenbaum, R. K., T. M. Bachmann, L. S. Gold, M. A. Huijbregts, O. Jolliet, R. Juraske, A. Koehler, *et al.* 2008. Usetox—the Unep-Setac Toxicity Model: Recommended Characterisation Factors for Human Toxicity and Freshwater Ecotoxicity in Life Cycle Impact Assessment. *The International Journal of Life Cycle Assessment* 13 (7): 532.

Royal, S. 2004. *Nanoscience and Nanotechnologies.* Royal Society, London.

Salieri, B., S. Righi, A. Pasteris, and S. I. Olsen. 2015. Freshwater Ecotoxicity Characterisation Factor for Metal Oxide Nanoparticles: A Case Study on Titanium Dioxide Nanoparticle. *Science of the Total Environment* 505: 494–502.

Salieri, B., D. A. Turner, B. Nowack, and R. Hischier. 2018. Life Cycle Assessment of Manufactured Nanomaterials: Where Are We? *NanoImpact* 10: 108–20.

Sánchez, C., M. Hortal, C. Aliaga, A. Devis, and V. Cloquell-Ballester. 2014. Recyclability Assessment of Nano-Reinforced Plastic Packaging. *Waste Management* 34 (12): 2647–55.

Savolainen, K., H. Alenius, H. Norppa, L. Pylkkänen, T. Tuomi, and G. Kasper. 2010. Risk Assessment of Engineered Nanomaterials and Nanotechnologies—a Review. *Toxicology* 269 (2–3): 92–104.

Seager, T., E. Selinger, and A. Wiek. 2012. Sustainable Engineering Science for Resolving Wicked Problems. *Journal of Agricultural and Environmental Ethics* 25 (4): 467–84.

Şengül, H., T. L. Theis, and S. Ghosh. 2008. Toward Sustainable Nanoproducts: An Overview of Nanomanufacturing Methods. *Journal of Industrial Ecology* 12 (3): 329–59.

Shatkin, J. A. 2008. Informing Environmental Decision Making by Combining Life Cycle Assessment and Risk Analysis. *Journal of Industrial Ecology* 12 (3): 278–81.

Shatkin, J. A., and B. Kim. 2015. Cellulose Nanomaterials: Life Cycle Risk Assessment, and Environmental Health and Safety Roadmap. *Environmental Science: Nano* 2 (5): 477–99.

Sinha, A. K., K. Suzuki, M. Takahara, H. Azuma, T. Nonaka, and K. Fukumoto. 2007. Mesostructured Manganese Oxide/Gold Nanoparticle Composites for Extensive Air Purification. *Angewandte Chemie International Edition* 46 (16): 2891–94.

Skjolding, L. M., S. N. Sørensen, N. B. Hartmann, R. Hjorth, S. F. Hansen, and A. Baun. 2016. Aquatic Ecotoxicity Testing of Nanoparticles—the Quest to Disclose Nanoparticle Effects. *Angewandte Chemie International Edition* 55 (49): 15224–39.

Som, C., M. Berges, Q. Chaudhry, M. Dusinska, T. F. Fernandes, S. I. Olsen, and B. Nowack. 2010. The Importance of Life Cycle Concepts for the Development of Safe Nanoproducts. *Toxicology* 269 (2–3): 160–69.

Som, C., B. Nowack, H. F. Krug, and P. Wick. 2012. Toward the Development of Decision Supporting Tools That Can Be Used for Safe Production and Use of Nanomaterials. *Accounts of Chemical Research* 46 (3): 863–72.

Standard, B. 2006. Bs En Iso 14040: 2006. *Environmental Management—Life Cycle Assessment—Principles and Framework.* British Standard.

Su, C., H. Zhang, C. Cridge, and R. Liang. 2019. A Review of Multimedia Transport and Fate Models for Chemicals: Principles, Features and Applicability. *Science of the Total Environment* 668: 881–92.

Sun, T. Y., F. Gottschalk, K. Hungerbühler, and B. Nowack. 2014. Comprehensive Probabilistic Modelling of Environmental Emissions of Engineered Nanomaterials. *Environmental Pollution* 185: 69–76.

Sun, T. Y., D. M. Mitrano, N. A. Bornhöft, M. Scheringer, K. Hungerbühler, and B. Nowack. 2017. Envisioning Nano Release Dynamics in a Changing World: Using Dynamic Probabilistic Modeling to Assess Future Environmental Emissions of Engineered Nanomaterials. *Environmental Science & Technology* 51 (5): 2854–63.

Theron, J., J. Walker, and T. Cloete. 2008. Nanotechnology and Water Treatment: Applications and Emerging Opportunities. *Critical Reviews in Microbiology* 34 (1): 43–69.

Touati, N., M. Kaci, S. Bruzaud, and Y. Grohens. 2011. The Effects of Reprocessing Cycles on the Structure and Properties of Isotactic Polypropylene/Cloisite 15a Nanocomposites. *Polymer Degradation and Stability* 96 (6): 1064–73.

Tsai, S.-J., M. Hofmann, M. Hallock, E. Ada, J. Kong, and M. Ellenbecker. 2009. Characterization and Evaluation of Nanoparticle Release During the Synthesis of Single-Walled and Multiwalled Carbon Nanotubes by Chemical Vapor Deposition. Environmental Science *& Technology* 43 (15): 6017–23.

Türk, V., C. Kaiser, D. Vedder, C. Liedtke, H. G. Kastenholz, A. Köhler, H. Knowles, and V. Murray. 2005. Nanologue Background Paper: On Selected Nanotechnology Applications and Their Ethical, Legal and Social Implications European Commission, Brussels, Belgium.

Upreti, G., R. Dhingra, S. Naidu, I. Atuahene, and R. Sawhney. 2015. Life Cycle Assessment of Nanomaterials. *Green Processes for Nanotechnology*, 393–408: Springer, Cham, Switzerland.

Vance, M. E., and L. C. Marr. 2015. Exposure to Airborne Engineered Nanoparticles in the Indoor Environment. *Atmospheric Environment* 106: 503–09.

Vogel, U., K. Savolainen, Q. Wu, M. van Tongeren, D. Brouwer, and M. Berges. 2013. *Handbook of Nanosafety: Measurement, Exposure and Toxicology.* Elsevier.

Von der Kammer, F., P. L. Ferguson, P. A. Holden, A. Masion, K. R. Rogers, S. J. Klaine, A. A. Koelmans, N. Horne, and J. M. Unrine. 2012. Analysis of Engineered Nanomaterials in Complex Matrices (Environment and Biota): General Considerations and Conceptual Case Studies. *Environmental Toxicology and Chemistry* 31 (1): 32–49.

Wang, J., C. Asbach, H. Fissan, T. Hülser, T. A. Kuhlbusch, D. Thompson, and D. Y. Pui. 2011. How Can Nanobiotechnology Oversight Advance Science and Industry: Examples from Environmental, Health, and Safety Studies of Nanoparticles (Nano-Ehs). *Journal of Nanoparticle Research* 13 (4): 1373–87.

Wang, Y., L. Deng, A. Caballero-Guzman, and B. Nowack. 2016. Are Engineered Nano Iron Oxide Particles Safe? An Environmental Risk Assessment by Probabilistic Exposure, Effects and Risk Modeling. *Nanotoxicology* 10 (10): 1545–54.

Wardak, A., M. E. Gorman, N. Swami, and S. Deshpande. 2008. Identification of Risks in the Life Cycle of Nanotechnology-Based Products. *Journal of Industrial Ecology* 12 (3): 435–48.

Watlington, K. 2005. *Emerging Nanotechnologies for Site Remediation and Wastewater Treatment.* Environmental Protection Agency Washington DC.

Wiesner, M. R., G. V. Lowry, K. L. Jones, J. Hochella, Michael F, R. T. Di Giulio, E. Casman, and E. S. Bernhardt. 2009. *Decreasing Uncertainties in Assessing Environmental Exposure, Risk, and Ecological Implications of Nanomaterials.* ACS Publications.

Wohlleben, W., T. A. Kuhlbusch, J. Schnekenburger, and C.-M. Lehr. 2014. *Safety of Nanomaterials Along Their Lifecycle: Release, Exposure, and Human Hazards.* CRC Press.

Yang, N., W. Zhang, C. Ye, X. Chen, and S. Ling. 2019. Nanobiopolymers Fabrication and Their Life Cycle Assessments. *Biotechnology Journal* 14 (1): 1700754.

Yeganeh, B., C. M. Kull, M. S. Hull, and L. C. Marr. 2008. Characterization of Airborne Particles During Production of Carbonaceous Nanomaterials. *Environmental Science & Technology* 42 (12): 4600–06.

Zhang, Q.-Q., G.-G. Ying, C.-G. Pan, Y.-S. Liu, and J.-L. Zhao. 2015. Comprehensive Evaluation of Antibiotics Emission and Fate in the River Basins of China: Source Analysis, Multimedia Modeling, and Linkage to Bacterial Resistance. *Environmental Science & Technology* 49 (11): 6772–82.

10 Limitations in Commercialization of Green Polymeric Nanocomposites and Avenues for Rectification

Deepak Kumar Tripathi, Sharanya Sarkar, Mukesh Kumar Meher, and Krishna Mohan Poluri

CONTENTS

10.1 INTRODUCTION

The employment of natural resin for the fabrication of green composites has been demystified by researchers as a response to the increase in environmental concerns (Mohanty et al. 2005, Bhawani et al. 2018). Bionanocomposites are multifunctional materials obtained from biodegradable polymers and additives of inorganic and organic nature. The depletion of petroleum-based energy sources and compromised environmental conditions due to plastic pollutants disposed of by various industries encourage the use of biodegradable and renewable materials (Shchipunov 2012). With the increasing use of degradable materials, the field of degradable nanocomposites is evolving rapidly (Lalwani et al. 2013, Sridhar et al. 2013, Wang et al. 2011). Petroleum-based polymers used in earlier days contributed much to the generation of global waste, which has encouraged the development of green polymers as a better substitute for traditional polymers. The depletion of fossil fuels encourages the evolution of bio-based products, which reduce dependence on petroleum- based resources.

Green polymer nanocomposites are potential alternatives to petroleum-based polymers (Reig et al. 2014). Generally, polymers with biodegradable and renewable properties are called green polymers or environmentally friendly polymers. Green polymer nanocomposites are made by dispersing the nano-reinforcement (nanoparticle, nanocrystals, nanofibers) in green polymer matrices (Shchipunov 2012, Shahzad 2015). Nanocomposites can be easily decomposed and do not contribute to ecological contamination (Adeosun et al. 2012, Hamad et al. 2014). Integration of such nano-sized fillers improves the physiochemical and thermo-mechanical properties of bionanocomposites, including tensile strength, impact strength, permeability, and stability in its thermal traits (Dong et al. 2015). It also plays a significant role in revamping the biocompatibility and biodegradability of the materials, which proves them to be highly valued materials for the future (Sharma et al. 2019, Das et al. 2017). In the literature, various examples of green polymer nanocomposites carrying filler materials such as starch nanocrystals, cellulose nanofibers and nanocrystals, and nanoparticles derived from chitosan and other biodegradable materials are reported (Chen et al. 2008, Azeredo et al. 2009, Bilbao-Sáinz et al. 2010, de Moura et al. 2009, Podsiadlo et al. 2005). Enhancement of the aforementioned properties is allowing fillers and polymers to be used in the bionanocomposites, and henceforth they are finding a place in various fields such as medicine, electronics, the automotive industry, packaging, energy storage, the aerospace engineering, and the construction industry (Müller et al. 2017, Bajpai et al. 2014).

It can be seen that nanocomposites prepared using non-renewable synthetic sources are not completely degradable (Jamshidian et al. 2010, John and Thomas 2008). It was observed that synthetic materials such as polystyrene and polyester were being used extensively for nanocomposite production. However, use of these synthetic materials has various disadvantages, such as global warming, cross contamination, uneconomical costs, and various environmental concerns. The aforementioned synthetic materials also face the problems of the unavailability of organic compounds, the scarcity of oil and gases, and extremely high market prices (Leja and Lewandowicz 2010, Jamshidian et al. 2010). Recently, the use of degradable polymers obtained from natural sources such as lignin, starch, cellulose acetate, polylactic acid, polyhydroxylalkanoates (PHA), and polyhydroxylbutyrate (PHB), collectively known as biopolymers, have been used for the fabrication of green nanocomposites (Pandey et al. 2007). Use of renewable materials has received substantial attention both from academia and from industry because the utilization of these resources for the production of nanocomposites is more advantageous when compared to the use of synthetic resources. Research needs to be intensified into the innovation and development of products from renewable polymer composites which could be useful in attaining economic and ecologically sustainable technology (Okamoto 2004) (see Table 10.1).

Bionanocomposites have diverse fields of application owing to their improved properties. In spite of there being many benefits to the use of green composites, there is quite a long way to go in order to adapt this material. There are some broad challenges associated with the development of bionanocomposites (see Figure 10.1) that prevent them from being commercially available (Ramos et al. 2018). First, there

TABLE 10.1
Summary of Green Nanocomposites Application in Different Areas and Their Associated Disadvantages

Area of Application	Disadvantages	References
Food packaging	Migration of nanoparticles from packing material to food item, brittleness in wet state, and poor moisture barrier.	Ramos et al. (2018), Duncan (2014), Duncan and Pillai (2014), Koshy et al. (2015), Wihodo and Moraru (2013)
Medical	Release of toxic compounds, degradation of material, and lack of material response optimization.	Saini et al. (2017), Shukla et al. (2016), Nezakati et al. (2018)
Aerospace and automobile industry	Lack of durability, reduced shelf life, and high maintenance cost.	Naskar et al. (2016), Gopi et al. (2017)
Wastewater treatment	High functioning cost and limited range of application.	Ahmed and Kanchi (2018), Wihodo and Moraru (2013)

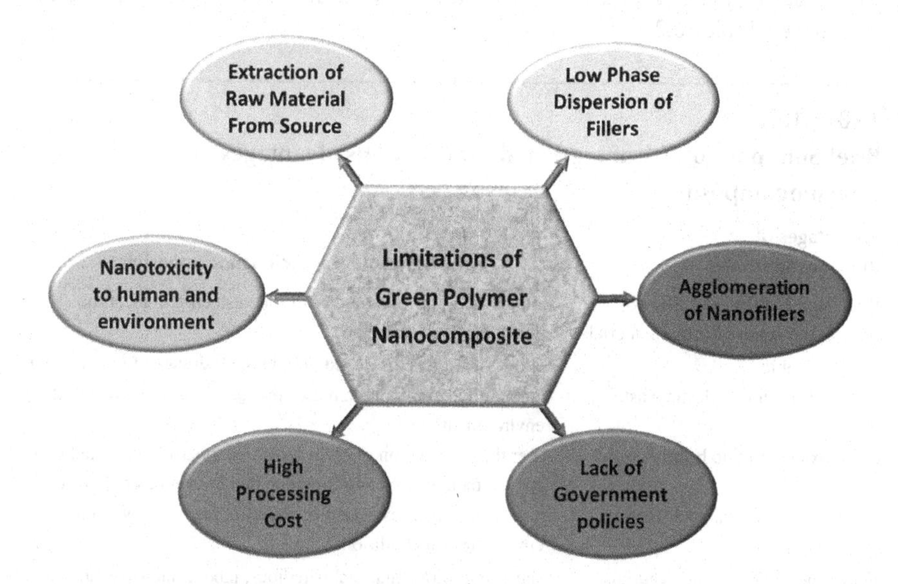

FIGURE 10.1 Limitations of green polymer nanocomposites that hinder the path of its commercialization.

are difficulties with the extraction of raw material from its source. Natural polymers such as starch, cellulose, chitosan etc., owing to their inherent properties, have limitations such as lack of stability, water resistance, solubility in specific solutions, and tedious and labor-intensive processing, which contribute to the increase in the cost of raw materials, thereby affecting the commercial aspects of bionanocomposites (Kumar Thakur and Kumari Thakur 2015, Sun et al. 2018, Lu et al. 2009, Encalada et al. 2018). In the future, on the grounds of there being a high demand for bio-based products, competition between food and cash crops may also be observed. Second, there exist complications with the production of bionanocomposites. In the course of the process, agglomeration of reinforced nanomaterial is seen due to the varying surface properties of polymer matrix (Kumar Thakur and Kumari Thakur 2015). In addition, along with the nature of nanoparticles, the development of an inappropriate interface between fillers and matrix due to improper dispersion of nanoparticle is a major obstacle that needs to be addressed. Likewise, hydrolysis of the material at the time of processing is also an issue that demands the prior modification of raw material so that it can withstand the harsh processing conditions at the time of composite fabrication (Shchipunov 2012, Müller et al. 2017). Third, durability or shelf life of bionanocomposites is also a crucial factor that needs to be considered. Due to the integral properties of these materials, they are reported to be sensitive to environmental changes, specifically heat and moisture, that cause permanent deterioration of the quality of the product (Bari et al. 2019, Jawaid et al. 2018, Cuilliere et al. 2015).

This current chapter provides insights into the challenges associated with the fabrication of green polymer nanocomposites (GNPC) that would govern the market uptake and commercial aspects of the synthesized green nanocomposites. Various techniques that can serve as possible solutions to the allied limitations are also discussed (see Table 10.2).

TABLE 10.2
Brief Summary of Advantages and Contrary Disadvantages of Bionanocomposites

Advantages of Bionanocomposites	Disadvantages of Bionanocomposites
Reduction in the use of petroleum-based non-renewable raw material	Competition between food crop and cash crop to meet the demand of raw materials
Biodegradable	Biodegradation is displayed only in the presence of O_2 and H_2O
Reduce the load of plastic waste	Nanoparticle emission increases the load of nanoparticle in the environment
Zero toxicity and no health hazard	After the degradation of biopolymer, nanoparticles emitted in the environment cause environmental and human health hazard
High thermo-mechanical strength	Change in properties is observed depending on the variation in environmental conditions
Easy availability of raw materials	Processing of raw material is tedious, labour intensive, and adds cost to the process

10.2 GPNCS AND THEIR MARKET

Green polymer nanocomposites are characterized by many intrinsic properties, including biodegradability, non-toxicity, thermo-mechanical resistance etc., which are the key factors behind the market demand for biocomposites. There are numerous areas of use of these eco-friendly composites in consumer products, which widens their range of applications and their market (Bhawani et al. 2018). There are a number of topical examples of success reported which emphasize the industrial applications of green composites in major areas such as biomedical, packaging, the automobile industry, construction, and energy storage. All these applications of green composites have displayed improvement when compared to conventional polymer composites. These improvements can be attributed to their functional performances and their economical traits (Georgios et al. 2016, Fowler et al. 2006). In the past few decades, consumer awareness has led to the acceptance of the wider use of green nanocomposites. Even if these products are more expensive in comparison to pre-existing products, buyers are interested because of the rising demand for sustainability in lifestyles. Consumers are greatly attracted to biodegradable solutions and services because of environmental sensitivity (Sharma et al. 2019). With appropriate improvement, biocomposites have the potential to enter the new-age market with its growing demand for environmentally friendly and sustainable products. Therefore, many old as well as new organizations are putting intense effort into developing a chain of such products to display their ecological credentials and claim a reduction in environmental deterioration (Fowler et al. 2006, Lu and Ozcan 2015).

From a commercial standpoint, conventional corporations as well as various start-ups are putting in hard work to make products economically and environmentally sustainable, along with good profit generation. Taking consumer products into consideration, NEC Corporation, a Japan-based organization, introduced composites made of kenaf fiber and PLA. This was used in many electronic devices and in housing industries in the fiscal year 2005; subsequently, this PLA/kenaf fiber composite was reinforced with flexibilizers, and in 2006 an "eco" mobile phone was introduced into the Japanese market. Following that, in 2010, NEC introduced a computer in which 75% of the parts are made up of plant-based material. In order to make it a lucrative business venture, NEC introduced cellulose-based bioplastic in 2016, targeting the production of electronic devices and other products. Bionanocomposites, due to their intrinsic properties, also find application in the textile and furniture industries; recent advances in this area were displayed in the exhibition BioFurniture in London 2013 (Georgios et al. 2016). Apart from these applications, automobile parts used by top car manufacturing companies are also preferably made up of the bio-based composites, thus opening up the commercialization of green nanocomposites. Moreover, the biomedical industry is developing products such as implants, surgical instruments, nanomedicine, and constituents of drug delivery systems etc. using green nanocomposites, owing to their biocompatibility.

The market for bionanocomposites is differentially categorized based upon its types and applications. On the basis of type, the green composite market is categorized into structural or non-structural. As of 2017, the structural market contributed more than 50% of the total generated revenue. Among the wide range of its applications, structural biocomposites in housing and the construction industry contributed to a larger share because of their exceptional load-bearing capacities. With reference to their applications, bionanocomposites can be classified into automobile, packaging, medical, aerospace, and so forth. The automobile industry is gradually shifting toward green composites, as natural polymers blended with nanoparticles retain considerable toughness and are comparatively cost-effective. Owing to this, green nanocomposites are used preferentially in the automobile industry and this has strengthened the growth of the global market. Moreover, because green nanocomposites with some specific reinforcement hold good rigidity and are easily moldable, their employment in marine engineering is anticipated. In terms of contribution, the market was previously dominated by the application of bionanocomposites in the automobile and biomedical industries, but according to a report published in January 2019 by "Research and Market", biocomposites are expected to observe the fastest growth in the field of aerospace engineering, with an expected annual growth rate of 13.6% by 2023. This is because the production of aircraft is anticipated to increase owing to the surge in air travel and tourism (www.researchandmarkets.com/research/jg447x/global_8_4_bn?w=12).

According to the report published by "Allied Market Research" (AMR), the market for green composites has had an immense growth rate over recent years, and it is expected to have similar growth value in the coming years. The major driving force behind the green composite market is the environmental consciousness that aims to produce sustainable products while reducing the levels of pollution. Moreover, bionanocomposites developed by the amalgamation of natural polymers and nanoparticles can lead to a decline in the dependence on petroleum-based products. This will then unleash considerable growth prospects for the universal market for green nanocomposites (www.marketresearchfuture.com/reports/green-composites-market3992?fbclid=IwAR1mk43WAiOdFrp_dh1LDfZUj36E894ga29hULF1syUa4CV70yyJyZlhQlY)

10.3 STAGES OF DEVELOPMENT

10.3.1 Raw Materials: Advantages and Limitations

Biopolymers or biodegradable polymers are those that can be degraded by the naturally occurring organisms resulting in the production of compounds such as water, biomass, salts, and gases such as carbon dioxide and nitrogen that do not deteriorate the environmental conditions (Rhim et al. 2013). Most of them, namely starch, cellulose, chitosan, PLA, PHA, etc., that are obtained from natural sources such as plants, animals, and microorganisms, are significantly employed for the fabrication of green polymer composites. They confer the improved properties of nanocomposites in the sense that their structural compatibility, functions, and their biodegradability are enhanced. These renewable and natural polymer composite sources are more advantageous when compared to the synthetic sources, notably serving as an ideal solution to environmental concerns about the surge in the burden of plastic waste. Use of biopolymers for the fabrication of nanocomposites not only provides sustainability in development technology but it can also make the economy green (Okamoto 2004). There are a number of challenges associated with the bionanoreinforcement of biopolymers; these limitations, which are associated with individual raw materials being used as nanofillers or matrix in the course of green nanocomposite fabrication, add to the cost of the process, thereby leading to obstacles in the way of commercialization of the final products.

10.3.1.1 Starch

Starch, a semi-crystalline biopolymer, possesses versatile properties that support its use in the synthesis of green nanocomposites (Lu et al. 2009). Starch-based eco-friendly nanocomposite products reveal numerous disadvantages which are mainly accredited to the highly hydrophilic nature of starch. Starch can be used as both filler and matrix. Many starch-based green polymers are currently reported, and researchers have used plasticized starch-based matrix fortified with cellulose nanocrystals and observed that they have profound effects on the thermo-mechanical attributes of the composite (Cao et al. 2008, Kaushik et al. 2010, Mathew et al. 2008). On a similar note, use of starch-based nanoparticles/nanocrystals as fillers has been successfully manifested in biopolymers such as pullulan (Kristo and Biliaderis 2007), PLA (Yu et al. 2008), PVA (Chen et al. 2008), poly(butylene succinate) (PBS) (Lin et al. 2011), soy protein isolate (SPI) (Zheng et al. 2009), and starch (García et al. 2011). Nevertheless, up to now, starch-based eco-friendly polymers have not been widely accepted because of certain limitations. Generally, starch is insoluble in cold water because of the strong hydrogen bonds that hold the chain of starch together. However, the crystalline structure of starch is broken due to the reinforcement of starch in the composites, which facilitates the interaction of water molecules with the hydroxyl group leading to starch granule solubilization. This deliquescent character of starch triggers low water stability, limiting the establishment of starch-based nanocomposites. It generates composites with compromised process ability and mechanical properties (Ribba et al. 2017). With regard to biodegradability, it is crucial to look into the reality; despite the fact that starch-based bionanocomposites are good alternatives for solving environment-related issues, their thermo-mechanical properties are often contrary to their degradability (Encalada et al. 2018).

Taking into consideration the fact that the properties of polymers can be customized by modifying the molecular structure of the polymer, starch can be modified physically, chemically, and enzymatically. It is important to ensure that these physicochemical modifications fall within the ambit of green chemistry and no environment harming chemicals are used. Techniques such as starch modification, incorporation with renewable polymers, and composting are majorly employed for the rectification of the disadvantages associated with starch as a raw material for the development of green nanocomposites. The use of chemically modified starch to serve the purpose of green material synthesis is reported by a number of research groups (Gutiérrez et al. 2015, Bonacucina et al. 2006, Seker and Hanna 2006). Contrarily, various methods of physical modification of starch have been demonstrated. Deep-freezing and thawing is the basic technique employed for the changes in the crystal structure of starch (Szymońska et al. 2003). Extrusion heating (EH) and fluidized bed heating (FBH) can also be used for the modification of crystalline structure (González et al. 2007). Moreover, crystalline structure alteration can be performed by micronization or mechanical activation (Che et al. 2007). Gelatinization transition temperature is also implied to serve the physical modifications via an instantaneous decrement in the pressure (Maache-Rezzoug et al. 2009). Enzymes such as glycogen branching enzyme (SmGBE) and Cyclomaltodextrinase have been used for the enzymatic modification of starch to meet the need for sustainability in the fabrication of starch-based bionanocomposites (Auh et al. 2006). Compatibility of starch-based biopolymers is another crucial issue, which can be resolved by using compatibilizers such as carboxylic acid (Martins and Santana 2016), maleic anhydride and acrylic acid (Huang et al. 2005), Polyolefin elastomer (Zhang et al. 2015) etc., which in general enhance the thermodynamic miscibility. Processing of starch as a biopolymer is more complex and strenuous than of conventional petroleum-based synthetic polymers. Modification in the conventional processing strategies, appropriate use of additives, and vigilantly controlled monitoring of the processing conditions have been employed to tackle the drawbacks associated with the processing of starch-based biopolymers.

10.3.1.2 Cellulose

Cellulose is one of the abundantly available biodegradable polymers; its use as reinforcement material for the fabrication of nanocomposites is an emerging field in nanotechnology (Moon et al. 2011, Lee et al. 2014, Shakeri et al. 2012, Siqueira et al. 2010). However, there are still some drawbacks associated with its employment. First, extraction of cellulose from its natural source is a tedious process and its processing techniques are not that well demonstrated, which implies that more research needs to be done in the field in order to increase it to industrial scale (Oksman et al. 2006). Second, a nano or micro form of cellulose is obtained by either chemical treatment or mechanical processing. Treatment processes include chemical treatment such as alkalization using NaOH solution, acid hydrolysis using H_2SO_4, and $NaClO_2$ bleaching followed by ultrasonic mechanical treatment to transform cellulose fibers into micro or nano size (Kovacs et al. 2010). The production requires a huge amount of energy and water. Moreover, it is a time-consuming process with very low yields. The processing step adds non-essential costs to the production, which in turn affects the commercial facet of the product (Ismail et al. 2018). Third, a cellulose chain has strong hydrogen bonds that restrict the application of cellulose-based nanofibers to the production of water-soluble polymers. During the process, huge amounts of water are also used and wasted because even after the cellulose nanofibers have been obtained via chemo-mechanical treatment, they are stored in water (Dufresne et al. 2000). Cellulose-based polymers have not been substantially used in conventional thermoplastics. Therefore, to widen the range of bionanocomposite applications, it is necessary to tackle the problems and improve their dissipation in solid-phase polymer matrix by surface modification of the nanofibers without degrading their capability of reinforcement. It is reported that the modification of the cellulose-based polymer makes it compatible with bio-based reinforcement (Gousse et al. 2004, Mohanty et al. 2005).

10.3.1.3 Chitosan

Chitosan is a natural polymer that has great potential to be used in the fabrication of green composites. Chitosan has both amine and hydroxyl groups, which make it increasingly more open for the modification of its polyfunctional characteristics such as biocompatibility, non-toxicity and biodegradability. This broadens its application not only in the field of polymer technology, but also in the field of nanotechnology for the development of eco-friendly green polymer nanocomposites (Ahmad et al. 2017). For green polymer composites prepared by either using chitosan as a filler or as reinforcement, it is observed in various studies that it improves the tensile strength of the synthesized composites and it also widens the range of applications of the products generated (Uddin et al. 2012, Chang et al. 2010, Li et al. 2009, Fernandes et al. 2009). Along with the pragmatic characteristics of chitosan that have been discussed, there are some challenges associated with chitosan that present difficulties in the final fabrication of the product.

Chitosan is an inadequately soluble compound; it is only soluble in acidic solutions, which increases the difficulty of the deacetylataion of chitosan during processing (Balázs and Sipos 2007). Since chitosan is obtained in abundant amounts from animal sources which show seasonal inconsistencies in the supply of raw material, this variation leads to differences in the physiochemical characteristics of the final product obtained (Kannan et al. 2010, Croisier and Jérôme 2013). Chitosan, being a natural polymer compound, might have organic and inorganic impurities associated with it, which need to be processed before chitosan is used as a matrix or a filler during the fabrication of green composites. This extra purification step leads to an increase in the processing cost and, therefore, the commercial value of the product.

10.3.1.4 Polyhydroxyalkanoates

PHAs belong to the class of biopolymers that are synthesized intracellularly by microorganisms from varieties of carbon sources and exist as energy storage granules. Plastics synthesized using PHAs are observed to be completely biodegradable and thus help in keeping the environment free of unwanted loads of non-degradable plastics. Major polymers of this class are PHB and its copolymer poly (hydroxybutyrate-co-hydroxyhexanoate) (PHBV) (Shan et al. 2011, Zribi-Maaloul et al. 2013, Yu et al. 2014). These polymers have potential applications in various sectors such as agriculture, medicine, and packaging and have viable commercial possibilities associated with them. Similar to most of the other natural polymers, they also hold some deficiencies such as high gas permeability, low electrical and thermal conductivity, low mechanical properties etc. thereby limiting their range of applications (Díez-Pascual and Díez-Vicente 2014, Peelman et al. 2013, Sridhar et al. 2013). Industrial fabrication of PHA is being practiced nowadays, but its cost is very high since it uses pure microbial cultures. Therefore, the final product is more expensive in comparison to traditional nanocomposites. Consequently, approaches are being made to produce PHA using mixed cultures that make both process and product more economical (Fradinho et al. 2013). In various studies, it is witnessed that this polymer has a high degree of crystallinity, low elongation, low impact strength, and limited processing range. These factors affect the processability of PHA and thus generate a hindrance to making profits from nanocomposites.

10.3.1.5 Poly Lactic Acid

Poly lactic acid (PLA) has attracted the interest of researchers as the alternative to conventional petroleum-based polymers because it displays characteristics such as biodegradability, better process ability, low toxicity, crystallinity, and extraordinary mechanical attributes. PLA is a product of lactic acid derived from natural sources such as corn, sugarcane etc. (Swaroop and Shukla 2018, Sun et al. 2018, Pal et al. 2017, Pinto et al. 2013). Numerous studies have shown that the reinforcement of nanofillers to PLA polymer shows enhancement in its properties, hence widening the range of its application (Huang et al. 2018, Vasile et al. 2017). Though PLA can be considered as a renewable biomaterial possessing extraordinary properties as a polymer, it also holds many evident constraints that can lower the commercial value of the product when employed in certain applications: (1) the time period of its degradation through hydrolysis ranges from months to years, thereby hindering its application in food packaging and the biomedical field; (2) PLA is very brittle in its native form, thus being unsuitable for mechanical applications; (3) the strong hydrophobic nature of PLA limits its use in tissue engineering, as it stimulates the inflammatory response from the host when used as a tissue-engineering material; (4) PLA has inadequate gas barrier properties which restrict the application of PLA in the packaging industry; (5) the cost of PLA is comparatively high, which does not make it feasible for industrial sectors. Nevertheless, various approaches of modification such as integration of other constituents, control monitoring of surface energy and surface charge in PLA have been made to rectify these issues (Hassan et al. 2013, Xiao et al. 2012).

10.3.1.6 Miscellaneous

Lignin is a plant-based biopolymer that is being utilized for the synthesis of bionanocomposites along with cellulose and hemicellulose. Though lignin displays a number of advantages over the other available reinforcements, complexity in the properties of lignin limits its successful utilization. Incompatibility of lignin with polymer matrices and fillers is the major issue that generates hindrances in its commercial utilization. Therefore, considerable hard work is required to modify the surface of lignin in such a way that it becomes compatible with other polymers and fillers (Thakur et al. 2014).

Pectin, an anionic biodegradable polymer generally present in all higher plant cells, and abundantly found in citrus fruits like lemon, oranges etc., has considerable potential as a constituent of eco-friendly nanocomposites. Conversely, extraction of pectin from its source is an incompetent process that leads to the degradation of pectin. Consequently, stabilizers such as EDTA are used to extract pectin from its source sustainably, thereby incurring an additional cost to the final product. In addition, hydrophilicity and brittleness are the major constraints associated with pectin as a raw material for the fabrication of bionanocomposites (Nešić et al. 2018, Kumar et al. 2017). Gorrasi et al. conveyed in their study that in spite of pectin being acknowledged as a biopolymer with sustainable nanocomposite-forming abilities, generated biocomposites exhibit a deprived moisture barrier, low thermal stability, fragile mechanical attributes etc. Modification of the above-mentioned properties is required to enhance the quality of pectin as a raw material. (Gorrasi et al. 2012).

Alginate is a natural polymer chiefly comprising (1,4)-linked β-D-mannuronic acid units and α-L-guluronic acid units (Deepa et al. 2016). Mainly, alginate is present in the cell wall of brown algae, and to serve a commercial purpose it is extracted from seaweed. Along with properties encouraging its employment in the fabrication of bionanocomposites, it exhibits some limitations as a raw material such as high rigidity and fragility with compromised elasticity and mechanical attributes (Wang et al. 2019). It is also observed to have poor water resistance which limits its application, principally in the presence of water and moisture (Abdollahi et al. 2013).

10.3.2 Production of GPNCs

It has been two decades since the term nanocomposite was first used; in due course, numerous techniques, as discussed previously, have evolved for the fabrication of green polymer nanocomposites and for their production on an industrial scale (Mann et al. 2018). One among many problems when processing green nanocomposites is the agglomeration of nanoparticles, which is followed by their inadequate dissipation in the expected formulations. The issue of agglomeration is raised due to the specific surface area and volume effect (Kango et al. 2013). Unfortunately, most of the polymer nanocomposites possess strong interparticle van der Waals interactions, which facilitate agglomeration. Such aggregation of the fillers affects the interaction between the polymer and the fillers, thus leading to deteriorating effects on the properties of the composite (Schaefer and Justice 2007, Ganesan 2008). Micronized clusters of the fillers are created owing to agglomeration. Most of the time, air is trapped inside the aggregates, which, in turn, generates voids in the final product, ultimately leading to a decline in its properties (Alateyah et al. 2013). This problem can be analyzed by using techniques such as a scanning electron microscope (SEM), a transmission electron microscope (TEM), infrared spectroscopy (IS), and X-ray diffraction (XRD), but these techniques are limited to laboratory scale and are difficult to apply in industrial-scale productions.

Manufacturers attempt to achieve homogeneous dispersion of nanoparticles in the polymer matrix of the composites in order to unleash the full prospective thermomechanical characteristics, but still face certain challenges with the dispersion procedures. Further, the interaction between the matrix and the nanoparticles also plays a significant role; the interaction controls the efficiency of stress transfer in polymer nanocomposites. Introduction of functional groups into the fillers leads to the strengthening of the interface between fillers and the polymer matrix (Hirsch 2002). Certainly, it is observed in the studies that an exceptionally high surface area promotes the modification in a macromolecular state of nanoparticles such as crystallinity, composition gradient etc. leading to modification in the overall behavior of the materials (Bugnicourt et al. 2016, Bugnicourt 2005, Azman et al. 2016). Numerous physiochemical strategies have been employed to enhance the dispersion quality. Surface modification is one the major approaches implied to enhance the dispersion by resolving the compatibility of fillers and polymer matrix. With regard to physical methods, along with the use of mechanical mixing methods, ultrasonic vibration has also been applied for the effective dispersion of nanoparticle reinforcement into polymer matrix in both solution and melt polymer forms (Karci et al. 2019, Xia and Wang 2003).

The compatibility between the hydrophilic fibers and hydrophobic polymer matrix has been a challenge associated with the fabrication of nanocomposites over past decades. Incompatibility results in the non-uniform dissemination of fibers within the matrix which in turn, leads to poor mechanical properties of the composite (John and Thomas 2008, Ashori 2008). Numerous steps have been taken to deal with this limitation associated with nanocomposites, including the use of different chemical coupling agents or compatibilizers. Unfortunately, all these measures have failed to produce any significant result (Pandey and Singh 2005, Qu et al. 2010). Currently, major research is being done on the interfacial interaction between the matrix and fibers and also efforts are being made on the molecular structure of matrix and fibers to attain any significant result.

One of the many challenges associated with naturally occurring polymer matrices are that they have poor thermo-mechanical strength. Green materials such as starch generally have poor water resistance and they are highly brittle, with abysmal tensile strengths. To deal with the problem a large amount of compatibilizers are being employed which enhance the properties of these biopolymers to withstand an environment of high pressure and heat (Ke and Sun 2001, Zabihzadeh 2009). Unusually, PLA has recently been considered the most favorable biopolymer to be in use as compared to others. It has great mechanical strength and low toxicity, though some of its properties, such as weak thermal stability and poor ductility, may limit the scope of its application (Jamshidian et al. 2010). Other biopolymers drawn from sources such as cellulose, gelatine, chitosan, and other plant-based oils have significant shortcomings such as insufficiency of sources as well as laborious and expensive fabrication processes.

Regarding the fabrication processes and desired dimensions for nanocomposites, they must be produced between a wide range of approximately 1 nm in height and 500 nm in width. Moreover, the acclimation of nanoparticles are also significantly associated with the enhancement of the thermo-mechanical properties of composite materials. Furthermore, aggregation of the particles, which leads to the formation of lumps, is a processing failure that needs to be taken into consideration while implementing the preventive measures (Tunç and Duman 2011, Kaushik et al. 2010). New approaches toward exploring and evolving the techniques of nanocomposite fabrication throughout product development, from raw material source selection, manufacturing, to process technology are now well manifested. Considerable effort has been made in obtaining an enhanced performance composite reinforced with biodegradable fillers. Forthwith, there are numerous perspectives and strategies that are being used for the fabrication of ideally optimal nanocomposites. They include (a) the melt extrusion technique – in which the speed of extrusion plays a critical role in governing the thermo-mechanical property of the composites (Majdzadeh-Ardakani et al. 2010, Guan and Hanna 2006); (b) the film stacking method – in which a stack of polymer film and fibers is compressed which prominently influences the tensile strength of composites (Bodros et al. 2007); (c) extrusion followed by an injection-molding approach for the fabrication of nanocomposites giving technically prime manifestation of fibers in the matrix (Averous and Boquillon 2004, Averous and Le Digabel 2006) etc. Many other preparation strategies such as melt compounding followed by compression molding (Lei et al. 2007), direct melting (Huskić and Žigon 2007), injection molding, and gelatinization before casting (Liu et al. 2010) have also been tested. These methods exhibit advantages such as easy handling and improved performance of the composites. All these systems modify functional groups in such a way that the interconnection between matrix polymers and nanofillers is strengthened (Zou et al. 2008, Wei et al. 2010). Along with improvement in the nanocomposite performance, sustainability toward the environment is a major concern which needs to be addressed throughout the product development trajectory.

10.3.3 Shelf-Life of GPNCs

Nanocomposites can undergo a variety of environmental effects which can retard their mechanical properties and consequently limit their shelf life (Jayaram, n.d.). Although previous research work has shown that incorporation of nanomaterial considerably increases the physical, thermo-mechanical properties of the composites, there are associated problems, particularly with the natural polymers that are employed in the manufacture of such composites (Njuguna et al. 2008, Wu et al. 2016, Dittenber et al. 2012, Eshraghi et al. 2013, Debelak and Lafdi 2007, Yasmin and Daniel 2004). The polymer matrices are prone to degradation by moisture, temperature, UV light, and microbial attack, which restricts the durability of the composites. For example, the shelf life of cellulose-fiber-reinforced composites was found to be reduced in environments of high alkalinity (Marouani et al. 2012). The degradation depends upon chemical composition and the inherent properties of the natural fibers (John and Thomas 2008).

Natural polymers such as hemicellulose have a major role in water absorption and consequent biodegradation. This is due to the presence of voids in the polymer structure. Moisture absorption can lead to the swelling of the composite, degradation of the polymer surface owing to corrosion by water movement, and also lead to decreased interfacial bonding in the composite, thus attenuating its mechanical properties (Nkurunziza et al. 2005).

Temperature usually has two kinds of effect on polymers – short-term, where the polymer undergoes temporary physical changes that revert on returning to the original temperature and long-term, where the polymer experiences permanent chemical change. While certain polymers have been reported to decompose at very high temperatures of around 300 to 500 °C they are observed to soften or distort at lower temperatures of 100 to 200 °C (Mouritz and Gibson 2007). Recent studies suggest that increased temperatures can attenuate mechanical properties of the polymer such as compression strength, tensile strength etc. (Reis 2012, Zhang et al. 2016, Ou et al. 2016). Moreover, composites containing such polymers are extremely susceptible to catching fire. The property of flammability thus possesses a great risk to the shelf life of such composites. Stringent safety regulations must be passed and proper flame-retardant treatment must be performed on the composites to ensure public safety as well as to enhance their shelf life (Das et al. 2019).

Some natural polymers such as lignin are highly susceptible to photo-degradation by UV light. Since absorption of UV light by lignin leads to the formation of chromospheres, yellowing of the surface is a common phenomenon. The covalent bonds in the polymer are weakened, leading to deterioration in the mechanical properties of the composite-enabled product (Azwa et al. 2013).

Natural polymers are also vulnerable to microbial attack as the microbes secrete exo-enzymes that have the ability to degrade them into monomeric units or smaller fragments. For example, cellulose in a composite is highly susceptible to fungal attack under moist conditions (Catto et al. 2016, Naumann et al. 2012). Bacterial degradation of the polymers usually ends up in erosion, tunneling, and cavity formation in the composite (Blanchette et al. 1990).

To summarize, the composites may be subjected to structural damage due to factors such as moisture, thermal stress, fire, microbes, fatigue etc. The damage impact in a composite is often manifested as internal delamination. Moreover, owing to the fact that the composite is non-ductile and the damage often begins at a non-impacted surface unlike a metal, damage in the composite is generally detected at a later stage (Al-Mosawi and Hatif 2012). The defects in composites, including porosity, voids, cracks etc. usually occur at the structure interfaces (Vaara and Leinonen 2012).

10.3.4 Applications

The packing material chosen for the packing of food products should ensure that the food quality is maintained during transport and storage under appropriate physical and chemical conditions. It should also enhance the shelf life of the food materials (Yam et al. 2005, Kumar et al. 2017). Bio-based polymer nanocomposites with certain properties have been shown to be potential packing material in the food packaging industry (Souza et al. 2018, Rhim et al. 2013). However, prior to their commercialization, the durability, strength, and degree of the migration of nanoparticles from the packing material to the food item should be examined properly. The bio-based packing material must retain its stability and should function appropriately from storage until its disposal (Ramos et al. 2018, Youssef and El-Sayed 2018). Duncan and Pillai demonstrated in their study that the rigorous use and abuse of nanocomposites leads to the release of filler materials that might, in turn, have detrimental effects on environmental and human health (Peelman et al. 2013, Duncan and Pillai 2014, Huang et al. 2015, Honarvar et al. 2016). As of now, very limited studies have been carried out in the field, but findings assure us that the migration of nanoparticles from packing material into food has a very limited effect on human health as well as on the environment (Othman and Procedia 2014, Silvestre et al. 2011). However, due to other challenges, such as the intrinsic hydrophilic property of the polymer used in nanocomposites, the material displays major drawbacks such as brittleness in wet conditions and the poor nature of humidity barriers. Many efforts, including physical, chemical, and enzymatic treatment, have been made to minimize this effect (Wihodo and Moraru 2013, Guerrero et al. 2011). Still, the compounds that are used as reinforcement may have some cytotoxicity that limits their use in food packaging. Therefore, other viable methods for the enhancement of properties of nanocomposites need to be employed.

Green nanocomposites have very high biocompatibility and are degraded mainly through hydrolysis or processes mediated by metabolism. Moreover, there is dimensional similarity between nanostructured material and components of natural tissues, because of which the composites can be potential sustainable alternatives to the current biomedical technologies (Dubey et al. 2019). Various materials have been reported to be employed in orthopedic procedures (Daramola et al. 2019). Among them, hydroxyapatite (HAP) polymer nanocomposites have been used as osteoconductive biocompatible agents for bone repair and replacement. However, HAP powder migrates from the implant site to other parts of the body which is not acceptable from a safety viewpoint (Bianco et al. 2011, Armentano et al. 2018, Jahangirian et al. 2018). Moreover, green composites have crucial morphologies that support regulated growth and migration of cells that can successively direct tissue regeneration. Cellulose-based nanowhiskers (CNWs), starch-based nanocomposites etc. have been shown to be promising candidates but lack of rigidity adds complications to the processing of these materials. The production range is narrowed down and consequently, the commercialization of this product is hampered (Gomes et al. 2002, Jahangirian et al. 2018, Zakaria et al. 2017). On a similar note, silk-based polymers are also being extensively explored for tissue engineering as they possess excellent biological and mechanical strength, biodegradability, as well as biocompatibility (Wang et al. 2015). Stability of the composites in formulation is also an aspect that needs to be studied well before moving on to the commercialization of the product. Further, the role of bionanocomposites of carrageenan has been explored in drug delivery systems. Carrageenan possesses surface charge and has good adhesiveness that offers the advantage of prolonging drug release (Kianfar et al. 2013, Ghanam and Kleinebudde 2011, Kalsoom Khan et al. 2017). Mahadavinia and coworkers reported in their study that drug release from carrageenan-based bionanocomposites is dependent upon the pH and magnetic field of the environment where the drug has be delivered (Mahdavinia et al. 2014). Due to the limitation of specific pH and magnetic field for the optimal performance of the formulation, it will cover a very small fraction of the market. Therefore, before commercializing a product, more research is required to widen the range of its applications.

The automobile industry is a material-assisted industry and therefore the current global urgency of environmental degradation has required the employment of green polymer composites. Use of nanocomposites in the automotive industry has considerably increased because of the numerous advantages they offer, such as light weight, extraordinary thermo-mechanical characteristics, durability, high resistance to corrosion, and comparatively low cost (Naskar et al. 2016, Ramli et al. 2018, Koronis and Silva 2018). Currently most of the top automobile industries are also taking an interest in universal sustainability. Major car manufacturers, for instance, BMW, Audi, Ford, Volvo, Mercedes, Toyota, and Volkswagen, are using bio-based composites instead of traditional composites for manufacturing parts such as air filter boxes and interior trim parts, dashboards, door panels, cabin linings etc. (Koronis et al. 2013, Arjmandi et al. 2017). Although these materials have advantages with respect to their mechanical properties and environmental issues, there are some challenges that need to be analyzed before the commercialization of such products takes place. These challenges include the durability of the product, the final cost of the product, and its shelf life. In many cases, use of wood fibers has been suggested. This may look fascinating initially, but with increasing demand for such fibers, a long-term impact on natural vegetation can be expected. Consequently, the cost of the product will also increase and would create a hindrance in the commercial facets of the product (Koronis and Silva 2018).

10.3.5 Environmental Effects and Nano-Toxicology

It is crucial to understand the environmental effects of green polymer nanocomposites and their possible toxicological scenarios in order to unravel the potential risks of commercial products that are fabricated from them. The green polymers that are employed in these composites are often natural polymers. Derived from sources occurring in nature itself, these polymers are environmentally friendly and are ultimately subjected to metabolic degradation (Aravamudhan et al. 2014). Moreover, owing to their similarity with components of the extracellular matrix, they are unlikely to cause toxicity or adverse immunological reactions in humans (Mano et al. 2007). However, there might be some toxicity associated with green polymers that are not naturally occurring. For example, poly-lactic acid (PLA), a man-made polymer synthesized using green chemistry methods, although non-toxic in solid form, can have a potential for toxicity in the molten or vaporized forms (www.ampolymer.com/SDS/PolylacticAcidSDS.html). A tool known as "Green Toxicology" can predict the potential toxicity associated with such green materials, utilizing omics, in-silico, and in-vitro methods (Zhang and Cue 2012). Green Toxicology may also be employed to assess the environmental safety of such green products, although there has not yet been any considerable progress in this regard (Crawford et al. 2017).

However, the primary toxicity potential and the environmental effects of green polymer nanocomposites, in effect, do not arise from the polymer, but from the constituent nanomaterials that may be released into the surrounding media during the use or abuse of the product (nanocomposite). Nanomaterials can be released into the external milieu by two paradigms (Duncan and Pillai 2014, Duncan 2014). The first release mechanism includes diffusion, desorption, or dissolution of the nanomaterials into an external liquid media and does not involve the alteration of the polymer matrix integrity. This mechanism is pertinent to human exposure from nanocomposite-based products that experience liquid environments for considerable periods of time, such as medical equipment and food packaging. The second mechanism entails the degradation of the host matrix (polymer) for the release of the engineered nanomaterials from the composite. The matrix may be degraded either mechanically, photo-chemically, thermally, or hydrolytically. In the case of mechanical abrasion, the nanomaterials are released when the degradation products are aerosolized. These released nanomaterials thus pose occupational hazards via inhalation or have environmental effects via processes such as sedimentation. The other types of matrix degradation either directly release the nanomaterials by exposing them from the core of the matrix to its surface or indirectly via weakened diffusion properties. The mode of degradation of host matrix usually depends on the arena where the nanocomposite-based commercial product is used. For example, hydrolytic degradation becomes relevant for nanocomposites that end up in landfill, whereas photodegradation is pertinent for construction-site-based products that experience outdoor environments (see Figure 10.2).

Once the nanomaterials are released into the environment, they assemble in different environmental media, namely soil, water, air (Iavicoli et al. 2014). However, owing to the fact that there are many unknown factors regarding the released nanomaterials (environmental concentrations, accurate identification, ineffective limits of detection etc.), their environmental impact cannot be completely unraveled (Kabir et al. 2018). Even a slight chemical change can radically convert an otherwise non-toxic nanomaterial into a toxic substance. However, a few studies have attempted

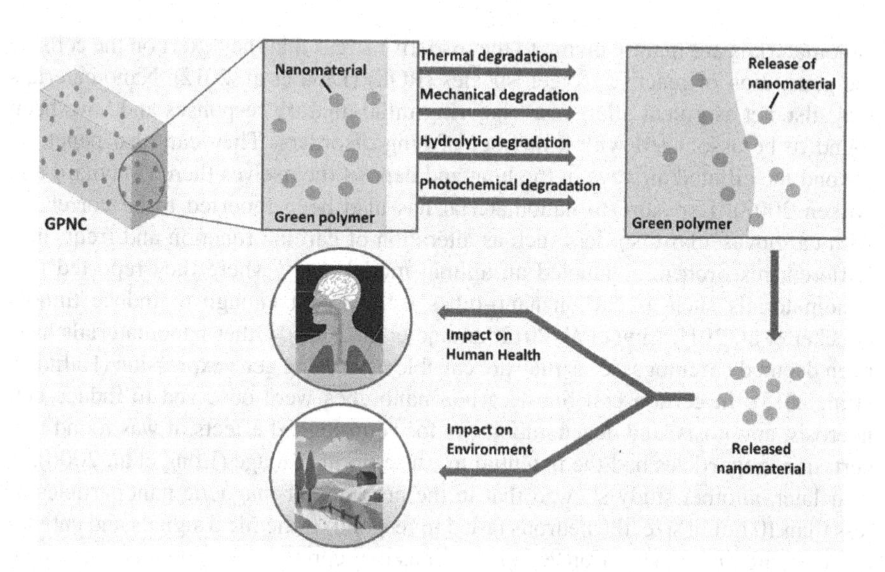

FIGURE 10.2 Migration of nanomaterial from polymer nanocomposites after degradation and its impact on human and environmental health.

to delineate the effect of some nanomaterials in different environmental matrices. For nanomaterials released into the soil, it was observed that plants were capable of translocating them from the soil. This, in turn, affected plant growth and germination via mechanisms such as gene up-regulation, induction of the plant defense system, genotoxic effects, DNA damage etc. (Ghosh et al. 2010, Wang et al. 2017, Hong et al. 2014, Khodakovskaya et al. 2009). Moreover, the distribution of the nanoparticles leading to their accumulation in the leaves presented a possibility of trophic transfer through the food chain. Apart from plants, nanomaterials in the soil were also found to exert adverse effects on the microbiome (Ge et al. 2011). Release of nanomaterials into the aquatic media had toxic impacts on the organisms residing in that environment. A host of effects were reported, including oxidative damage, mortality, growth reduction etc. (Baker et al. 2014, Grillo et al. 2015, Rocha et al. 2015). Cytotoxic effects were not only reported for marine animals such as *Daphnia magna* (Gao et al. 2018) and freshwater crab (Khalid et al. 2016), but also for algae (Li et al. 2017). Nanomaterials released into the air were also reported to have detrimental effects, as they have a pivotal role in dust-cloud formation (Turkevich et al. 2015).

Released nanomaterial from green polymer nanocomposites have been concluded to have adverse effects on human beings depending on the toxicity of the bulk substance from where they are synthesized. Nanomaterials may be hazardous because they can easily penetrate the cell membranes and enter the cells owing to their small size and shape. As a result, they can either destroy organelle integrity or impair other cellular functions (Bystrzejewska-Piotrowska et al. 2009). Nanomaterials may also enter cells via endocytosis or phagocytosis (Mallapragada et al. 2015). Humans may be exposed to nanomaterial release via oral, dermal, inhalation, and/or gastrointestinal routes while using commercial products based on such composites (Ma et al. 2015, Mackevica and Foss Hansen 2016). The antagonistic effects exhibited by nanomaterials are mainly owing to the oxidative stress that they exert on the cells by the generation of reactive oxygen species (ROS) (Liou et al. 2012). Nanomaterials may also act as potent allergens, triggering inflammatory responses and have been found to be associated with a number of lung disorders. They can also penetrate beyond the ciliated airways in the lung and deposit themselves there (Maynard and Aitken 2007). Exposure to nanomaterial has also been reported to be correlated with cardiovascular disorders such as alteration of cardiac function and frequency. Different researchers conducted an animal model study where they reported that nanomaterials such as carbon nano-tubes were potent enough to induce tumors (Becker et al. 2011, Liou et al. 2012, Araujo et al. 2009). Other nanomaterials have been deemed carcinogenic, as they are capable of altering gene expression (Haliullin et al. 2015). In certain cell lines, carbon nanotubes were observed to induce cell necrosis, apoptosis, and degeneration. As for neurological effects, it was found that certain nanoparticles had the potential to cause brain damage (Long et al. 2006). A year later, another study showed that in the presence of magnetic nanoparticles of less than 10 nm in size, the neurons failed to respond to chemical signals and entered a latent state instead (Johnson 2007). As far as the reproductive system is concerned, a study reported that nanoparticles had the ability to traverse the blood testes barrier and suggested it as a potent source of toxicity to sperm cells (McAuliffe and Perry 2007). Therefore, the release of nanomaterials from the composites can pose a great threat to both environmental and human health, thus marking a great limitation in its commercialization.

10.3.6 Limitations in Commercialization of GPNC

Apart from the methodological improvements required at the various junctures of fabricating green polymer nanocomposites, there are many areas in the field where upgradation is needed. Until now, most of green nanotechnology research has been limited to the laboratory scale or in certain cases, to the pilot scale. Among those which are commercialized, their area of application is very narrow, and needs to be widened. Initiatives are needed to study the probability of existing technologies becoming accustomed to commercial-scale production (Mann et al. 2018, Bari et al. 2016).

All of the challenges associated with bionanocomposites can be categorized as technical, economic, or environmental. Technical challenges to the commercial acceptance of green composites include the physical limitations of raw material and restriction in its processing. In particular, the lack of processability of the raw material is a significant hindrance to product commercialization. Quality and uniformity of the resources are crucial hurdles to the commercial utilization of green nanocomposites (Kumar Thakur and Kumari Thakur 2015, Sharma et al. 2019). Moreover, reliability of the supply of resources is a major issue, which needs to be considered. Green polymer nanocomposites have a diversified field of application (Bari et al. 2016). However, development and application of green polymer nanocomposites are two entirely different aspects. Collective efforts from all disciplines such as physics, chemistry, materials science, biomedicine, engineering, and several others are required to make the process effective and sustainable at the same time. This combined approach will be helpful in mobilizing the area ahead in a way that stimulates positive feedback between the disciplines and in turn will encourage the development and commercialization of new product, claiming application and a market in novel areas.

Since the market for bionanocomposites is based on their environmental credentials, it is crucial to get it authenticated by life cycle assessment (LCA) (as discussed in the previous chapter of this book) or other dependable methods. In addition to this, the standards that are used for the assessment of materials display slight positive bias toward pre-existing products (Hervy et al. 2015, Luzi et al. 2019). Accreditation of a new product is a costly process which therefore presents a hurdle for the adoption of these new materials. This paves the way for the requirement of intensive research in this new technological field to mitigate these hindrances and make commercialization of sustainable bionanocomposites a reality. Despite the great interest in the application of bionanocomposites, their testing and verification for biodegradability and compostability were issues when the field was in its early phase. Researchers should form an alliance with private sectors for the fruitful commercialization of green nanotechnology. Moreover, government should provide financial aid to further accelerate the process of commercialization (Matus et al. 2012). Currently, many governments and independent organizations are taking an interest in verifying the biodegradability of the final product and providing the certification that allows corporations to claim environmental credibility. Still, there is scope for improvement in the field, Lack of government interest, deficiency of government policies, and less participation by governmental organizations make the field dull, in turn affecting the path of commercialization of such products.

In the last few years it has been observed that the implementation of environmental protection guidelines have had crucial control over petroleum-based industries. Almost all the rules and regulations place stress on maintaining sustainability during the process of nanocomposite fabrication to maintain the health of environment (Lu and Ozcan 2015). Consequently many organizations have to use their resources accordingly, and have to take green material into consideration, and this is where bionanocomposites come into the picture. Similarly, for effective commercialization of green nanotechnology, the government should introduce appropriate policies and regulations.

10.4 CONCLUDING REMARKS

Green polymer nanocomposite has been found to have similar structural properties to conventional nanocomposites. The surge in the importance of GPNCs is due to the decline in non-renewable sources and an increase in environmental pollution. There are promising prospects in the field of green polymer nanocomposites. Bionanocomposites need to be fabricated with enhanced mechanical, biocompatible, thermal, and biodegradable properties. Furthermore, they display other functionalities such as resistance to microbes, energy storage capacity, aerospace engineering etc. Green polymer nanocomposites find their way into daily life, despite the fact that most of the materials used consist of metals as well as petroleum-based synthetic plastics and composites.

This chapter aimed to provide a comprehensive summary of the challenges associated with the bionanocomposites which affect its commercial aspects. Commercialization is not yet successful owing to factors such as attenuated adhesion properties between the nanomaterial and matrix, poor compatibility, ineffective distribution of nanomaterial in the matrix etc. In addition to this, production cost is one of the principle obstacles to the industrial use of bionanocomposites. Moreover, the final product faces the problem of the migration of nanofillers from the matrix, which exerts harmful effects on the environment as well as on human health. Likewise, the bionanocomposites behave differentially when exposed to the varying environmental conditions that affect the durability of the composite material; in turn shelf life is compromised, and this hinders the commercial path of biodegradable polymer nanocomposite products. Based upon the wide expansion of both fundamental understanding and industrial innovation, recent advances and future avenues hopefully assure the continuous growth and development of bio-based polymer nanocomposites materials research, thus rectifying the challenges associated with the commercialization of green polymer nanocomposites.

ACKNOWLEDGMENTS

DKT and MKM acknowledge the MHRD fellowship from IIT-Roorkee for pursuing a PhD. KMP acknowledges the research support of DBT-IYBA fellowship, research grants from NMCG-MoWR, SERB-DST, and MHRD-IITR from Govt. of India, and BEST Pvt. Ltd.

REFERENCES

Abdollahi, M., M. Alboofetileh, R. Behrooz, M. Rezaei, and R. Miraki. 2013. Reducing Water Sensitivity of Alginate Bio-Nanocomposite Film Using Cellulose Nanoparticles. *International Journal of Biological Macromolecules* 54: 166–73.

Adeosun, S. O., G. Lawal, S. A. Balogun, and E. I. Akpan. 2012. Review of Green Polymer Nanocomposites. *Journal of Minerals and Materials Characterization and Engineering* 11 (04): 385.

Ahmad, M., K. Manzoor, S. Singh, and S. Ikram. 2017. Chitosan Centered Bio-nanocomposites for Medical Specialty and Curative Applications: A Review. *International Journal of Pharmaceutics* 529 (1–2): 200–17.

Ahmed, S., and S. Kanchi. 2018. *Handbook of Bio-nanocomposites*. Pan Stanford.

Al-Mosawi, A. I., and A. H. Hatif. 2012. Reinforcing by Palms-Kevlar Hybrid Fibers and Effected on Mechanical Properties of Polymeric Composite Material. *Journal of University of Babylon* 20 (1): 188–93.

Alateyah, A., H. Dhakal, and Z. Zhang. 2013. Processing, Properties, and Applications of Polymer Nanocomposites Based on Layer Silicates: A Review. *Advances in Polymer Technology* 32(4): 36.

Araujo, J. A., and A. E. Nel. 2009. Particulate Matter and Atherosclerosis: Role of Particle Size, Composition and Oxidative Stress. *Particle and Fibre Toxicology* 6 (1): 24.

Aravamudhan, A., D. M. Ramos, A. A. Nada, and S. G. Kumbar. 2014. Natural Polymers: Polysaccharides and Their Derivatives for Biomedical Applications. *Natural and Synthetic Biomedical Polymers*, 67–89: Elsevier.

Arjmandi, R., A. Hassan, and Z. Zakaria. 2017. Polylactic Acid Green Nanocomposites for Automotive Applications. *Green Biocomposites*, 193–208: Springer.

Armentano, I., D. Puglia, F. Luzi, C. Arciola, F. Morena, S. Martino, and L. Torre. 2018. Nanocomposites Based on Biodegradable Polymers. *Materials* 11 (5): 795.

Ashori, A. J. 2008. Wood–Plastic Composites as Promising Green-Composites for Automotive Industries! *Bioresource Technology* 99 (11): 4661–67.

Auh, J.-H., H. Y. Chae, Y.-R. Kim, K.-H. Shim, S.-H. Yoo, and K.-H. Park. 2006. Modification of Rice Starch by Selective Degradation of Amylose Using Alkalophilic Bacillus Cyclomaltodextrinase. *Journal of Agricultural and Food Chemistry* 54 (6): 2314–19.

Averous, L., and N. Boquillon. 2004. Biocomposites Based on Plasticized Starch: Thermal and Mechanical Behaviours. *Carbohydrate Polymers* 56 (2): 111–22.

Averous, L., and F. Le Digabel. 2006. Properties of Biocomposites Based on Lignocellulosic Fillers. *Carbohydrate Polymers* 66 (4): 480–93.

Azeredo, H. M., L. H. C. Mattoso, D. Wood, T. G. Williams, R. J. Avena-Bustillos, and T. H. McHugh. 2009. Nanocomposite Edible Films from Mango Puree Reinforced with Cellulose Nanofibers. *Journal of Food Science* 74 (5): N31–N35.

Azman, I. A. B., R. B. M. Salleh, S. B. M. Alauddin, and M. I. B. Shueb. 2016. Dispersion Techniques of Nanoparticles. *International Journal of Applied Chemistry* 12 (1): 51–56.

Azwa, Z., B. Yousif, A. Manalo, and W. Karunasena. 2013. A Review on the Degradability of Polymeric Composites Based on Natural Fibres. *Materials & Design* 47: 424–42.

Bajpai, P. K., I. Singh, and J. Madaan. 2014. Development and Characterization of Pla-Based Green Composites: A Review. *Journal of Thermoplastic Composite Materials* 27 (1): 52–81.

Baker, T. J., C. R. Tyler, and T. S. Galloway. 2014. Impacts of Metal and Metal Oxide Nanoparticles on Marine Organisms. *Environmental Pollution* 186: 257–71.

Balázs, N., and P. Sipos. 2007. Limitations of Ph-Potentiometric Titration for the Determination of the Degree of Deacetylation of Chitosan. *Carbohydrate Research* 342 (1): 124–30.

Bari, E., J. J. Morrell, and A. Sistani. 2019. Durability of Natural/Synthetic/Biomass Fiber–Based Polymeric Composites: Laboratory and Field Tests. *Durability and Life Prediction in Biocomposites, Fibre-Reinforced Composites and Hybrid Composites*, 15–26: Elsevier.

Bari, S. S., A. Chatterjee, and S. Mishra. 2016. Biodegradable Polymer Nanocomposites: An Overview. *Polymer Reviews* 56 (2): 287–328.

Becker, H., F. Herzberg, A. Schulte, and M. Kolossa-Gehring. 2011. The Carcinogenic Potential of Nanomaterials, Their Release from Products and Options for Regulating Them. *International Journal of Hygiene and Environmental Health* 214 (3): 231–38.

Bhawani, S., A. Bhat, F. Ahmad, and M. Ibrahim. 2018. Green Polymer Nanocomposites and Their Environmental Applications. *Polymer-Based Nanocomposites for Energy and Environmental Applications*, 617–33: Elsevier.

Bianco, A., B. M. Bozzo, C. Del Gaudio, I. Cacciotti, I. Armentano, M. Dottori, F. D'Angelo, *et al.* 2011. Poly (L-Lactic Acid)/Calcium-Deficient Nanohydroxyapatite Electrospun Mats for Bone Marrow Stem Cell Cultures. *Journal of Bioactive and Compatible Polymers* 26 (3): 225–41.

Bilbao-Sáinz, C., R. J. Avena-Bustillos, D. F. Wood, T. G. Williams, and T. H. McHugh. 2010. Composite Edible Films Based on Hydroxypropyl Methylcellulose Reinforced with Microcrystalline Cellulose Nanoparticles. *Journal of Agricultural and Food Chemistry* 58 (6): 3753–60.

Blanchette, R. A., T. Nilsson, G. Daniel, and A. Abad. 1990. *Biological Degradation of Wood*. ACS Publications.

Bodros, E., I. Pillin, N. Montrelay, and C. Baley2007. Could Biopolymers Reinforced by Randomly Scattered Flax Fibre Be Used in Structural Applications? *Composites Science and Technology* 67 (3–4): 462–70.

Bonacucina, G., P. Di Martino, M. Piombetti, A. Colombo, F. Roversi, and G. F. Palmieri. 2006. Effect of Plasticizers on Properties of Pregelatinised Starch Acetate (Amprac 01) Free Films. *International Journal of Pharmaceutics* 313 (1–2): 72–77.

Bugnicourt, E. Development of Sub-Micro Structured Composites Based on an Epoxy Matrix and Pyrogenic Silica: Mechanical Behavior Related to the Interactions and Morphology at Multi-Scale. Lyon, INSA, 2005.

Bugnicourt, E., T. Kehoe, M. Latorre, C. Serrano, S. Philippe, and M. Schmid. 2016. Recent Prospects in the Inline Monitoring of Nanocomposites and Nanocoatings by Optical Technologies. *Nanomaterials* 6 (8): 150.

Bystrzejewska-Piotrowska, G., J. Golimowski, and P. L. Urban. 2009. Nanoparticles: Their Potential Toxicity, Waste and Environmental Management. *Waste Management* 29 (9): 2587–95.

Cao, X., Y. Chen, P. Chang, A. Muir, and G. Falk. 2008. Starch-Based Nanocomposites Reinforced with Flax Cellulose Nanocrystals. *Express Polymer Letters* 2 (7): 502–10.

Catto, A. L., L. S. Montagna, S. H. Almeida, R. M. Silveira, and R. M. Santana. 2016. Wood Plastic Composites Weathering: Effects of Compatibilization on Biodegradation in Soil and Fungal Decay. *International Biodeterioration & Biodegradation* 109: 11–22.

Chang, P. R., R. Jian, J. Yu, and X. Ma. 2010. Fabrication and Characterisation of Chitosan Nanoparticles/Plasticised-Starch Composites. *Food Chemistry* 120 (3): 736–40.

Che, L.-M., D. Li, L.-J. Wang, X. Dong Chen, and Z.-H. Mao. 2007. Micronization and Hydrophobic Modification of Cassava Starch. *International Journal of Food Properties* 10 (3): 527–36.

Chen, Y., X. Cao, P. R. Chang, and M. A. Huneault. 2008. Comparative Study on the Films of Poly (Vinyl Alcohol)/Pea Starch Nanocrystals and Poly (Vinyl Alcohol)/Native Pea Starch. *Carbohydrate Polymers* 73 (1): 8–17.

Crawford, S. E., T. Hartung, H. Hollert, B. Mathes, B. van Ravenzwaay, T. Steger-Hartmann, C. Studer, and H. F. Krug. 2017. Green Toxicology: A Strategy for Sustainable Chemical and Material Development. *Environmental Sciences Europe* 29 (1): 16.

Croisier, F., and C. Jérôme. 2013. Chitosan-Based Biomaterials for Tissue Engineering. *European Polymer Journal* 49 (4): 780–92.

Cuilliere, J.-C., L. Toubal, K. Bensalem, V. François, and P.-B. Gning. 2015. *Water Uptake and Its Effect on Biocomposites*. SPE PRO.

Daramola, O. O., J. L. Olajide, S. C. Agwuncha, M. J. Mochane, and E. R. Sadiku. 2019. Nanostructured Green Biopolymer Composites for Orthopedic Application. *Green Biopolymers and Their Nanocomposites*, 159–90: Springer.

Das, O., N. Kim, and D. Bhattacharyya. 2017. The Mechanics of Biocomposites. *Biomedical Composites*, 375–411: Woodhead Publishing .

Das, O., N. K. Kim, M. S. Hedenqvist, and D. Bhattacharyya. 2019. The Flammability of Biocomposites. *Durability and Life Prediction in Biocomposites, Fibre-Reinforced Composites and Hybrid Composites*, 335–65: Elsevier.

de Moura, M. R., F. A. Aouada, R. J. Avena-Bustillos, T. H. McHugh, J. M. Krochta, and L. H. Mattoso. 2009. Improved Barrier and Mechanical Properties of Novel Hydroxypropyl Methylcellulose Edible Films with Chitosan/Tripolyphosphate Nanoparticles. *Journal of Food Engineering* 92 (4): 448–53.

Debelak, B., and K. Lafdi. 2007. Use of Exfoliated Graphite Filler to Enhance Polymer Physical Properties. *Carbon* 45 (9): 1727–34.

Deepa, B., E. Abraham, L. Pothan, N. Cordeiro, M. Faria, and S. Thomas. 2016. Biodegradable Nanocomposite Films Based on Sodium Alginate and Cellulose Nanofibrils. *Materials* 9 (1): 50.

Díez-Pascual, A., and A. Díez-Vicente. 2014. Poly (3-Hydroxybutyrate)/Zno Bio-nanocomposites with Improved Mechanical, Barrier and Antibacterial Properties. *International Journal of Molecular Sciences* 15 (6): 10950–73.

Dittenber, D. B., and H. V. GangaRao. 2012. Critical Review of Recent Publications on Use of Natural Composites in Infrastructure. *Composites Part A: Applied Science and Manufacturing* 43 (8): 1419–29.

Dong, P., R. Prasanth, F. Xu, X. Wang, B. Li, and R. Shankar. 2015. Eco-Friendly Polymer Nanocomposite—Properties and Processing. *Eco-Friendly Polymer Nanocomposites*, 1–15: Springer.

Dubey, N., C. S. Kushwaha, and S. Shukla. 2019. A Review on Electrically Conducting Polymer Bio-nanocomposites for Biomedical and Other Applications. *International Journal of Polymeric Materials and Polymeric Biomaterials* 1–19.

Dufresne, A., D. Dupeyre, and M. R. Vignon. 2000. Cellulose Microfibrils from Potato Tuber Cells: Processing and Characterization of Starch–Cellulose Microfibril Composites. *Journal of Applied Polymer Science* 76 (14): 2080–92.

Duncan, T. V., and K. Pillai. 2014. Release of Engineered Nanomaterials from Polymer Nanocomposites: Diffusion, Dissolution, and Desorption. *ACS Applied materials & Interfaces* 7 (1): 2–19.

Duncan, T. V. 2014. Release of Engineered Nanomaterials from Polymer Nanocomposites: The Effect of Matrix Degradation. *ACS Applied Materials & Interfaces* 7 (1): 20–39.

Encalada, K., M. B. Aldás, E. Proaño, and V. Valle. 2018. An Overview of Starch-Based Biopolymers and Their Biodegradability. *Ciencia e Ingeniería* 39 (3): 245–58.

Eshraghi, A., H. Khademieslam, I. Ghasemi, and M. Talaiepoor. 2013. Effect of Weathering on the Properties of Hybrid Composite Based on Polyethylene, Woodflour, and Nanoclay. *BioResources* 8 (1): 201–10.

Fernandes, S. C., L. Oliveira, C. S. Freire, A. J. Silvestre, C. P. Neto, A. Gandini, and J. Desbriéres. 2009. Novel Transparent Nanocomposite Films Based on Chitosan and Bacterial Cellulose. *Green Chemistry* 11 (12): 2023–29.

Fowler, P. A., J. M. Hughes, and R. M. Elias. 2006. Biocomposites: Technology, Environmental Credentials and Market Forces. *Journal of the Science of Food and Agriculture* 86 (12): 1781–89.

Fradinho, J., J. Domingos, G. Carvalho, A. Oehmen, and M. Reis. 2013. Polyhydroxyalkanoates Production by a Mixed Photosynthetic Consortium of Bacteria and Algae. *Bioresource Technology* 132: 146–53.

Ganesan, V. 2008. Some Issues in Polymer Nanocomposites: Theoretical and Modeling Opportunities for Polymer Physics. *Journal of Polymer Science, Part B: Polymer Physics* 46 (24): 2666–71.

Gao, M., Z. Zhang, M. Lv, W. Song, and Y. Lv. 2018. Toxic Effects of Nanomaterial-Adsorbed Cadmium on Daphnia Magna. *Ecotoxicology and Environmental Safety* 148: 261–68.

García, N. L., L. Ribba, A. Dufresne, M. Aranguren, and S. Goyanes. 2011. Effect of Glycerol on the Morphology of Nanocomposites Made from Thermoplastic Starch and Starch Nanocrystals. Carbohydrate Polymers 84 (1): 203–10.

Ge, Y., J. P. Schimel, and P. A. Holden. 2011. Evidence for Negative Effects of Tio2 and Zno Nanoparticles on Soil Bacterial Communities. Environmental Science & *Technology* 45 (4): 1659–64.

Georgios, K., A. Silva, and S. Furtado. 2016. Applications of Green Composite Materials. *Biodegradeable Green Composites* 16: 312.

Ghanam, D., and P. Kleinebudde. 2011. Suitability of K-Carrageenan Pellets for the Formulation of Multiparticulate Tablets with Modified Release. *International Journal of Pharmaceutics* 409 (1–2): 9–18.

Ghosh, M., M. Bandyopadhyay, and A. Mukherjee. 2010. Genotoxicity of Titanium Dioxide (Tio2) Nanoparticles at Two Trophic Levels: Plant and Human Lymphocytes. *Chemosphere* 81 (10): 1253–62.

Gomes, M. E., J. Godinho, D. Tchalamov, A. Cunha, and R. Reis. 2002. Alternative Tissue Engineering Scaffolds Based on Starch: Processing Methodologies, Morphology, Degradation and Mechanical Properties. *Materials Science and Engineering: C* 20 (1–2): 19–26.

González, R., C. Carrara, E. Tosi, M. Añón, and A. Pilosof. 2007. Amaranth Starch-Rich Fraction Properties Modified by Extrusion and Fluidized Bed Heating. *LWT-Food Science and Technology* 40 (1): 136–43.

Gopi, S., P. Balakrishnan, M. Sreekala, A. Pius, and S. Thomas. 2017. Green Materials for Aerospace Industries. *Biocomposites for High-Performance Applications*, 307–18: Elsevier.

Gorrasi, G., V. Bugatti, and V. Vittoria. 2012. Pectins Filled with Ldh-Antimicrobial Molecules: Preparation, Characterization and Physical Properties. *Carbohydrate Polymers* 89 (1): 132–37.

Gousse, C., H. Chanzy, M. Cerrada, and E. Fleury. 2004. Surface Silylation of Cellulose Microfibrils: Preparation and Rheological Properties. *Polymer* 45 (5): 1569–75.

Grillo, R., A. H. Rosa, and L. F. Fraceto. 2015. Engineered Nanoparticles and Organic Matter: A Review of the State-of-the-Art. *Chemosphere* 119: 608–19.

Guan, J., and M. A. Hanna. 2006. Selected Morphological and Functional Properties of Extruded Acetylated Starch–Cellulose Foams. *Bioresource Technology* 97 (14): 1716–26.

Guerrero, P., Z. N. Hanani, J. Kerry, and K. De La Caba. 2011. Characterization of Soy Protein-Based Films Prepared with Acids and Oils by Compression. *Journal of Food Engineering* 107 (1): 41–49.

Gutiérrez, T. J., M. S. Tapia, E. Pérez, and L. Famá. 2015. Structural and Mechanical Properties of Edible Films Made from Native and Modified Cush-Cush Yam and Cassava Starch. *Food Hydrocolloids* 45: 211–17.

Haliullin, T., R. Zalyalov, A. Shvedova, and A. Tkachov. 2015. Hygienic Evaluation of Multilayer Carbon Nanotubes. *Meditsina truda i promyshlennaia ekologiia* 7: 37–42.

Hamad, K., M. Kaseem, Y. G. Ko, and F. Deri. 2014. Biodegradable Polymer Blends and Composites: An Overview. *Polymer Science Series A* 56 (6): 812–29.

Hassan, A., H. Balakrishnan, and A. Akbari. 2013. Polylactic Acid Based Blends, Composites and Nanocomposites. *Advances in Natural Polymers*, 361–96: Springer.

Hervy, M., S. Evangelisti, P. Lettieri, and K.-Y. Lee. 2015. Life Cycle Assessment of Nanocellulose-Reinforced Advanced Fibre Composites. *Composites Science and Technology* 118: 154–62.

Hirsch, A. 2002. Functionalization of Single-Walled Carbon Nanotubes. *Angewandte Chemie International Edition* 41 (11): 1853–59.

Honarvar, Z., Z. Hadian, and M. Mashayekh. 2016. Nanocomposites in Food Packaging Applications and Their Risk Assessment for Health. *Electronic Physician* 8 (6): 2531.

Hong, J., J. R. Peralta-Videa, C. Rico, S. Sahi, M. N. Viveros, J. Bartonjo, L. Zhao, and J. L. Gardea-Torresdey. 2014. Evidence of Translocation and Physiological Impacts of Foliar Applied Ceo_2 Nanoparticles on Cucumber (Cucumis Sativus) Plants. *Environmental Science & Technology* 48 (8): 4376–85.

Huang, C.-Y., M.-L. Roan, M.-C. Kuo, and W.-L. Lu. 2005. Effect of Compatibiliser on the Biodegradation and Mechanical Properties of High-Content Starch/Low-Density Polyethylene Blends. *Polymer Degradation and Stability* 90 (1): 95–105.

Huang, J.-Y., X. Li, and W. Zhou. 2015. Safety Assessment of Nanocomposite for Food Packaging Application. *Trends in Food Science & Technology* 45 (2): 187–99.

Huang, Y., L. Mei, X. Chen, and Q. Wang. 2018. Recent Developments in Food Packaging Based on Nanomaterials. *Nanomaterials* 8 (10): 830.

Huskić, M., and M. Žigon. 2007. Pmma/Mmt Nanocomposites Prepared by One-Step in Situ Intercalative Solution Polymerization. *European Polymer Journal* 43 (12): 4891–97.

Iavicoli, I., V. Leso, W. Ricciardi, L. L. Hodson, and M. D. Hoover. 2014. Opportunities and Challenges of Nanotechnology in the Green Economy. *Environmental Health* 13 (1): 78.

Ismail, I., D. Sa'adiyah, P. Rahajeng, A. Suprayitno, and R. Andiana. 2018. Extraction of Cellulose Microcrystalline from Galam Wood for Biopolymer. Paper presented at the AIP Conference Proceedings.

Jahangirian, H., E. G. Lemraski, R. Rafiee-Moghaddam, and T. J. Webster. 2018. A Review of Using Green Chemistry Methods for Biomaterials in Tissue Engineering. *International Journal of Nanomedicine* 13: 5953.

Jamshidian, M., E. A. Tehrany, M. Imran, M. Jacquot, and S. Desobry. 2010. Poly-Lactic Acid: Production, Applications, Nanocomposites, and Release Studies. *Comprehensive Reviews in Food Science and Food Safety* 9 (5): 552–71.

Jawaid, M., M. Thariq, and N. Saba. 2018. *Durability and Life Prediction in Biocomposites, Fibre-Reinforced Composites and Hybrid Composites.* Woodhead Publishing.

Jayaram, S. H. Impingement of Environmental Factors That Defines a System on Composites Performance. *Civil Engineering Portal.*

John, M. J., and S. Thomas. 2008. Biofibres and Biocomposites. *Carbohydrate Polymers* 71 (3): 343–64.

Johnson, R. C. 2007. Studies Warn of Nanoparticle Health Effects. *EE Times* 4: 13.

Kabir, E., V. Kumar, K.-H. Kim, A. C. Yip, and J. Sohn. 2018. Environmental Impacts of Nanomaterials. *Journal of Environmental Management* 225: 261–71.

Kalsoom Khan, A., A. U. Saba, S. Nawazish, F. Akhtar, R. Rashid, S. Mir, B. Nasir, *et al.* 2017. Carrageenan Based Bio-nanocomposites as Drug Delivery Tool with Special Emphasis on the Influence of Ferromagnetic Nanoparticles. *Oxidative Medicine and Cellular Longevity* 2017.

Kango, S., S. Kalia, A. Celli, J. Njuguna, Y. Habibi, and R. Kumar. 2013. Surface Modification of Inorganic Nanoparticles for Development of Organic–Inorganic Nanocomposites—a Review. *Progress in Polymer Science* 38 (8): 1232–61.

Kannan, M., M. Nesakumari, K. Rajarathinam, and A. Singh. 2010. Production and Characterization of Mushroom Chitosan under Solid-State Fermentation Conditions. *Advanced Biomedical Research* 4 (1): 10–13.

Karci, M., N. Demir, and S. Yazman. 2019. Evaluation of Flexural Strength of Different Denture Base Materials Reinforced with Different Nanoparticles. *Journal of Prosthodontics* 28 (5): 572–79.

Kaushik, A., M. Singh, and G. Verma. 2010. Green Nanocomposites Based on Thermoplastic Starch and Steam Exploded Cellulose Nanofibrils from Wheat Straw. *Carbohydrate Polymers* 82 (2): 337–45.

Ke, T., and X. J. Sun. 2001. Effects of Moisture Content and Heat Treatment on the Physical Properties of Starch and Poly (Lactic Acid) Blends. *Journal of Applied Polymer Science* 81 (12): 3069–82.

Khalid, P., V. B. Suman, M. A. Hussain, and A. B. Arun. 2016. Toxicology of Carbon Nanotubes - a Review. *International Journal of Applied Engineering Research* 11 (1): 148–57.

Khodakovskaya, M., E. Dervishi, M. Mahmood, Y. Xu, Z. Li, F. Watanabe, and A. S. Biris. 2009. Carbon Nanotubes Are Able to Penetrate Plant Seed Coat and Dramatically Affect Seed Germination and Plant Growth. *ACS nano* 3 (10): 3221–27.

Kianfar, F., M. Antonijevic, B. Chowdhry, and J. S. Boateng. 2013. Lyophilized Wafers Comprising Carrageenan and Pluronic Acid for Buccal Drug Delivery Using Model Soluble and Insoluble Drugs. *Colloids and Surfaces B: Biointerfaces* 103: 99–106.

Koronis, G., and A. Silva. 2018. *Green Composites for Automotive Applications.* Woodhead Publishing.

Koronis, G., and A. Silva. 2018. Green Composites Reinforced with Plant-Based Fabrics: Cost and Eco-Impact Assessment. *Journal of Composites Science* 2 (1): 8.

Koronis, G., A. Silva, and M. Fontul. 2013. Green Composites: A Review of Adequate Materials for Automotive Applications. *Composites Part B: Engineering* 44 (1): 120–27.

Koshy, R. R., S. K. Mary, L. A. Pothan, and S. Thomas. 2015. Soy Protein-and Starch-Based Green Composites/Nanocomposites: Preparation, Properties, and Applications. *Eco-Friendly Polymer Nanocomposites*, 433–67: Springer.

Kovacs, T., V. Naish, B. O'Connor, C. Blaise, F. Gagné, L. Hall, V. Trudeau, and P. Martel. 2010. An Ecotoxicological Characterization of Nanocrystalline Cellulose (Ncc). *Nanotoxicology* 4 (3): 255–70.

Kristo, E., and C. G. Biliaderis. 2007. Physical Properties of Starch Nanocrystal-Reinforced Pullulan Films. *Carbohydrate Polymers* 68 (1): 146–58.

Kumar, N., P. Kaur, and S. Bhatia. 2017. Advances in Bio-Nanocomposite Materials for Food Packaging: A Review. *Nutrition & Food Science* 47 (4): 591–606.

Kumar Thakur, V., and M. Kumari Thakur. 2015. *Eco-Friendly Polymer Nanocomposites: Processing and Properties*. Springer, 579p.

Lalwani, G., A. M. Henslee, B. Farshid, L. Lin, F. K. Kasper, Y.-X. Qin, A. G. Mikos, and B. Sitharaman. 2013. Two-Dimensional Nanostructure-Reinforced Biodegradable Polymeric Nanocomposites for Bone Tissue Engineering. *Biomacromolecules* 14 (3): 900–09.

Lee, K.-Y., Y. Aitomäki, L. A. Berglund, K. Oksman, and A. Bismarck. 2014. On the Use of Nanocellulose as Reinforcement in Polymer Matrix Composites. *Composites Science and Technology* 105: 15–27.

Lei, Y., Q. Wu, F. Yao, and Y. Xu. 2007. Preparation and Properties of Recycled Hdpe/Natural Fiber Composites. *Composites Part A: Applied Science and Manufacturing* 38 (7): 1664–74.

Leja, K., and G. J. P. Lewandowicz. 2010. Polymer Biodegradation and Biodegradable Polymers: A Review. *Polish Journal of Environmental Studies* 19 (2): 255–66.

Li, J., S. Schiavo, G. Rametta, M. L. Miglietta, V. La Ferrara, C. Wu, and S. Manzo. 2017. Comparative Toxicity of Nano Zno and Bulk Zno Towards Marine Algae Tetraselmis Suecica and Phaeodactylum Tricornutum. *Environmental Science and Pollution Research* 24 (7): 6543–53.

Li, Q., J. Zhou, and L. Zhang. 2009. Structure and Properties of the Nanocomposite Films of Chitosan Reinforced with Cellulose Whiskers. *Journal of Polymer Science Part B: Polymer Physics* 47 (11): 1069–77.

Lin, N., J. Yu, P. R. Chang, J. Li, and J. Huang. 2011. Poly (Butylene Succinate)-Based Biocomposites Filled with Polysaccharide Nanocrystals: Structure and Properties. *Polymer Composites* 32 (3): 472–82.

Liou, S.-H., T.-C. Tsou, S.-L. Wang, L.-A. Li, H.-C. Chiang, W.-F. Li, P.-P. Lin, *et al.* 2012. Epidemiological Study of Health Hazards among Workers Handling Engineered Nanomaterials. *Journal of Nanoparticle Research* 14 (8): 878.

Liu, D., T. Zhong, P. R. Chang, K. Li, and Q. Wu. 2010. Starch Composites Reinforced by Bamboo Cellulosic Crystals. *Bioresource Technology* 101 (7): 2529–36.

Long, T. C., N. Saleh, R. D. Tilton, G. V. Lowry, and B. Veronesi. 2006. Titanium Dioxide (P25) Produces Reactive Oxygen Species in Immortalized Brain Microglia (Bv2): Implications for Nanoparticle Neurotoxicity. *Environmental Science & Technology* 40 (14): 4346–52.

Lu, D., C. Xiao, and S. Xu. 2009. Starch-Based Completely Biodegradable Polymer Materials. *Express Polymer Letters* 3 (6): 366–75.

Lu, Y., and S. Ozcan. 2015. Green Nanomaterials: On Track for a Sustainable Future. *Nano Today* 10 (4): 417–20.

Luzi, F., L. Torre, J. M. Kenny, and D. Puglia. 2019. Bio-and Fossil-Based Polymeric Blends and Nanocomposites for Packaging: Structure–Property Relationship. *Materials* 12 (3): 471.

Ma, J., R. R. Mercer, M. Barger, D. Schwegler-Berry, J. M. Cohen, P. Demokritou, and V. Castranova. 2015. Effects of Amorphous Silica Coating on Cerium Oxide Nanoparticles Induced Pulmonary Responses. *Toxicology and Applied Pharmacology* 288 (1): 63–73.

Maache-Rezzoug, Z., T. Maugard, I. Zarguili, E. Bezzine, M.-N. El Marzouki, and C. Loisel. 2009. Effect of Instantaneous Controlled Pressure Drop (Dic) on Physicochemical Properties of Wheat, Waxy and Standard Maize Starches. *Journal of Cereal Science* 49 (3): 346–53.

Mackevica, A., and S. Foss Hansen. 2016. Release of Nanomaterials from Solid Nanocomposites and Consumer Exposure Assessment–a Forward-Looking Review. *Nanotoxicology* 10 (6): 641–53.

Mahdavinia, G. R., and H. Etemadi. 2014. In Situ Synthesis of Magnetic Carapva Ipn Nanocomposite Hydrogels and Controlled Drug Release. *Materials Science and Engineering: C* 45: 250–60.

Majdzadeh-Ardakani, K., and S. Sadeghi-Ardakani. 2010. Experimental Investigation of Mechanical Properties of Starch/Natural Rubber/Clay Nanocomposites. *Digest Journal of Nanomaterials & Biostructures* 5 (2): 307–16.

Mallapragada, S. K., T. M. Brenza, J. M. McMillan, B. Narasimhan, D. S. Sakaguchi, A. D. Sharma, S. Zbarska, and H. E. Gendelman. 2015. Enabling Nanomaterial, Nanofabrication and Cellular Technologies for Nanoneuromedicines. *Nanomedicine: Nanotechnology, Biology and Medicine* 11 (3): 715–29.

Mann, G. S., L. P. Singh, P. Kumar, and S. Singh. 2018. Green Composites: A Review of Processing Technologies and Recent Applications. *Journal of Thermoplastic Composite Materials*. doi: 10.1177/0892705718816354.

Mano, J., G. Silva, H. S. Azevedo, P. Malafaya, R. Sousa, S. S. Silva, L. Boesel, *et al.* 2007. Natural Origin Biodegradable Systems in Tissue Engineering and Regenerative Medicine: Present Status and Some Moving Trends. *Journal of the Royal Society Interface* 4 (17): 999–1030.

Marouani, S., L. Curtil, and P. Hamelin. 2012. Ageing of Carbon/Epoxy and Carbon/ Vinylester Composites Used in the Reinforcement and/or the Repair of Civil Engineering Structures. *Composites Part B: Engineering* 43 (4): 2020–30.

Martins, A. B., and R. M. C. Santana. 2016. Effect of Carboxylic Acids as Compatibilizer Agent on Mechanical Properties of Thermoplastic Starch and Polypropylene Blends. *Carbohydrate Polymers* 135: 79–85.

Mathew, A. P., W. Thielemans, and A. Dufresne. 2008. Mechanical Properties of Nanocomposites from Sorbitol Plasticized Starch and Tunicin Whiskers. *Journal of Applied Polymer Science* 109 (6): 4065–74.

Matus, K. J., W. C. Clark, P. T. Anastas, J. B. Zimmerman. 2012. Barriers to the Implementation of Green Chemistry in the United States. *Environmental Science & Technology* 46 (20): 10892–99.

Maynard, A. D., and R. J. Aitken. 2007. Assessing Exposure to Airborne Nanomaterials: Current Abilities and Future Requirements. *Nanotoxicology* 1 (1): 26–41.

McAuliffe, M. E., and M. J. Perry. 2007. Are Nanoparticles Potential Male Reproductive Toxicants? A Literature Review. *Nanotoxicology* 1 (3): 204–10.

Mohanty, A. K., M. Misra, and L. T. Drzal. 2005. *Natural Fibers, Biopolymers, and Biocomposites*. CRC Press.

Moon, R. J., A. Martini, J. Nairn, J. Simonsen, and J. Youngblood. 2011. Cellulose Nanomaterials Review: Structure, Properties and Nanocomposites. *Chemical Society Reviews* 40 (7): 3941–94.

Mouritz, A. P., and A. G. Gibson. 2007. *Fire Properties of Polymer Composite Materials*. Vol. 143: Springer Science & Business Media.

Müller, K., E. Bugnicourt, M. Latorre, M. Jorda, Y. Echegoyen Sanz, J. Lagaron, O. Miesbauer, *et al.* 2017. Review on the Processing and Properties of Polymer Nanocomposites and Nanocoatings and Their Applications in the Packaging, Automotive and Solar Energy Fields. *Nanomaterials* 7 (4): 74.

Naskar, A. K., J. K. Keum, and R. G. Boeman. 2016. Polymer Matrix Nanocomposites for Automotive Structural Components. *Nature Nanotechnology* 11 (12): 1026.

Naumann, A., I. Stephan, M. Noll, and Biodegradation. 2012. Material Resistance of Weathered Wood-Plastic Composites against Fungal Decay. *International Biodeterioration & Biodegradation* 75: 28–35.

Nešić, A., M. Gordić, S. Davidović, Ž. Radovanović, J. Nedeljković, I. Smirnova, and P. Gurikov. 2018. Pectin-Based Nanocomposite Aerogels for Potential Insulated Food Packaging Application. *Carbohydrate Polymers* 195: 128–35.

Nezakati, T., A. Seifalian, A. Tan, and A. M. Seifalian. 2018. Conductive Polymers: Opportunities and Challenges in Biomedical Applications. *Chemical Reviews* 118 (14): 6766–843.

Njuguna, J., K. Pielichowski, and S. Desai. 2008. Nanofiller-Reinforced Polymer Nanocomposites. *Polymers for Advanced Technologies* 19 (8): 947–59.

Nkurunziza, G., A. Debaiky, P. Cousin, and B. Benmokrane. 2005. Durability of Gfrp Bars: A Critical Review of the Literature. *Progress in Structural Engineering and Materials* 7 (4): 194–209.

Okamoto, M. J. 2004. Biodegradable Polymer/Layered Silicate Nanocomposites: A Review. *Journal of Industrial and Engineering Chemistry* 10 (7): 1156–81.

Oksman, K., A. P. Mathew, D. Bondeson, I. Kvien, and technology. 2006. Manufacturing Process of Cellulose Whiskers/Polylactic Acid Nanocomposites. *Composites Science and Technology* 66 (15): 2776–84.

Othman, S. H. 2014. Bio-Nanocomposite Materials for Food Packaging Applications: Types of Biopolymer and Nano-Sized Filler. *Agriculture and Agricultural Science Procedia* 2: 296–303.

Ou, Y., D. Zhu, H. Zhang, L. Huang, Y. Yao, G. Li, and B. Mobasher. 2016. Mechanical Characterization of the Tensile Properties of Glass Fiber and Its Reinforced Polymer (Gfrp) Composite under Varying Strain Rates and Temperatures. *Polymers* 8 (5): 196.

Pal, N., P. Dubey, P. Gopinath, and K. Pal. 2017. Combined Effect of Cellulose Nanocrystal and Reduced Graphene Oxide into Poly-Lactic Acid Matrix Nanocomposite as a Scaffold and Its Anti-Bacterial Activity. *International Journal of Biological Macromolecules* 95: 94–105.

Pandey, J. K., W.-S. Chu, C. S. Lee, and S.-H. Ahn. 2007 Preparation Characterization and Performance Evaluation of Nanocomposites from Natural Fiber Reinforced Biodegradable Polymer Matrix for Automotive Applications. Paper presented at the International Symposium on Polymers and the Environment: Emerging Technology and Science, BioEnvironmental Polymer Society (BEPS).

Pandey, J. K., and R. P. Singh. 2005. Green Nanocomposites from Renewable Resources: Effect of Plasticizer on the Structure and Material Properties of Clay-Filled Starch. *Starch-Stärke* 57 (1): 8–15.

Peelman, N., P. Ragaert, B. De Meulenaer, D. Adons, R. Peeters, L. Cardon, F. Van Impe, and F. Devlieghere. 2013. Application of Bioplastics for Food Packaging. *Trends in Food Science & Technology* 32 (2): 128–41.

Pinto, A. M., J. Cabral, D. A. P. Tanaka, A. M. Mendes, and F. D. Magalhães. 2013. Effect of Incorporation of Graphene Oxide and Graphene Nanoplatelets on Mechanical and Gas Permeability Properties of Poly (Lactic Acid) Films. *Polymer International* 62 (1): 33–40.

Podsiadlo, P., S.-Y. Choi, B. Shim, J. Lee, M. Cuddihy, and N. A. Kotov. 2005. Molecularly Engineered Nanocomposites: Layer-by-Layer Assembly of Cellulose Nanocrystals. *Biomacromolecules* 6 (6): 2914–18.

Qu, P., Y. Gao, G. Wu, and L. Zhang. 2010. Nanocomposites of Poly (Lactic Acid) Reinforced with Cellulose Nanofibrils. *BioResources* 5 (3): 1811–23.

Ramli, N., N. Mazlan, Y. Ando, Z. Leman, K. Abdan, A. Aziz, and N. Sairy. 2018. Natural Fiber for Green Technology in Automotive Industry: A Brief Review. Paper presented at the IOP Conference Series: Materials Science and Engineering.

Ramos, Ó. L., R. N. Pereira, M. A. Cerqueira, J. R. Martins, J. A. Teixeira, F. X. Malcata, and A. A. Vicente. 2018. Bio-Based Nanocomposites for Food Packaging and Their Effect in Food Quality and Safety. *Food Packaging and Preservation*, 271–306: Elsevier.

Reig, C. S., A. D. Lopez, M. H. Ramos, and V. A. C. Ballester. 2014. Nanomaterials: A Map for Their Selection in Food Packaging Applications. *Packaging Technology and Science* 27 (11): 839–66.

Reis, J. M. L. d. 2012. Effect of Temperature on the Mechanical Properties of Polymer Mortars. *Materials Research* 15 (4): 645–49.

Rhim, J.-W., H.-M. Park, and C.-S. Ha. 2013. Bio-Nanocomposites for Food Packaging Applications. *Progress in Polymer Science* 38 (10–11): 1629–52.

Ribba, L., N. L. Garcia, N. D'Accorso, and S. Goyanes. 2017. Disadvantages of Starch-Based Materials, Feasible Alternatives in Order to Overcome These Limitations. *Starch-Based Materials in Food Packaging*, 37–76: Elsevier.

Rocha, T. L., T. Gomes, V. S. Sousa, N. C. Mestre, and M. J. Bebianno. 2015. Ecotoxicological Impact of Engineered Nanomaterials in Bivalve Molluscs: An Overview. *Marine Environmental Research* 111: 74–88.

Saini, R. K., A. K. Bajpai, and E. Jain, P. 2017. 13 Advances in Bio-nanocomposites for Biomedical Applications. *Biocompatible Polymer Composites: Processing, and Applications* 379: Woodhead Publishing.

Schaefer, D. W., and R. S. Justice. 2007. How Nano Are Nanocomposites? *Macromolecules* 40 (24): 8501–17.

Seker, M., and M. A. Hanna. 2006. Sodium Hydroxide and Trimetaphosphate Levels Affect Properties of Starch Extrudates. *Industrial Crops and Products* 23 (3): 249–55.

Shahzad, A. 2015. Mechanical Properties of Eco-Friendly Polymer Nanocomposites. *Eco-Friendly Polymer Nanocomposites*, 527–59: Springer.

Shakeri, A., A. P. Mathew, and K. Oksman. 2012. Self-Reinforced Nanocomposite by Partial Dissolution of Cellulose Microfibrils in Ionic Liquid. *Journal of Composite Materials* 46 (11): 1305–11.

Shan, G.-F., X. Gong, W.-P. Chen, L. Chen, and M.-F. Zhu. 2011. Effect of Multi-Walled Carbon Nanotubes on Crystallization Behavior of Poly (3-Hydroxybutyrate-Co-3-Hydroxyvalerate). *Colloid and Polymer Science* 289 (9): 1005–14.

Sharma, A., M. Thakur, M. Bhattacharya, T. Mandal, and S. Goswami. 2019. Commercial Application of Cellulose Nano-Composites-a Review. *Biotechnology Reports* 21: e00316.

Sharma, H., U. K. Komal, I. Singh, J. P. Misra, and P. K. Rakesh. 2019. Introduction to Green Composites. *Processing of Green Composites*, 1–13: Springer.

Shchipunov, Y. 2012. Bio-nanocomposites: Green Sustainable Materials for the near Future. *Pure and Applied Chemistry* 84 (12): 2579–607.

Shukla, S., S. K. Shukla, P. P. Govender, and N. Giri. 2016. Biodegradable Polymeric Nanostructures in Therapeutic Applications: Opportunities and Challenges. *RSC Advances* 6 (97): 94325–51.

Silvestre, C., D. Duraccio, and S. Cimmino. 2011. Food Packaging Based on Polymer Nanomaterials. *Progress in Polymer Science* 36 (12): 1766–82.

Siqueira, G., J. Bras, and A. Dufresne. 2010. Cellulosic Bio-nanocomposites: A Review of Preparation, Properties and Applications. *Polymers* 2 (4): 728–65.

Souza, V., J. Pires, É. Vieira, I. Coelhoso, M. Duarte, and A. Fernando. 2018. Shelf Life Assessment of Fresh Poultry Meat Packaged in Novel Bio-nanocomposite of Chitosan/Montmorillonite Incorporated with Ginger Essential Oil. *Coatings* 8 (5): 177.

Sridhar, V., I. Lee, H. Chun, and H. Park. 2013. Graphene Reinforced Biodegradable Poly (3-Hydroxybutyrate-Co-4-Hydroxybutyrate) Nano-Composites. *Express Polymer* Letters 7 (4): 320–28

Sun, J., J. Shen, S. Chen, M. Cooper, H. Fu, D. Wu, and Z. Yang. 2018. Nanofiller Reinforced Biodegradable Pla/Pha Composites: Current Status and Future Trends. *Polymers* 10 (5): 505.

Swaroop, C., and M. Shukla. 2018. Nano-Magnesium Oxide Reinforced Polylactic Acid Biofilms for Food Packaging Applications. *International Journal of Biological Macromolecules* 113: 729–36.

Szymońska, J., F. Krok, E. Komorowska-Czepirska, and K. Rebilas. 2003. Modification of Granular Potato Starch by Multiple Deep-Freezing and Thawing. *Carbohydrate Polymers* 52 (1): 1–10.

Thakur, V. K., M. K. Thakur, P. Raghavan, and M. R. Kessler. 2014. Progress in Green Polymer Composites from Lignin for Multifunctional Applications: A Review. *ACS Sustainable Chemistry & Engineering* 2 (5): 1072–92.

Tunç, S., and O. Duman 2011. Preparation of Active Antimicrobial Methyl Cellulose/ Carvacrol/Montmorillonite Nanocomposite Films and Investigation of Carvacrol Release. *LWT-Food Science and Technology* 44 (2): 465–72.

Turkevich, L. A., A. G. Dastidar, Z. Hachmeister, and M. Lim. 2015. Potential Explosion Hazard of Carbonaceous Nanoparticles: Explosion Parameters of Selected Materials. *Journal of Hazardous Materials* 295: 97–103.

Uddin, A. J., M. Fujie, S. Sembo, and Y. Gotoh. 2012. Outstanding Reinforcing Effect of Highly Oriented Chitin Whiskers in Pva Nanocomposites. *Carbohydrate Polymers* 87 (1): 799–805.

Vaara, P., and J. Leinonen. 2012. Technology Survey on Ndt of Carbon-Fiber Composites.

Vasile, C., M. Râpă, M. Ştefan, M. Stan, S. Macavei, R. Darie-Niţă, L. Barbu-Tudoran, *et al.* 2017. New Pla/Zno: Cu/Ag Bio-nanocomposites for Food Packaging. *eXPRESS Polymer Letters* 11 (7).

Wang, B., Y. Wan, Y. Zheng, X. Lee, T. Liu, Z. Yu, J. Huang, *et al.* 2019. Alginate-Based Composites for Environmental Applications: A Critical Review. *Critical Reviews in Environmental Science and Technology* 49 (4): 318–56.

Wang, L., C. Lu, Y. Li, F. Wu, B. Zhao, and X. J. Dong. 2015. Green Fabrication of Porous Silk Fibroin/Graphene Oxide Hybrid Scaffolds for Bone Tissue Engineering. *RSC Advances* 5 (96): 78660–68.

Wang, P., E. Lombi, S. Sun, K. G. Scheckel, A. Malysheva, B. A. McKenna, N. W. Menzies, F.-J. Zhao, and P. M. Kopittke. 2017. Characterizing the Uptake, Accumulation and Toxicity of Silver Sulfide Nanoparticles in Plants. *Environmental Science: Nano* 4 (2): 448–60.

Wang, X., H. Yang, L. Song, Y. Hu, W. Xing, and H. Lu. 2011. Morphology, Mechanical and Thermal Properties of Graphene-Reinforced Poly (Butylene Succinate) Nanocomposites. *Composites Science and Technology* 72 (1): 1–6.

Wei, L., N. Hu, and Y. Zhang. 2010. Synthesis of Polymer—Mesoporous Silica Nanocomposites. *Materials* 3 (7): 4066–79.

Wihodo, M., and C. I. Moraru. 2013. Physical and Chemical Methods Used to Enhance the Structure and Mechanical Properties of Protein Films: A Review. *Journal of Food Engineering* 114 (3): 292–302.

Wu, G., L. Ma, Y. Wang, L. Liu, and Y. Huang. 2016. Improvements in Interfacial and Heat-Resistant Properties of Carbon Fiber/Methylphenylsilicone Resins Composites by Incorporating Silica-Coated Multi-Walled Carbon Nanotubes. *Journal of adhesion science and Technology* 30 (2): 117–30.

Xia, H., and Q. Wang. 2003. Preparation of Conductive Polyaniline/Nanosilica Particle Composites through Ultrasonic Irradiation. *Journal of Applied Polymer Science* 87 (11): 1811–17.

Xiao, L., B. Wang, G. Yang, and M. Gauthier. 2012. Poly (Lactic Acid)-Based Biomaterials: Synthesis, Modification and Applications. *Biomedical Science, Engineering and Technology* 11: 247–82.

Yam, K. L., P. T. Takhistov, and J. Miltz. 2005. Intelligent Packaging: Concepts and Applications. *Journal of Food Science* 70 (1): R1–R10.

Yasmin, A., and I. M. Daniel. 2004. Mechanical and Thermal Properties of Graphite Platelet/Epoxy Composites. *Polymer* 45 (24): 8211–19.

Youssef, A. M., and S. M. El-Sayed. 2018. Bio-nanocomposites Materials for Food Packaging Applications: Concepts and Future Outlook. *Carbohydrate Polymers* 193: 19–27.

Yu, H.-Y., Z.-Y. Qin, B. Sun, X.-G. Yang, and J.-M. Yao. 2014. Reinforcement of Transparent Poly (3-Hydroxybutyrate-Co-3-Hydroxyvalerate) by Incorporation of Functionalized Carbon Nanotubes as a Novel Bio-nanocomposite for Food Packaging. *Composites Science and Technology* 94: 96–104.

Yu, J., F. Ai, A. Dufresne, S. Gao, J. Huang, and P. R. Chang. 2008. Structure and Mechanical Properties of Poly (Lactic Acid) Filled with (Starch Nanocrystal)-Graft-Poly (E-Caprolactone). *Macromolecular Materials and Engineering* 293 (9): 763–70.

Zabihzadeh, S. M. 2009. Water Uptake and Flexural Properties of Natural Filler/Hdpe Composites. *BioResources* 5 (1): 316–232.

Zakaria, N., N. Muhammad, and M. Abdullah. 2017 Potential of Starch Nanocomposites for Biomedical Applications. Paper presented at the IOP Conference Series: Materials Science and Engineering.

Zhang, B., S. Dhital, B. M. Flanagan, P. Luckman, P. J. Halley, and M. J. Gidley. 2015. Extrusion Induced Low-Order Starch Matrices: Enzymic Hydrolysis and Structure. *Carbohydrate Polymers* 134: 485–96.

Zhang, H., Y. Yao, D. Zhu, B. Mobasher, and L. Huang. 2016. Tensile Mechanical Properties of Basalt Fiber Reinforced Polymer Composite under Varying Strain Rates and Temperatures. *Polymer Testing* 51: 29–39.

Zhang, W., and B. W. Cue. 2012. *Green Techniques for Organic Synthesis and Medicinal Chemistry*. John Wiley & Sons.

Zheng, H., F. Ai, P. R. Chang, J. Huang, and A. Dufresne. 2009. Structure and Properties of Starch Nanocrystal-Reinforced Soy Protein Plastics. *Polymer Composites* 30 (4): 474–80.

Zou, H., S. Wu, and J. Shen. 2008. Polymer/Silica Nanocomposites: Preparation, Characterization, Properties, and Applications. *Chemical Reviews* 108 (9): 3893–957.

Zribi-Maaloul, E., I. Trabelsi, L. Elleuch, H. Chouayekh, and R. B. Salah. 2013. Purification and Characterization of Two Polyhydroxyalcanoates from Bacillus Cereus. *International Journal of Biological Macromolecules* 61: 82–88.

Index

Printed in the United States
by Baker & Taylor Publisher Services